#73903-1

OUR WAR ON OURSELVES

Rethinking Science, Technology, and Economic Growth

Our approach to knowing and doing is based on delegating physical phenomena to physicists, biological phenomena to biologists, social phenomena to sociologists, economic phenomena to economists, and so on. This approach to knowledge and practice works well when one category of phenomena dominates (as in mechanical and technical systems), but does not work when many categories of phenomena make significant contributions (as in the biological and cultural spheres). As a result, our civilization succeeds in its scientific and technical endeavours yet fails in dealing with communities and ecosystems.

Following his groundbreaking *Labyrinth of Technology* and *Living in the Labyrinth of Technology*, Willem H. Vanderburg's *Our War on Ourselves* explores the type of war we have unleashed on our lives by emphasizing discipline-based processes. The work also illuminates how we can achieve a more balanced, livable, and sustainable future by combining both technical and cultural perspectives in our educational and institutional settings.

WILLEM H. VANDERBURG is Director of the Centre for Technology and Social Development at the University of Toronto.

WILLEM H. VANDERBURG

Our War on Ourselves

Rethinking Science, Technology, and Economic Growth

UNIVERSITY OF TORONTO PRESS
Toronto Buffalo London

© University of Toronto Press 2011
Toronto Buffalo London
www.utppublishing.com
Printed in Canada

ISBN 978-1-4426-4438-0 (cloth)
ISBN 978-1-4426-1261-7 (paper)

Printed on acid-free, 100% post-consumer recycled paper with vegetable-based inks.

Library and Archives Canada Cataloguing in Publication

Vanderburg, Willem H.
Our war on ourselves : rethinking science, technology, and economic growth / Willem H. Vanderburg.

Includes bibliographical references and index.
ISBN 978-1-4426-4438-0 (bound). ISBN 978-1-4426-1261-7 (pbk.)

1. Science – Social aspects. 2. Technology – Social aspects. 3. Decision making – Social aspects. 4. Economic development – Sociological aspects. I. Title.

T14.5.V35 2012 303.48'3 C2011-904803-5

This book has been published with the help of a grant from the Canadian Federation for the Humanities and Social Sciences, through the Aid to Scholarly Publications Program, using funds provided by the Social Sciences and Humanities Research Council of Canada.

University of Toronto Press acknowledges the financial assistance to its publishing program of the Canada Council for the Arts and the Ontario Arts Council.

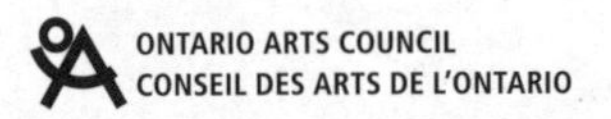

University of Toronto Press acknowledges the financial support of the Government of Canada through the Canada Book Fund for its publishing activities.

For Rita

Contents

Preface

The purpose of this work is straightforward but far-reaching. I am proposing that we consider modifying the discipline-based intellectual and professional division of labour to endow specialists with the capacity to make decisions that have a better ratio of desired to undesired effects. My confidence that this is possible is grounded in a diagnosis tracing the spectacular successes of our civilization as well as its equally spectacular failures to this division of labour. Simply put, our present form of specialization does not permit the practitioners of disciplines and specialties to see the consequences of their decisions because these fall mostly beyond their areas of expertise. Hence, the negative consequences must be dealt with by end-of-pipe approaches, which have made the system so top-heavy that it is no longer economical, socially viable, or environmentally sustainable.

This proposal for deep structural reforms in our approaches to scientific knowing and technical doing hopefully comes at a good time. It is likely that we are finally emerging from decades of economic fundamentalism, which has ignored the fact that industrial civilization has gone from one economic crisis to another, except during the period when Keynesian economics was practised. Perhaps there is a growing willingness to face the deep structural roots of the present situation. If we can reform our universities to educate practitioners whose decisions will have a significantly better ratio of desired to undesired effects, we may be able to change course and slowly restore a level of context appropriateness to our technology, create an economy that serves people, rather than the other way around, and at the same time reduce our impact on all life.

For many of my readers this will undoubtedly sound like an impossibility. How can I promise scientific and technical advances that will lead to a more viable economy, and do all this with much higher social, health, labour, and environmental protection? How can we redress some of the terrible injustices of the current system, which is making the wealthy even richer and the poor and the vulnerable weaker? The answer is simple, at least in principle. If the discipline-based organization of contemporary ways of life were gradually shifted to enable practitioners to understand better the consequences of their decisions and to use this understanding to adjust these decisions to achieve the desired results, but at the same time prevent or greatly minimize harmful effects, a little thinking can achieve what now must be done by expensive end-of-pipe technologies or social, health, labour, and environmental services. In other words, the practices of these specialists would gradually begin to take on a more preventive orientation as opposed to the current end-of-pipe one. It would allow our university and college students to regain a vision of a more liveable and sustainable future, as well as an ability to live that vision in their education and subsequent practices. It may provide governments with political options that are now closed. It may brighten the hopes of our children and grandchildren.

I know this goes against common sense. It is true that under current discipline-based approaches all this would be impossible. If we deal with our problems in an end-of-pipe fashion, doing less harm means the addition of all kinds of end-of-pipe technologies and services. There is no future in this business, other than lobbying governments to postpone the inevitable. Twice in my life I convinced senior policymaking bodies that what I am proposing was a viable and practical option. Twice, the election of a new government, which believed that the Market provided a better alternative, killed the initiatives. It is time to try it again, because the events in 2008 demonstrated clearly what that kind of economic system will do time and time again. What is required is neither a conservative approach, nor a liberal, democratic approach, nor a republican approach. What is required is a coalition government dedicated to the public good, which recognizes that there are no political solutions to our deep structural crises other than to rebuild our knowledge bases on which everything in our contemporary societies depend. Such is the agenda of this book.

Although my previous works will help the reader fill in many details, they are not necessary to follow the present argument. However, there is one important thing to keep in mind. I was born in the Netherlands and, for the first sixteen years of my life, grew up as any other Dutchman. Had I been born in China, Chile, Egypt, or any other country I would have grown up very differently. Surely, much the same kind of argument can be made for whatever 'intellectual culture' we bring to this work. I was intellectually 'born' as an engineer, and this profoundly shaped my thinking. Surely this is true for all of us. Whatever discipline or school of thought into which we were intellectually born will shape our thinking. If my reader is willing to entertain the possibility that we are all people of our time, place, and culture and that we are all shaped by the disciplines we have studied and practised, then let us attempt to journey together as far as possible. I am not attempting to convert you to my interpretation of what is happening. When I first sought to transcend my intellectual discipline because of the questions raised by the social and environmental crises, I was profoundly shaken by reading a book that implicitly presented the most accurate description of how my mindset worked and which then tied this to our crises. It forced me to rethink everything. I am asking no more from my reader than to allow for the possibility of a similar journey. All living systems are in decline, and this is the consequence of our discipline-based approaches to knowing and doing. None of us can possibly remain the same if we seriously create an ecology of knowledge capable of understanding our current dilemma in a more comprehensive way. Such an ecology can point the way towards exercising a genuine human responsibility for the future of ourselves and our children.

I wish to thank my anonymous reviewers for their helpful suggestions, which resulted in several footnotes that build bridges and clarify relations with other thinkers. I wish to thank my wife, Rita, who again has been my best in-house editor. I also wish to thank my daughter, Esther, whose assistance made it possible for me to complete this work with greater speed. I also wish to thank my students, who year after year encourage me to continue this intellectual struggle for their future. May this work make a small contribution to the celebration and protection of life on this planet.

Toronto, 2011

OUR WAR ON OURSELVES

Rethinking Science, Technology, and Economic Growth

Introduction: A Civilization of Non-sense?

Failure to Connect the Dots?

Most of us attempt to make sense of our lives and the world. Doing so may be likened to 'connecting the dots' in order to symbolically represent the fabric of relations within which everything, including ourselves, adapts and evolves in relation to everything else. In the past each community did so very differently, by its members symbolizing their experiences via an acquired culture that endowed everything with meanings and values.

Our civilization stands apart by insisting that this task is best left to science for knowing ourselves and the world and to technology for providing the means to help us make better adaptations. This completely transforms how we 'connect the dots.' For example, physicists connect the dots between physical phenomena, but not between them and everything else, and the same holds for all other scientific disciplines. Similarly, professionals connect the dots only between what lies within their areas of specialization and not with anything beyond them. Does this represent an anti-life bias?

To varying degrees many people recognize that everything living is in decline: from organisms to ecosystems and from human lives to societies. It has become increasingly difficult to believe that these kinds of issues are temporary because tomorrow's scientific discoveries and technical developments will overcome them. There is an emerging awareness that our economies have been turned into anti-economies that extract rather than create wealth. The decline of all life may well be a symptom of this extraction. We may have to confront the possibility that our economic, social, and environmental crises have a common root: we are no longer connecting the dots to ensure that what we achieve in one area (as measured by certain criteria) is not undermined by its consequences in other areas (as measured by different criteria).

Efforts to connect the dots better with approaches such as interdisciplinarity, multidisciplinarity, transdisciplinarity, wholism, and systems theory have achieved very little. The development of the scientific approach to knowing and the technical approach to doing and the attempts to remedy their weaknesses were all premised on human life, society, and the biosphere being separable and divisible without any significant loss in our understanding or a reduction in the effectiveness of our doing. Ironically, these forms of science and technology heightened our awareness of how everything is related to everything else. Concepts such as inseparability and indivisibility became thinkable, but they have had little impact. This is the bad news.

The good news is that there may be a possibility of recognizing that our current forms of science and technology are ill adapted to the study and manipulation of a reality that turns out to be much less separable and divisible than was originally believed. Changes in the current intellectual and professional division of labour may well become necessary, and this will change our discipline-based approaches. Our approaches to knowing and doing must be adapted to a living world in order that we do not undermine it. We shall begin with an embryonic sketch of this diagnosis, to be developed further in the first four chapters. Next, an embryonic prescription will be presented, to be fully developed in the last chapter.[1]

Science as Biased Knowing

Western civilization gradually developed a unique approach to knowing ourselves and our world. It did away with any attempts to tackle these complexities, with their numerous aspects and relationships. The task of knowing was greatly simplified by parcelling it out to various disciplines. Physical phenomena were the business of physicists, chemical phenomena of chemists, biological phenomena of biologists, social phenomena of sociologists, political phenomena of political scientists, and so on. Western science holds that everything can be known by examining one category of phenomena at a time. When difficulties arose, this discipline-based approach was refined to create hybrid disciplines such as physical chemistry, biochemistry, social psychology, and sociobiology.

Imagine for a moment if we attempted to understand our daily lives in this way. How many of our daily-life activities can be better known by consulting a single discipline? How many others appear to be constituted by the intermingling of several categories of phenomena, thus requiring the consultation of a number of disciplines and the integration of their findings? In the absence of a 'science of the sciences' capable of *scientifically* integrating the findings, how are we to proceed in these cases?

The difficulties that confront us when we attempt to understand our daily lives better through science do not end here. Many of these activities directly or indirectly involve technologies of one kind or another and their products. These technological phenomena in our lives are excluded from the 'one category of phenomena at a time' approach of the social sciences. A mirror-image exclusion occurs in the professions, whose disciplines examine categories of phenomena other than those dealt with by the social sciences. All of this becomes evident when we consult the index of an introductory textbook to any discipline for entries on categories of phenomena other than the one constituting its focus.

The above kind of exercise immediately reveals the limitations of the discipline-based approach to knowing. It works extremely well for situations in which the influence of one category of phenomena dwarfs the influences of all others, so that these can be neglected. It does not work very well for those situations in which several categories of phenomena make important contributions, so that these cannot be understood one category of phenomena at a time. The reason that physics became the model discipline and that many other disciplines failed to replicate its success also becomes apparent. Physics happens to study situations in which physical phenomena dwarf all others, as is the case for the 'big bang' and the subatomic 'world.' It shares this characteristic with some other disciplines, including chemistry and molecular biology.

The social sciences were not so fortunate. Their founding thinkers took the intermingling of many categories of phenomena in human life and society as an indication that whatever category they focused on had to be studied against the background of all the others. Such is clearly the case in the works of Adam Smith, Karl Marx, Max Weber, and many others. It is also implicit in historical works such as those of Arnold Toynbee, which examine the ways in which many categories of phenomena help constitute a society or civilization and how it grows, declines, and eventually collapses.

The success of physics is directly linked to its limitations. When we began to study physics in high school, we were told that initially certain aspects such as friction, air resistance, and inertia had to be neglected, reducing complex situations to simple ones to which Newton's laws of motion could be applied in an elementary fashion. It was understood that as we advanced we would gradually learn to include these and other aspects, with the result that these situations would approximate the real world ever more closely. This is true in one respect, and entirely false in another, because the intellectual domain of physics excludes all but physical phenomena. Physicists know this very well and would not dream of attempting to solve the appropriate differential equations to help them ride their bikes or play squash. Doing this is impossible because physical phenomena are intermingled with many others, which determine the boundary conditions required to solve the equations. This is also why high school physics does not build on our prior experiences of physical phenomena implied in our ability to make sense of and skilfully carry out a range of activities that are dependent on physical phenomena as well as others, from crawling to walking to tree climbing to manipulating toys to much more skilful activities. There are good reasons that our high school physics teachers did not build on this experience but instead started us off in a mathematical domain populated exclusively by very simple physical phenomena.

The same kinds of issues emerge when discipline-based knowledge is applied to the creation of all manner of devices. For example, designing and building a nuclear bomb is predominantly a matter of exploiting physical phenomena; however, the moment it is exploded, the after-effects soon mingle with every other imaginable category of phenomena. Although the consequences of most devices are much less powerful, they nevertheless transform everything they touch.

Studying human life and the world, one category of phenomena at a time, also has important and far-reaching implications for the conduct of experiments. Direct empirical investigation is impossible in all cases in which the category of phenomena being studied is mingled with a great many others whose influences cannot be neglected. Hence, special domains must be designed and built, which we call *laboratories*. These imitate the domains of a particular discipline as opposed to our own world. It is a highly simplified environment designed to study a small set of variables related to one category of phenomena, preferably by modifying one variable at a time. Other categories of phenomena must be kept at bay so that their influence on the experiment is negligibly small. Once again, the results are extremely useful for advancing our understanding of situations dominated by one category of phenomena, and much less so for studying all the others.

It would appear that discipline-based science is biased in favour of situations in which one category of phenomena dwarfs all others, and consequently is biased against those situations that cannot be understood in this way. The symbolization of experience by means of a culture does not have this bias, as it seeks to understand everything in relation to everything else.

Technology as Biased Doing

During the closing decades of the nineteenth century the discipline-based approach to knowing became the model for organized doing, especially in technology and industry. Until then, technological knowing and doing had been embedded in experience because they were learned by apprentices under the watchful eye of someone with a great deal of experience. This dependence on symbolizing experience first ran into difficulties in the chemical and electrical industries. What little can be observed of chemical processes does not correlate very well with what is actually happening, and nothing at all can be observed regarding electrical circuits when they function normally. The discipline-based approach to knowing was ideally suited to these industries because they arranged their artefacts and processes to take advantage of a single category of phenomena. A chemical plant is organized in terms of reactor vessels into which different chemical compounds are introduced. The product that is created from this reaction is piped to the next reactor vessel, where the next chemical reaction takes place. This continues until the final product emerges. Similarly, electrical circuits exclude all but electrical phenomena, which are intellectually modelled in mathematical domains and experimentally verified in laboratories. Germany, the first nation to develop this discipline-based approach to doing in industry, rapidly became the leading industrial power and for decades dominated the patents in these industries.

Simply put, technological artefacts, processes, and systems are built up from separate but interacting domains in which one category of phenomena is endlessly repeated to contribute one sub-function to the larger entity. An electronic device is based on a circuit in which various components use a particular kind of electrical phenomenon to produce a sub-function. These are connected by the circuit to create the overall desired effect. An internal combustion engine is based on a sequence of four different processes (intake of air, injection of fuel and combustion, expansion of combustion gases, and exhaust of this mixture) in the space between the engine block and the piston head. Any pressure on the piston head is translated into mechanical forces. These are transmitted through a linkage that rotates the crankshaft. In this way, distinct and separate domains produce the necessary sub-functions by exploiting a single category of phenomena or by a sequence of distinct phenomena. Almost without exception, every technological artefact, process, or system is designed to operate in this manner. Gilbert Simondon has argued that a more advanced approach ought to make use of multiple categories of phenomena at the same time, but his ideas have not been developed.[2]

As a result, discipline-based knowing and doing are ideally suited to technology and industry built up in this way. They are also well suited to anything reorganized in the image of a classical or information machine. For example, no mechanization could occur until work was reorganized in such a way that a machine could perform the same functions, by what is called the technical division of labour. Once this was accomplished, any production step could be assigned either to a machine or to a human being, but the latter had to work as if he or she were a machine.

Opposing Biases

Our previous discussion suggests that it is essential to reflect on the differences between living entities and machines. Because of the way the latter are organized they thrive on repetition, while living organisms are destroyed by it. A machine is built up from separate domains in which the repetition of a particular instance of one category of phenomena contributes a sub-function to another domain, and this is continued until the final result is achieved. Consequently each domain can do nothing but endlessly repeat the production of its sub-function.

If we attempted to organize daily-life activities in this way, we would encounter insurmountable obstacles. For example, the ones related to ordering a meal in a restaurant may be impinged on in many different ways. We may glimpse a friend whom we have been attempting to contact all day pass in front of the restaurant. While apologizing to the waitress, we get up and wave to attract his attention. Alternatively, the waitress may notice someone trying to leave without paying, causing her to dash off. Other kinds of interruptions may result from patrons asking the waitress a question, someone having a heart attack, a burning smell coming from the kitchen, people singing 'Happy Birthday' at another table, or the fire alarm going off. The possibilities are almost endless, with the result that it is impossible to treat the activities related to the ordering of the meal in a restaurant as a kind of domain characterized by a script. The intervening activities may cause the activities of ordering a meal to evolve in very different ways. Each situation enfolds something of a way of life. When a great many different categories of phenomena intermingle, each one adapts and adjusts to the others, with the result that when the diversity is significant, it is highly unlikely that any situation will be repeated in quite the same way. There is no possibility of coping with such situations on the basis of repetition. Everything needs to be designed to adapt to, and evolve in, surroundings in which nothing repeats itself.

In other words, in a 'world' designed and built in the image of the machine there is nothing but performance measured in terms of output-input ratios such as efficiency, productivity, and profitability. It is by these outputs and inputs that the different domains are connected together. Increases in performance of the machine thus depend on the performance of its constituent domains and the design for their integration. Everything is accomplished by exploiting one instance of one category of phenomena at a time.

However, in situations in which many different categories of phenomena mingle, with the result that each and every one contributes to what is happening, there is no possibility of anything recurring in quite the same way. There are simply too many reciprocal adaptations to too many other categories of phenomena, which means that the probability of anything ever occurring in exactly the same way is so small as to be negligible. There is order and disorder, predictability and chaos, and all this is rooted in the indivisible character of life in the world, where everything evolves in relation to everything else. Nothing is ever entirely separable from everything else without some impairment and loss of understanding. It is the very design, construction, and operation of machines that is fundamentally at odds with anything that depends on the positive and negative synergies of a great many phenomena, all making their unique contributions, with the result that repetition is out of the question. On some level we all know this, but our civilization has organized everything in such a way that this has become practically invisible. It is possible to improve the performance of any living entity by repeating one or more aspects, but this comes at a great cost to its integrality and its ability to adapt and evolve within a living world.

All this is deeply rooted in the way living beings come into existence. We are not assembled out of separate parts in our mother's womb. If this were the case, we could not adapt and evolve to everything around us. Instead, embryos develop by progressive cell differentiation. The biological whole is represented by the DNA in each and every stem cell, with the result that within the embryo each one of these cells can specialize to become a unique expression of that whole, which permits it to see, to hear, to clean the blood, and to perform all the other functions that are required to sustain the whole. There are no 'parts' in the sense of a machine. Each cell is both whole and part. It is both internally and externally connected to all the others and, via them, to the whole, which allows for ways of participation that are impossible in a machine. Moreover, all the cells in our bodies are constantly being repaired and replaced, with the exception of our brain cells, which are only repaired. All the others have a lifespan ranging from days to about seven years, with the result that, except for our brains, our bodies 'turn over' many times during our lives, and yet they remain our own. As an expression of the whole, each cell sustains, and is in turn sustained by, this whole.

Much the same development occurs on the sociocultural level. Following birth, human communities act as a social womb in which we become people of our time, place, and culture. The limited organizations of the brain-minds, with which people are born, progressively differentiate as a result of the symbolization of their experiences by means of neural and synaptic changes to these organizations. This development of our brain-minds by progressive differentiation implies that we internalize the cultural design of our community for making sense of, and living in, the world. The organizations of our brain-minds symbolically map this cultural design through the experiences of learning to participate in it. Once again, there is no question of our becoming 'parts' of a social mechanism. Growing up makes us into individually unique manifestations of a cultural whole, analogous to the relationship between ourselves and our bodies. The organizations of our brain-minds are the equivalent of the DNA of the sociocultural level. All aspects of our being a person of our time, place, and culture evolve in relation to all others, with the result that all of them sustain, and are sustained by, the community.

As we adapt to our social and physical surroundings, these surroundings are, in large measure, the result of other people also living their lives. This includes their interaction with, and modification of, our physical surroundings. As others sustain our lives we sustain theirs, and to the extent that we fail to do this we all suffer.

Much the same is true for all non-symbolic species. As unique manifestations of the biosphere, their lives are sustained by the whole represented as the DNA. The niches in a local ecosystem are the result of all other species doing the same. In the biosphere, with its enormous diversity of situations created by the mingling of numerous phenomena, nothing ever repeats itself in quite the same way. There is a reciprocal adaptation and evolution of all species to all others, and it is this reciprocity that sustains all life. It accounts for the incredible resilience of all organisms and living entities. Each and every life form is constantly sustaining all the others by adjusting to them, as they do to it. No engineered system can come anywhere close to having this kind of resilience; such a system is in the business of repetition, which is the exact opposite of adaptation and evolution.

We are now beginning to discover that our civilization has taken all this too much for granted during the last two centuries. The situation is somewhat analogous to people losing their short-term memories, which interferes with each experience being symbolized by neural and synaptic changes to the organizations of their brain-minds. Each experience can no longer be symbolically related to all the others and thus become a moment of their lives. Their lives can no longer symbolically evolve, because the organizations of their brain-minds cannot adapt. Consequently, these people can no longer participate in a conversation, because they cannot remember what was said before. Nor are they able to make sense of a story they read or a movie they watch. Their lives are disconnected in space and time as well. When these people are taken into a building, they will not remember how they got there, with the result that unless they knew the building from before the onset of the disease they will be totally lost. In the same vein, they have no idea of the time when they last ate, or did anything else, for that matter. With the onset of short-term memory loss, the limited ability of these people's lives working in the background becomes painfully obvious.

Much the same argument can be made for our interference in the processes of the biosphere. We experience the results as a loss in the biosphere's capacities to sustain all life. However, we generally treat the environmental crisis as a phenomenon in its own right rather than treating it as a symptom of the malfunction of our discipline-based approach to knowing and doing. This is particularly evident in the way our universities deal with it and in the influence this has on governments and corporations.

Some Consequences

Since our civilization values the discipline-based approaches to knowing and doing over all alternatives, its successes and failures can readily be explained. Much of individual and collective human life is made up of situations to which many different categories of phenomena make non-negligible contributions. When any discipline-based approach is applied to these kinds of situations, it will abstract those phenomena that belong to the category in which it specializes in order to place them in a domain exclusively populated by this one category of phenomena. It thereby gains the advantages of comparative approaches at the expense of getting a grip on how these phenomena contribute to the adaptation and evolution of the situations in which they participate. As a result, the application of any discipline-based knowledge will necessarily multiply the tensions in any living world (natural or cultural), thereby revealing its anti-life bias. Discipline-based approaches will not produce these problems wherever individual and collective human life has been reorganized in the image of classical or information machines. In other words, discipline-based approaches can improve what a particular phenomenon does in comparison to all other phenomena of the same kind without being able to evaluate how this affects the context to which this particular phenomenon contributes. Without intending to do so, our civilization has ended up with approaches that deliver performance by tearing the fabric of relations integral to all life.

From the outset, our civilization was unaware of the anti-life bias of its discipline-based approaches, and this largely remains the case. Science was not perceived to have any limitations in the domain of knowing, nor technology in the domain of doing. Culture-based designs for making sense of and living in the world were replaced by countless decisions supported by discipline-based knowing and doing. Customs and traditions all but disappeared. The consequences have been devastating. For example, when economists study the economic phenomena within human life and society, they necessarily behave as if these phenomena dominated all of life, which implies that, in essence, life is economic. All the other phenomena can be neglected, or understood in terms of economic phenomena. They will inevitably be biased against situations in which economic phenomena contribute very little or where other phenomena are much more important. The pinnacle of this bias has been reached by the Chicago School of Economics. It has essentially decreed that markets, rates of unemployment, and *homo economicus* are 'natural' and thus beyond our responsibility.

In contrast, Keynesian economics essentially softened the anti-life bias by insisting that economies ought to serve their communities, not the other way around. It sought to understand how economic phenomena contributed to individual and collective human life and how these phenomena should serve the public good.

When Keynesian economics was displaced by monetarism, many governments inflicted devastating effects on their people. Many citizens were needlessly denied meaningful participation in their (economically oriented) societies, and a great deal of wealth was transferred from the many to the few. Trapped in their disciplinary silos, economists had no answers to one of the greatest recent financial crises and failed to notice the mutation of economies into anti-economies. As a consequence of the discipline-based approach to economic phenomena, tensions between these phenomena and a living world multiplied. Rapidly rising costs quickly outstripped wealth creation, with the result that net wealth has remained flat for many decades or has substantially declined depending on where people were located in the socio-economic hierarchy. Although the influence of monetarism is weakening somewhat, many governments continue to behave as if human life were all about economic growth. This idea would have been totally absurd to many people were it not explained to them by the mass media.

In the same vein, we are endlessly told that technology is neutral and that its influence on human life, society, and the biosphere is the result of its use, as opposed to its structure and its intermingling with other phenomena. From the above brief exploration it would appear that such a position is untenable. Science, technology, and the economic growth achieved with them are hardly neutral and certainly not objective. Our ways of life have introduced a bias in favour of everything technological. We succeed brilliantly in improving the performance of everything that has been reorganized in the image of the machine and fail equally spectacularly in ensuring that it is adapted to, and able to evolve with, everything else. All this has become very destructive of all life.

The discipline-based technical approach has all but replaced the symbolic cultural approach in almost every sphere of human life. It proceeds by treating everything we wish to improve as a domain constituted of a limited number of variables and by neglecting all the others. However, doing so is scientifically acceptable only when everything else remains the same (in which case it does not evolve and is therefore assumed to be dead), when everything else repeats itself (in which case it is assumed to be a dead machine or a technical system), or when everything else may be neglected (in which case it is devalued to the point that it might as well not exist). Its widespread use is unscientific in most cases, with the result that these assumptions reveal the anti-life bias of our modern ways of life and the nihilistic orientation of our civilization. Max Weber examined the embryonic beginnings of this process, which he called *rationality*. Decades later, Jacques Ellul examined it as the phenomenon and system of technique.[3] These developments are characterized by the drive for efficiency based on reorganizing everything in mechanistic or informational terms by using a discipline-based approach. The technical approach builds a technical order separated from the cultural order that has evolved on the basis of experience and culture.

This technical order of non-sense is destructive of life in four important ways. First, discipline-based expertise is separated from human life and the world through a triple abstraction. As an example, consider what happens in our hospitals. Since many different phenomena are involved in their operations, there is no discipline that corresponds to hospitals. To bring discipline-based knowing and doing to bear on their operations, hospitals must be abstracted from the world, which is replaced with the 'inputs' of sick or injured people received and the 'outputs' of discharged patients returned to it. Before anyone can participate in the healing process that transforms these inputs into outputs, they must further abstract from it those aspects commensurate with their disciplines or specialties. Doctors, nurses, physiotherapists, nutritionists, psychiatrists, social workers, information systems specialists, administrators, accountants, maintenance engineers, public relations specialists, and security consultants all know different aspects according to their disciplines. Since these disciplines exclude how the running of a hospital interacts with everything else, specialists cannot decide between alternative courses of action in response to an issue on the basis of what is best for human life, society and the biosphere. They cannot use the values of their culture from which they have abstracted all they could – namely, only the part of the healing process related to their discipline and whatever interventions they return to this process – and this can be measured only in terms of output-input ratios abstracted from human values. As a result, the discipline-based division of labour in hospitals carries on as if these hospitals were organized in terms of separate and distinct domains in which one category of phenomena, corresponding to a single discipline or specialty, contributes a subfunction first to the healing process and then, via it, to the functioning of hospitals. Everything these discipline-based specialists do is thus separated from human life and society via a triple abstraction, and their collective efforts build a technical order that evolves without making any reference to sense.

A second important consequence of discipline-based approaches results from the fact that specialists suspended in this triple abstraction cannot intellectually 'see,' let alone deal with, most of the negative consequences of their decisions because these fall beyond their domains of expertise. They cannot symbolize their professional experiences in relation to everything else, with the result that they are unable to see out of their disciplines (justly referred to as *silos*) in order to adjust their decision making to achieve their goals and at the same time prevent or greatly minimize undesired and harmful effects. The daily-life equivalent would be to train people to drive their cars by having them concentrate on their performance as indicated by the gauges on the dashboard and only occasionally glance out of their windows when they hear a loud noise. Consequently, the many undesired and harmful consequences of discipline-based decision making must be dealt with in an end-of-pipe fashion. Additional goods and services must be created to compensate or mitigate these effects, and this has become so expensive that we have all but given up on effective regulation. It has also transformed wealth creation into wealth extraction.

The third consequence makes matters even worse, because all that the discipline-based approach to knowing and doing can do is improve the performance of everything. When genuine solutions require prevention by better adaptation and evolution, this approach to knowing and doing is structurally incapable of providing them. For example, the gridlock in many cities will not be resolved by endlessly optimizing the carrying capacities of our transportation systems. In addition to 'supply side' approaches that increase carrying capacities, 'demand-side' approaches are essential to reduce our need for mobility. These transcend the usual disciplines. In the meantime, most discipline-based solutions lead to further deterioration in the compatibility between people, their transportation needs, urban forms, and the biosphere.

Finally, individual specialists and their collective efforts through the current intellectual and professional division of labour behave as if human life and the world were organized in the image of conventional or information machines, meaning that they are built up from distinct and separate domains, in which one category of phenomena contributes a sub-function. If we ever needed evidence that human life and the world are not organized in this way, it is surely furnished by the mushrooming human, social, and environmental crises of our time – provided we understand that they are the result of the common bias in discipline-based knowing and doing.

Where Do We Go from Here?

If the above embryonic diagnosis of how we have backed ourselves into a corner during the last two hundred years has any merit (and all of this will be developed in subsequent chapters), what should we do? It goes without saying that we cannot abandon specialization. It is a question of inventing a different *form* of specialization (other than interdisciplinarity, multidisciplinarity, transdisciplinarity, wholism, and systems thinking). The search for a viable alternative to a discipline-based intellectual and professional division of labour will be explored on two levels. First, it is essential to recognize how we continue to depend on two parallel and interdependent modes of knowing and doing, even though we have completely devalued the mode based on symbolization, experience, and culture to the benefit of the discipline-based approaches. Since the strengths and weaknesses of the approaches to knowing and doing based on symbolization are opposite to those of the approaches based on disciplines, a synergy between the two can be created in almost any discipline or specialty. The strengths of one kind of approaches will then cover for the weaknesses of the other kind. Second, discipline-based approaches can be modified by examining the typical undesired consequences that flow from the practice of a particular discipline or specialty, consulting the disciplines that study these effects, importing some of this knowledge into the original disciplines or specialties, and using it in a negative feedback mode to adjust decision making to obtain the desired results, but at the same time preventing and greatly minimizing harmful effects.

There have been two precedents that sought to implement this kind of reform in Canada. Under the former Premier's Council of Ontario, a round table was organized to explore the potential of preventive approaches by modifying the education of practitioners, based on the above diagnosis. It was terminated in its early phase by a newly elected government believing in Market solutions. On the federal level, in 2003 the Natural Sciences and Engineering Research Council and the Social

Sciences and Humanities Research Council of Canada proposed Society, Technology, Science 21 (STS21) to redirect research and teaching in Canadian universities, in part based on the above diagnosis and prescription. Once again, the winds of economic fundamentalism helped sweep it all away. As the absurdity of monetarism is gradually dawning on more and more people, the time will hopefully come soon when we will be ready for a change in direction. However, recent developments in the United States ought to make us very cautious about the prospect of genuine change.

Even when significant opportunities present themselves, the road to genuine change is a rocky one. We conduct ourselves as if we lived in a secular age, that is, as if the forces of morality and religion no longer presented any obstacles. The secular religious forces that replaced the traditional ones have thus become invisible, but they have done the same kind of devastating work in the background. We live as if science were omnipotent in the domain of knowing. Otherwise, we could have recognized its limitations decades ago and searched for alternatives. The same is true for discipline-based doing. These approaches to knowing and doing have not been recognized for what they are: human inventions that are good for certain things, harmful to others, and irrelevant to still others. If we had known, we would not have backed ourselves into our present corner. In the meantime we do not allow parallel modes of knowing and doing to flourish in our universities, and certainly not in our courts. There is nothing new under the sun. While congratulating ourselves on having come of age, we misinterpreted the deposing of traditional gods, who were replaced by more powerful ones.

How We Shall Proceed

The chapters that follow will argue that our civilization is undermining what until now, in the human story, has made us what we are: a symbolic species. Simply put, to symbolize human experiences by means of a culture is to dialectically organize the ways in which everything is related to everything else in the world, and desymbolization is the undermining of these abilities. Hence, human journeys guided by cultures with low levels of desymbolization result in everything they create being highly compatible with everything else, just as we observed in the evolution of the biosphere. In contrast, human journeys that are guided by highly desymbolized cultures result in everything being much less compatible with everything else, the symptoms of which are inappropriate technologies and unsustainable ways of life. When we acknowledge that our cultures are highly desymbolized, it will be difficult to continue to behave as if phenomena such as global warming, peak oil, pollution, and resource depletion are disturbing and possibly life threatening but solvable. Jointly they are the symptoms of a path of development that has been permeated with an anti-life bias. We continue to proceed as if these issues can be solved one at a time by more science and technology. As a result, there continues to be a proliferation of '-isms' that further manifest the high levels of desymbolization of our cultures. We are simply unable to connect the dots and treat these issues as inseparable symptoms of our anti-life bias. It produces the ultimate technical bluff: that what our contemporary ways of life have to offer us is worth giving up what has made us cultural beings.

With hindsight, it is apparent that desymbolization and its effects on human life have had precursors that resulted in warnings being issued by a number of scholars. Some of these have turned out to be prophetic. Adam Smith warned that the technical division of labour would produce a new wealth of nations but that it would also make human beings as stupid as they could possibly become.[4] Karl Marx showed that the capitalist system enslaved the rich and the poor alike.[5] Max Weber warned that humanity was shutting itself into an iron cage.[6] John Kenneth Galbraith deplored the fact that we were serving the system we created to serve us.[7] Jacques Ellul warned us against the autonomy of what he called the system of technique;[8] by this he meant that the influence of this system on people and communities had begun to outweigh the influence of people and communities on it.

These warnings can readily be understood provided we recognize that being a symbolic species transforms our relations with our surroundings (both physical and social). These relations become reciprocal because each and every experience of these surroundings modifies the organizations of our brain-minds, as it is symbolically placed within our lives. Hence, as we affect our surroundings, they simultaneously affect us. We are thus internally and externally connected to these surroundings as a consequence of being a symbolic species. However, although we directly experience how we affect our surroundings, we do not directly experience how these in turn influence us. Paying critical attention to this latter interaction will fundamentally change our perception of the former interaction.

The concerns of the above authors are shared by the present study. Human slavery (the equivalent of sin in the Jewish and Christian traditions and of alienation in the social sciences) is an unacceptable form of human life. Our freedom is threatened when the influence we have on our surroundings is dwarfed by the influence these surroundings have on us. There can be no pretence of objectivity in this matter. The tensions between freedom and alienation are at the heart of our analysis of our being a symbolic species and of our desymbolizing activities.

In the first chapter, we shall revisit the issue of humanity being a symbolic species, since this is fundamental to understanding what little remains of cultural designs for making sense of and living in the world. No matter how highly desymbolized these designs have become, babies and children continue to be utterly dependent on them in order to become members of our symbolic species. To a significant degree, we all continue to go about our lives in the biosphere in a manner unlike that of the animals. All human experiences and relations are symbolized by means of a culture. Such a culture includes everything that a community creates and passes on from generation to generation and which is not the result of the processes by which the remainder of the biosphere adapts and evolves. From the perspective we are developing, not all cultures were born equal. Most of them created ways of life that gave meaning, purpose, and direction to communities without threatening local ecosystems. Their languages had no concepts equivalent to 'appropriate technologies' or 'sustainable ways of life,' because all this could be taken for granted most of the time. Being a symbolic species is therefore not likely to be the underlying issue. Instead, we may be faced with fundamental differences between our cultures and those that preceded us. However, we shall begin with an examination of what we continue to hold in common, before embarking on a study of desymbolizing processes in later chapters. In this way, it will become apparent that our cultures are likely to be among the most highly desymbolized ones on which humanity has ever relied.

In the second chapter we shall begin to explore the desymbolization of contemporary cultures through some of the experiences of babies, children, and teenagers in contemporary mass societies. We shall discover that the anti-life biases of our civilization have many undesired and unintended consequences, resulting in desymbolization. The chapter begins with the development of a conceptual framework for examining the desymbolizing effects on human life as a consequence of relations that restrict the process of resymbolization. It then turns to examining the reciprocal relations that children have with television and with computers. It concludes by examining what happens when teenagers are introduced to scientific disciplines in high school. In each case, the desymbolization advances a little further, thus socializing each new generation into what, in the following chapters, will be seen as highly desymbolized cultures.

The third and fourth chapters will examine the forces of desymbolization unleashed on human life by the process of industrialization. One of the roots of the present technical order goes back to the reorganization of human work in the image of the machine by means of the technical division of labour. It marked the beginning of the reorganization of human life in the image of non-life. Another root goes back to the necessity of creating an economic order capable of organizing the growing throughput of matter and energy and their composites required by industrialization. Until then, the exchanges of matter and energy necessary to sustain the way of life of a society were usually enfolded into the patterns of kinship and other social bonds that constituted the social fabric of a community. The new forms of economic organization began to constitute orders of non-life because they depended on the Market (made up of many markets), ruled not by a culture but by mechanisms of supply and demand jointly acting as an invisible hand. The Market was humanity's first universal institution operating at arm's length from the cultural order of any society.

The economic orders of the industrializing societies could not be easily integrated into their cultural orders or the natural orders of their ecosystems, because these were the orders of life. The tensions between them were further aggravated when these economic orders began to run into serious limitations that necessitated their mutations, first into rational orders and subsequently into technical ones. These orders could not fit into the cultural orders any better than could their economic predecessors, for the same reasons.

These transformations were driven by attempts to overcome the limitations of culture-based approaches to knowing and doing and by the desymbolizing pressures that industrialization put on all cultures. The successful response to overcoming these limitations was an entirely new approach to knowing and doing, based on disciplines. These disciplines took elements from a living world and transformed them in a way that made no reference to sense, thus creating an order that lay outside the cultural order but could not operate without it.

The discipline-based approach to knowing and doing helped build a universal technical order, which became increasingly dependent on technical information. The result was an information explosion that, in turn, created a need for, and the diffusion of, a new kind of information machine. In order to use computers, intellectual work had to be reorganized by means of a technical division of labour, with some tasks delegated to these machines and others to people. However, people would have to do their work in the image of these machines, as had been the case with mechanization and industrialization. In addition, these machines had to be connected to an overall information system, creating a kind of two-dimensional (both horizontal and vertical) assembly line tended by information workers. These developments further increased the tensions between the orders of life and non-life.

Everything in our contemporary world testifies to this opposition. Virtually all our accomplishments are in the domain of increasing the power and performance of any process or activity, while almost all our failures are related to our inability to ensure that all these more efficient and more powerful elements are individually and jointly compatible with human life, society, and the biosphere.

Computer-based information and communication technologies have greatly strengthened the technical order at the expense of the cultural and natural orders. All the operations within a computer are built up from zeros and ones, and each and every process would fail to accomplish its mission if it could be affected by everything stored in what we refer to as its 'memory.' In this respect, the computer represents the ultimate triumph of performance over context. Anything dialectical (by being both zero and one, for example) or enfolded into other entities (another way of being both zero and one) cannot be represented by a computer. Even when a computer is used to manipulate symbols, it does so without any regard to the dialectical context that makes them what they are, thereby destroying their symbolic capacities. Moreover, any work-related activity carried out with the aid of a computer must first be reorganized by means of the technical division of labour, which transforms it into the image of this information machine. As such, these new technologies have been instrumental in extending the enormous influence that the transformation of human work in the image of the machine has had on the economy and, via it, on human life, society, and the biosphere.

Biotechnology and nanotechnology represent a kind of ultimate experiment: seeing to what extent discipline-based approaches to knowing and doing can incorporate elements of living orders into a technical order or include parts of a technical kind into living orders to produce new efficiencies. There is little doubt that this mixing of living orders with non-living orders will produce the same kinds of results that we have seen during the last two centuries. There will be spectacular gains in efficiency and equally devastating pollution of matter and the DNA pool. By this definition, pollutants are foreign bodies that cannot be absorbed into the order into which they are launched and in relation to which they have a disorganizing effect.

The perspective we shall develop in chapters 3 and 4 concerns the pollution of human cultures as a consequence of the emergence of an economic order followed by a technical order. These orders create 'foreign bodies' that literally belong nowhere and which have disordering effects on the living orders into which they are released. In a multitude of ways they interfere with the reproduction and evolution of order from order.

Kierkegaard criticized classical philosophers as people who think in certain categories but live in others. He likened this to building a large castle and living beside it in a hut.[9] Today, the same criticism could be addressed to all scientific and technical specialists, whose lives depend on two parallel modes of knowing and doing. The potential complementarity of these two modes has been destroyed by our contemporary secular myths.

The concluding chapter will examine how the relationship between the two parallel modes of knowing and doing can be reversed, leading to a transformation of various disciplines, and how in turn these transformations can be supported by university reforms. It must be emphasized that ultimately this involves a struggle against our secular myths, which continue to inflict so much harm on humanity. It will be a struggle to reclaim our symbolic abilities and to compensate for the limitations of these abilities by transforming discipline-based approaches to knowing and doing. It can be a contribution towards the ongoing struggle for a liveable and sustainable future.

1 Symbolization: Getting in Touch with Ourselves and the World

Living a Life

On some level we are all aware that we are able to live our lives – an awareness that implies a measure of inseparability and indivisibility of the moments of our lives. The extent to which this is the case is impossible to know with our current approaches to knowing and doing, which deal with these lives one category of phenomena at a time. A symbolic species cannot be adequately understood and sustained by means of a discipline-based intellectual and professional division of labour. The approaches to knowing and doing based on symbolization, however, imply that everything in human life and the world is related to everything else. Without this inseparability and indivisibility symbolization is impossible.

The indivisibility of a human life may be regarded as an evolving dialogue between all the experiences resulting from our relations with others and the world, our reflections on them, and our imagination of what might have been. In a society, such overlapping and interpenetrating dialogues are expressions of who we are as people of our time, place, and culture. This indivisibility of our lives is extremely fragile. To begin with, our lives embody many tensions and contradictions. Suppressing any of these prevents their participation in these dialogues and may even lead to mental illness. To prevent them from dissolving into chaos, every moment must be differentiated from all the others and integrated into the whole. Inadequate differentiation could blur differences that may be essential to our lives or to the community to which we belong, thereby curtailing the evolving dialogues. However, excessive differentiation could turn our lives and our world into unintelligible complexities, thereby making the living of our lives impossible. These dialogues act as interacting ecologies of experiences, feelings, emotions, intuitions, and thoughts and thus need to be protected from relativism, nihilism, and anomie by means of an absolute point of reference. Within the dialogues, any rules, algorithms, and everything built up with them must play a limited role because they represent sets of relations to the exclusion of all others, and thus to the exclusion of our lives lived as members of a symbolic species. We shall gradually show how many of the difficulties faced by our civilization are rooted in our failure to appreciate the potential complementarity between what is dialectical and what is rational in our lives and communities.

The indivisibility of our lives must therefore be sustained by the organizations of our brain-minds, where the dialogues between all our moments are symbolically represented. These dialogues are both conscious and metaconscious, individual and collective, cultural and historical. They are the common ground on which we stand and the cultural landscape within which human life evolves. They can sustain nations that take great pains to protect their most vulnerable and powerless members, and also those nations who do the exact opposite. Every aspect of our lives must be understood against this background. The present chapter will attempt to do so for individual and collective human life with low levels of desymbolization. These were common prior to industrialization.

Consider a moment of a person's life in which she responds to what someone else is saying to her. What she says to him involves processes in her brain-mind that are transmitted to her vocal cords. These generate pressure waves in the air that are detected by his ears and transmitted to his brain-mind, where other mental processes permit him to make sense of what she is saying. The mental processes of both people are dependent on the acquisition of a shared language and culture.

The verbal response is framed by what is detected by the other senses. The chain of phenomena associated with seeing contributes what is expressed by facial and body expressions, including an eye etiquette, a conversation distance, and gestures that help to convey certain emotional colourings. Smell, touch, and taste may contribute additional chains of phenomena associated with these senses, although these may be restricted to intimate relations. Hence, the response can involve up to five chains of phenomena associated with the senses, all of which connect the two people. While some of these phenomena operate within the reach of their conscious minds, most of them are non-conscious. The latter may be regarded as working in the background,[1] but their unconscious influence is very different from much of what was envisioned by Freudian psychology. Regarding the non-conscious components of human relations as the work of an adaptive unconscious[2] is closer to the mark, but it still invites many problems, rooted as this concept is in the information processing carried out by some kind of advanced computer.

These chains of phenomena cannot be divided into the links that lie within someone's conscious awareness and those that lie beyond it. Doing so would create a new dualism between the conscious and unconscious minds and between the mind and the body. Such a distinction would imply that an adaptive unconscious presents the conscious 'inner self' with experiences. Any kind of information processing by any type of computer has an 'output' that is presented to an external person. Without revisiting the debates in artificial intelligence and (before that) in philosophy, a change in perspective is proposed.

The previously mentioned problems may be avoided by recognizing that we are a symbolic species, with the result that our lives have a high measure of indivisibility. To begin with, the above five chains of phenomena corresponding to the five senses are symbolically interrelated in the brain-minds of the two people involved. In this way, what is being said takes on additional significance through what is detected by the other senses, and vice versa. Each of their senses thus contributes a dimension of experience. Furthermore, the 'internal worlds' of the two people may contribute additional aspects or dimensions to their experiences of the conversation. They may become aware of getting rather hungry or of their legs bothering them from standing too long. It is not simply a matter of many chains of phenomena working in the background but a matter of their bodies and lives doing so. Everything in their lives is symbolically interrelated to everything else by the organizations of their brain-minds. These interrelate all the chains of phenomena that connect these organizations to the two people's external and internal worlds. At the same time, these experiences are transformed into moments of their lives, because they are symbolized by neural and synaptic modifications to the organizations of their brain-minds, which includes the symbolization of all previous experiences. Each new experience is thus symbolically related to all others and placed into the life of a person, thereby transforming the experience into a moment of that life. Its meaning and value relative to all the other experiences, and hence for that life as a whole, is thus symbolically implied in the organization of the person's brain-mind. As a result, it is not only a person's body that is working in the background but a person's entire life lived as a member of a community in a particular time, place, and culture.

What we are suggesting is that when a human life symbolically and dialectically relates everything in that life to everything else, a complex fabric of relations is created that merges the inner world with the outer world. When a person's attention is drawn to something, some of this fabric of relations becomes the foreground of that life, and the remainder the background. Since this fabric of relations includes the social self of a person, there are no outputs presented by an adaptive unconscious, because all unconscious relations prepare for, and are implicit in, a person's experiences and a life lived in relation to a world. There is no separate memory of the kind we find in computers. Everything symbolized in a person's organization of the brain-mind evolves in relation to everything else because the details of the organization of the brain-mind symbolize particular experiences, while the whole symbolizes the person's entire life. There is no dualism in any foreground-background distinction or in multiple levels of symbolization, where each level comes closer to the living of a life.

The living of a human life thus transforms the world into a symbolic universe. Everything observed is symbolically mapped into a person's life, thereby revealing its meaning and value relative to everything else, and hence for that life. Before something useful can be made from anything in a person's surroundings, it must be symbolized as a constituent of these surroundings and as something that can be fitted into people's lives. Here, the evolutionary process of the human being diverts sharply from that of all the other animals. It is as much a question of a symbolic species adapting to its surroundings as it is of such a species adapting these surroundings and their constituents according to their meanings and values for people's lives. In this way, a sphere of human potential is symbolically inserted between a symbolic species and its surroundings, which distances the species from their surroundings to create a limited sphere of freedom. Insofar as people carry out symbolic acts in relation to other people, symbolization is inseparable from the creation and evolution of cultures. In all of this, nothing can be accomplished in a piecemeal fashion; everything is always related and evolved in relation to everything else, which includes the binding together of individual lives into communities.

Much of the living of a human life is metaconscious, in the sense that we can at best recollect specific experiences but we have no access to how the organizations of our brain-minds transcend all of them. Without going into any detail about how the brain functions biochemically and the mind symbolically, such a going beyond individual experiences to evolve the organizations of our brain-minds develops a great deal of metaconscious knowledge. Much of human behaviour may be understood by introducing the following analogy. Neural and synaptic changes resulting from the symbolization of individual daily-life activities relate to the organization of the brain-mind in a manner analogous to the way in which the data of a scientific experiment relate to the curve fitted through them. Just as our confidence in the experimental design is strengthened when we are able to fit such a curve, so it appears that our grip on our lives in the world is strengthened by the organization of our brain-minds working in the background. In other words, instead of rejecting the interpolation and extrapolation of the experimental data as an unscientific way of going beyond the evidence, we accept it as a confirmation of all our scientific experiences of the behaviour of our world. In the same way, the metaconscious knowledge implied in the organizations of our brain-minds confirms our lives in the world.

The metaconscious equivalent of the processes of interpolation and extrapolation is the symbolic processes of differentiation and integration, and the equivalent of the fitted curve is the metaconscious knowledge generated by the latter processes. A distinction may be made between the *metaconscious* knowledge enfolded in the organizations of our brain-minds and the *unconscious* knowledge enfolded in the organizations of our brains at birth and in the repressed experiences that no longer participate in the organizations of our brain-minds in the usual ways. I shall briefly revisit these models of the processes of differentiation, integration, and the generation of metaconscious knowledge, which have been developed in an earlier work.[3]

Before we get started, a brief note of caution is in order. Our world and our lives may be divided into two orders: the technical order, accommodated to machines in general and computers in particular, and the cultural order, created to sustain human life. The cultural order is so permeated by our experiences of the technical order that we have become quite confused about the differences between them and what happens in our lives. The very possibility of symbolization producing meanings and values for human lives has been dismissed by many as an unscientific idea. Furthermore, we use words such as *memory*, *information*, *knowledge*, and *expertise* for both machine functions and the human aspects of our daily lives. Yet, some differences are obvious.

When a computer performs a particular operation, whatever else is stored in its memory is not working in the background, while in the case of our lives it always is. When our memories can no longer work in the background, or do so partially, our lives become drastically transformed, as in the cases of people with short-term memory loss or the early phases of Alzheimer's disease. In contrast, if everything in our computer's memory worked in the background, the machine would be useless.

We bring a great deal of metaconscious knowledge to everything we think and do. It is one of the ways in which our lives work in the background. The role of metaconscious knowledge in human life has been widely recognized. For example, in science its presence has been identified by Michael Polanyi as tacit knowledge[4] and by Thomas Kuhn first as a paradigm and later as a disciplinary matrix.[5] Stuart Dreyfus has shown that human skill acquisition increasingly produces and relies on metaconscious knowledge, which he called *intuitive knowledge*.[6] My own research has identified many forms of metaconscious knowledge acquired by babies and children, including the conversation distance at which to stand when talking to others, an eye etiquette for keeping conversations on track, a design for arranging their lives in time and space, values for orienting every daily-life activity, their social selves, the selves of significant others, the social roles of strangers, a 'mental map' of the way of life of their society, and the myths that sustain the historical journey of their community.[7]

All this and much more is made possible by each moment of our lives being symbolized by neural and synaptic changes to the organizations of our brain-minds. As a result, our experiences are metaconsciously merged into our lives, permitting these lives to be enfolded into all our deeds and thoughts. The organizations of our brain-minds represent the equivalent of interpolating and extrapolating the experiential data of our interactions with one another in the world, thereby creating patterns or metaconscious knowledge that can be intuited. In this way, our 'memories' work in the background during each and every moment of our lives, and since the organizations of our brain-minds are built up with the experiences of our lives in a community, they bear the imprint that makes us individually unique and typical of our time, place, and culture.

In order to explore this reinterpretation of the role of the unconscious in human life, it is essential to transcend the discipline-based approach of the social sciences. Their findings must be reinterpreted as but one dimension of people living their lives. Much of the literature deals with laboratory experiments or situations described on surveys that have little or no meaning for the subjects. They are not relevant for the organizations of their brain-minds, thus permitting their lives to only minimally work in the background, which distorts the situation, creating the well-known 'garbage in, garbage out' scenarios. Little can be learned about the living of people's lives in that way. Those laboratory experiments or survey questions that do have significance for the subjects reveal something about human life, but the findings of the various disciplines must be interpreted in relation to one another and jointly in relation to human life in the world. This task is both difficult and heretical to many people, but we shall nevertheless attempt it, beginning with newborns.

Babies Getting in Touch with Themselves and the World

After birth, babies continue their development in the social womb of their community. They learn to eat its food, smile its happiness, cry its troubles, play in its world, communicate in its language, and live by its ways and spiritual orientation points. Given the considerable differences that have always existed between cultures, little of this is inborn. Babies must learn to transform the chaos initially detected by their senses into the cultural order of their community. Newborn babies confront a blur of stimuli from within and without, to which they can relate by means of very few innate abilities. However, doing so immediately involves an orientation of the brain, resulting from humanity having become a symbolic species.[8] What this means in practical terms is that human brains are capable of relating anything they detect by means of the senses and nerves to anything else detected at the same time as well as to anything detected previously. Whatever is detected is symbolically represented by neural and synaptic changes, thus gradually transforming human brains into brain-minds. The organizations of the brain-minds of babies include symbolic representations of everything they have related to in their lives in a manner that (symbolically) relates everything to everything else. As a result, anything from within or without that is merely a random stimulus cannot be given a place in their lives and thus cannot be experienced or lived. Otherwise, babies' lives would quickly be swamped by randomness and would descend into anomie. At this point in their development, the organizations of their brain-minds work in the background to protect them from being overwhelmed by the blur of stimuli from within and without and to place within their lives everything they have been able to make sense of and relate to. Their lives are sustained in such a way that they will not learn to live as if everything were simply noise and randomness without any possible meaning, direction, or purpose. Within the inner and outer blur of stimuli they begin to build their lives in relation to a world, even though in the very beginning this is exceedingly embryonic. All this is difficult to imagine and even more difficult to put into words. However, I shall venture to revisit a few details.[9]

After birth, babies are confronted with a blur of new stimuli from within and without, from which they draw those elements that can take on some meaning as a result of an innate ability to relate to them. Initially, the feelings that they will later name as being hungry or fed, being comfortable or dirty, being lonely or loved, and a great deal else come and go, and this coming and going is associated with changing visual blurs, noises, skin pressures, mouth movements, and odours. Owing to the ability of their brains to associate these rudimentary stimuli with each other, an experience with an equally rudimentary meaning can be derived from such associations. As a result, the rudimentary stimuli detected by the senses become signs of the rudimentary experience of 'something out there mostly making something else better.' The more often any pattern of rudimentary stimuli, feelings, and associations between them are detected, the more these signs take on a meaning related to 'something out there,' to the point that eventually babies can react to the sounds of something out there before anything else is detected by their other senses. Similarly, the 'something out there' becomes a sign of 'something else' that somehow can affect the sensory signs and the feeling signs. In this process, some signs are transformed from icons (having a direct correspondence to what is signified) into indices (that do not have direct correspondence to what is signified) on the way to symbol formation.

Everything that newborns discover about themselves and the world is relational. They learn something about the visual, oral, and tactile blurs of the stimuli of their mothers relative to the blurs of the stimuli of their inner world resulting from feeling an empty or full stomach, from feeling cold and wet or warm and dry, feeling bloated before being burped, and better after, and a general feeling of contentment when being played with or of discontentment when alone. At the same time, they learn something about these feelings of their inner world relative to the blurs of stimuli from the outer world. Although they as yet have no awareness of their physical selves and make no distinctions between these selves and their surroundings, this does not impede their ongoing differentiation of these kinds of embryonic experiences. What they are discovering are not factual details about themselves and the world but a dialectical ordering of these blurs of stimuli in relation to one another. Simply put, whatever the visual blur of stimuli of their mothers might mean is what all the other blurs of stimuli they have learned to differentiate from it do not mean. The same is true for all the other blurs of stimuli that their brain-minds have detected by differentiating them from all others. Out of a chaos of undifferentiated stimuli from within and without emerge blurs that are associated with one another in a way that has some meaning for the lives of these newborns. As their exposure to these blurs of stimuli is ongoing, their brain-minds continue to refine the ways they are differentiated from each other, on the way to becoming signs of something in their lives. In speaking of their brain-minds, there is the suggestion not that newborns are engaged in a kind of inborn information processing but instead that they have as yet no awareness of their physical or social selves.

In this way, the organizations of babies' brain-minds grow by means of the neural and synaptic changes that symbolize blurs of stimuli whose individual meanings are refined by differentiating them from all others, thereby creating a symbolic and dialectical order within their lives in the world. The expansion of this order constitutes a symbolic context within which new blurs of stimuli can begin to take on a meaning, thereby growing the organizations of their brain-minds, which in turn permits additional blurs of stimuli to do the same, and so on. As a result, the organizations of the brain-minds as well as what babies have learned to make sense of in terms of their inner and outer worlds co-evolve by a process of differentiation. New elements first appear out of the background of undifferentiated stimuli, and these elements may break up into one or more others when further distinctions are learned. In other words, the growth of the brain-minds of newborns is somewhat comparable to the preceding biological development of their embryos resulting from ongoing cell differentiation. All this will become clearer as we develop this working hypothesis.

Careful attention must be paid to how the ability of the human brain to relate everything that is detected to everything else transforms the physical embodiment of babies in the world. For example, their innate ability to follow movement does not become associated with their ability to focus the eyes until it is learned that within the visual blur of undifferentiated retinal stimuli there is something meaningful 'out there' on which to focus. Their innate ability to detect noise does not lead to a deliberate turning towards a source until it is learned that this source out there has meaning. Their innate ability to suck is transformed into more complex behaviour when it becomes a sign. In other words, innate abilities are launched towards symbolization as a consequence of being related to everything else. Owing to the physical embodiment of babies into the world, some of the visual blur detected as retinal stimuli eventually becomes the visual dimension of the symbol of mother. Similarly, from among all the noises they detect as aural stimuli, they gradually begin to differentiate the sounds that will become the aural dimension of the symbol of mother. Similar developments occur in the other dimensions of experience associated with touch, taste, and smell. Within the unknown, detected as 'noise' of undifferentiated stimuli in these five dimensions of experience, a foreground is created of interrelated signs of something out there that feeds, cleans up, hugs, and plays with 'something else.'

These developments are mentally related to many others. Although newborn babies spend a great deal of time sleeping in their cribs, when they are awake they may be wiggling their bodies without knowing what it is they are doing. They have as yet no awareness of their physical self that is able to wave arms, kick legs, make sounds, or turn their head or of any distinction between that self and the remainder of the world. Again, the ability of their brain to detect and associate various stimuli in different dimensions of experience launches all this towards symbolization. Differentiating the touching of the crib wall from touching themselves becomes possible because the former is associated with only one kind of tactile stimulus on a hand or foot, while the latter associates two kinds of tactile stimuli received from the two parts of the baby's body brought into contact with one another. As they focus on their moving arms, they may learn that this 'out there' can be moved somewhat at will, while the wall cannot be moved. Differentiating a great many of these kinds and other experiences gradually begins to imply a distinction between themselves and the external world. Once implicitly learned, this development paves the way for non-verbal communication with that external world by moving their arms, making sounds, and using their facial muscles. In other words, a more deliberate physical embodiment in the world paves the way for a greater social embodiment, not unlike the way two adults learn to communicate with each other when they do not speak each other's languages. They have to fall back on gesturing and making facial expressions in order to communicate. In this way, the baby learns to react to some of his mother's body languages and vice versa. Differentiating these experiences implies a pattern of different kinds of sounds made by the baby and the mother, each eliciting different reactions. Gradually, these developments begin to act as a metalanguage within which some sounds made by the mother can take on a meaning that evolves, in association with others, towards symbolization.

Words as Signs

Further light is shed on these developments when toddlers begin to master some words. For example, on a walk a mother may point to a bird flying over and call it *a bird*. A little later, a toddler may point to a plane and call it *a bird*. The mother then may respond that this is not a bird but a plane. Initially, this may make no sense to the toddler. Depending on the life she has thus far built up in the unknown, planes and birds may appear essentially alike: as things differentiated by everything else in her experience by moving above her head. She may now pay more attention to the sky and notice more birds and planes. Differentiating these experiences from each other, as well as from all other experiences, will lead to the experiences of birds being differentiated from each other as essentially alike but much less like planes, and the experiences of planes being essentially alike and much less like birds. She has learned something about birds relative to planes, and something about planes relative to birds. At the same time, she learns something about birds and planes relative to all other kinds of experiences, and vice versa.

On another occasion, the toddler may be taken to a zoo. If the family has a cat, the toddler may excitedly point to a tiger and express her amazement about the size of this pussycat. When her father corrects her, the toddler may look the tiger over more carefully and, by differentiating its various behaviours, learn something about tigers relative to cats and cats relative to tigers. Over time, tigers will appear less and less similar to cats, and cats to tigers, and the toddler will recognize that they are different animals. Once again, the differentiation of experiences of cats from those of tigers refines her knowledge of both animals relative to each other, while at the same time the differentiation of these animals from everything else in her life is confirmed and strengthened.

Had the toddler grown up on a hobby farm, her life and world might be full of experiences of cats, dogs, chickens, and horses. When taken to the zoo and observing a giraffe for the first time, she may ask what kind of funny horse it is with such a long neck. In other words, she has no difficulty recognizing that the giraffe is at best a very strange horse and more likely a different kind of animal altogether, in which case she can readily learn its proper name because it corresponds to the distinctions she has learned to make in her life and world.

Consider a toddler who has had many experiences with cats but few experiences with other four-footed animals. Encountering another family's dog, she may rush right over and begin to play with it as if it were a cat. She will quickly learn that dogs respond differently, and before long she will easily differentiate cats from dogs. Again, by means of differentiation, she learns something about cats relative to dogs, and dogs relative to cats, and these distinctions test and reinforce all others she has learned to make in her life and her world.

To sum up, babies gradually learn to differentiate what initially are undifferentiated stimuli in any dimension of experience related to their inner and outer worlds. Out of this noise gradually emerge different people, animals, plants, and objects. They learn to differentiate different people as essentially alike relative to animals, plants, and objects. Within each of these groups of experiences, they learn to differentiate different animals, plants, or objects. Each differentiation either reinforces or weakens others. In this way, young children create a limited and meaningful life and world out of an undifferentiated unknown. They are protected from meaninglessness noise by virtue of the fact that whatever cannot take its place in that limited life in the world by being meaningfully differentiated from everything else in that life will be excluded from it by remaining a part of the undifferentiated stimuli that represent the unknown. Babies and toddlers are protected from relativism and nihilism because their process of differentiation is temporarily open ended and thus supportive of a playful attitude towards the world. The organizations of their brain-minds sustain their lives in a largely unknown world by effortlessly differentiating everything from everything else they are able to relate to. These organizations of the brain-mind must not be confused with the lives they symbolize: what we live is not the same as what we have in our heads, and what we have in our heads is not what we have in our brain-minds. It is all part of the widespread confusion in our civilization between human life and how it is symbolically sustained, and between this symbolic sustaining and information processing.

The development of babies requires little else than a few innate abilities, of which the most important one is the orientation of the brain towards symbolization. There is no need to assume that babies' brains contain a rational soul or a logic engine to sort out their experiences. Nor is the brain the clean slate on which experiences are written. The brain is simply a part of the genetic endowment associated with being born as a member of a symbolic species.

Gradually, the distinctions implied in the limited life in the world of babies and toddlers begin to constitute a metalanguage that opens the door to symbolization, culture acquisition, and the children's entrance into the symbolic universe of the community. This may be illustrated by making a simple model that maps the acquired distinctions, as follows. Imagine a surface on which each dot represents an element of experience such as a person, an animal, a plant, or a thing. The dots are arranged in such a way that adjacent ones represent what, in the lives of babies or toddlers, are lived as the most alike and the least different from everything else. The distance between the dots increases according to the experienced level of difference.

Initially, the model would show a constellation of dots representing very weakly differentiated experiences that are the precursors of what (much later) will become those of being fed, being held, being changed, being picked up or put down, or being alone in the crib. Since there is no awareness as yet of a distinct physical self, of others, and of a world, any differentiation is based on differences between associations of blurs of stimuli from within or without. Even so, these weakly differentiated experiences form a context in which these blurs of stimuli can begin to take on rudimentary meanings. In turn, these meanings refine the experiences to which they belong, making them less alike. It is the beginning of a self-reinforcing process in which blurs of stimuli can take on a meaning in the context of the others. As they do so, they individually refine the experiences to which they belong, and collectively refine the context in which other blurs of stimuli can now begin to take on a meaning, and so on. In terms of the model, this means that the borders of the dots become less fuzzy and more defined. At the same time, the constellation of dots is undergoing constant changes as some of them move further apart while others move closer together to form temporary clusters. All this reflects the fact that some experiences in the lives of babies need to be lived as being essentially different, while others can be lived as being essentially alike. All this is evident from their behaviour when, for example, they no longer attempt to suck when they are being hugged or held. Eventually, the original constellation of dots is transformed into a constellation of clusters of dots representing the experiences of being fed, being held, being changed, and so on.

The appearance of the clusters of dots in the model also represents the formation of metaconscious knowledge. As noted, babies are differentiating clusters of experiences when they are wriggling in their cribs and touching them with their limbs, playing with a suspended rattle, or touching their own bodies. This contributes to a growing metaconscious knowledge of their physical selves as distinct from things in their surroundings. Intuiting this metaconscious knowledge helps to refine other distinctions, such as the one between their physical selves and the physical selves of the people who hold and play with them. All such developments test, refine, or modify the positions of individual dots and clusters of dots in relation to other dots and clusters.

Eventually, the metaconscious knowledge of the babies' physical selves and the selves of others opens up entirely new kinds of relationships. As noted, the awareness of the physical self opens the door to becoming a social self. By means of gestures and facial expressions, babies and toddlers learn to communicate in a manner somewhat analogous to adults communicating without a shared language. Gradually this communication is enhanced by some words taking on a meaning in the context of various experiences. Even a limited use of vocal signs will test, refine, or modify the differentiation of the experiences of the lives of babies and toddlers. Again, this is reflected in some dots in the model distancing themselves from others, causing them to move out of a cluster of dots or to confirm the cluster to which they belong. These patterns have been illustrated by our discussions of how toddlers learn to differentiate four-footed animals from each other and from everything else. It is important to note that the distinctions babies and toddlers learn to make are those that are essential to their lives. For example, what differentiates cats from dogs is what matters to their lives; it has nothing in common with analysis or taxonomy. However, with the onset of the use of vocal signs, and later with the use of language, there is a gradual convergence of the differences that matter to the lives of toddlers and those that matter to the members of their community, which embody the experiences of many generations. However, for this stage of development of babies and toddlers what is essentially alike or different is still mostly a question of their own lives as symbolized in the organizations of their brain-minds. As far as the model is concerned, the distances between dots and the way the dots are clustered are themselves symbolic of their lives.

The process of the development of babies and children becomes self-reinforcing in other ways. For example, the toddler growing up on a farm may have differentiated her experiences with a number of four-footed animals, represented in the model as distinct clusters of dots. These clusters form a context within which additional vocal signs can take on a meaning, thus testing, refining, or modifying what in her life appears to be essentially alike or different. After all, treating one kind of animal like another kind may have surprising and even unpleasant consequences. As a result, a cluster of dots representing the experiences of one kind of four-footed animal may break up into two sub-clusters when another kind of four-footed animal is encountered. Alternatively, the original cluster of four-footed animals may contain the experiences of several kinds of four-footed animals, provided that the experiences with them continue to be elementary and superficial. As children grow up, the root cluster of their encounters as toddlers will have broken up into a constellation of sub-clusters representing all the different kinds of four-footed animals they have learned to differentiate. Indirectly, this sub-constellation will affirm the differentiation of four-footed animals from people, plants, and objects. The pathways via which the constellation of sub-clusters forms in the model will vary from child to child according to how they encounter these animals in their lives. For example, if they grow to like dogs or cats, they may learn to differentiate many varieties, thus causing the sub-cluster of experiences of one of these animals to break up into many corresponding sub-clusters.

The metaconscious and symbolic differentiation of everything babies and toddlers have come to know and live is dialectical in character. Never do they establish what something *is*. Nor do they establish any 'facts' about their lives in the world. What they metaconsciously symbolize is what something is relative to everything else they have come to know and live. What makes something unique in their lives is dialectically related to what everything else is: what it *is* is essentially what everything else *is not*. The dialectical order thus created in their lives differentiates something directly from what most resembles it and, via these distinctions, from everything else. If we give a definition of anything in our lives, we can do so only in relation to other terms and, via these, to other terms, and so on. Doing so involves separating these terms from the dialectical order that relates everything to everything else in a person's life in the world. A closed definition without porous boundaries is impossible. This makes all definitions essentially contested because in our lives in the world everything is related to everything else, and we symbolize everything in relation to everything else.[10] All we know about what something is depends dialectically on everything else from which it is symbolically and metaconsciously differentiated. Human lives are dialectically suspended in an ultimately unknowable reality. There can be no such things as hard facts but only the relative meanings and values associated with this suspension.

For example, the ability to recognize and relate to a particular kind of animal is directly associated with a child's ability to deal with and relate to all other animals in her life. A particular animal is what the others are not, either by being similar and yet different or by being fundamentally different. To live one distinction, as when playing with a cat, is to indirectly live all the others. To change a distinction by recognizing that cats are fundamentally different from dogs indirectly refines all other distinctions. Jointly, the distinctions between different kinds of animals confirm or challenge the distinctions established between animals and other constituents of the child's world. In other words, the child's knowledge of the constituents of her world is relative. What she knows about dogs via her senses is their distinctiveness from all other animals and vice versa, thereby constituting what makes dogs essentially alike. What they are not helps to form what they are. A playful relationship with the unknown as a source of new discoveries continuously confirms or challenges all of the distinctions a child has learned to make and live in the world as drawn from undifferentiated stimuli. What any constituent is in that world remains open to that unknown and a constant stream of new discoveries that flow from it. What babies and children know about anything does not come from any attempt to define what it is but from metaconsciously situating it in relation to everything in their lives, which makes it what it is.

Let us return to our simple model of how all relations in the lives of babies and children are associated as being essentially alike or different. Those relations that are experienced as essentially alike form a cluster that has emerged within a larger, previously established cluster, which essentially broke up into this cluster as well as others. For example, everything that moves in a baby's surroundings may have first been differentiated from the background blur of undifferentiated stimuli. Depending on circumstances and the way of life of the community (including how it deals with children), this cluster may have broken up into a sub-cluster that later became that of people, and into another related to objects. The latter may have later split into one related to plants and another related to things. Dialectically associating the distinctions between the relations of a person's life in terms of whether they are lived as essentially alike or different can be expanded indefinitely in a manner that tests, confirms, or modifies all previously made distinctions. In this way, all relations lived as being essentially alike are integrated into a cluster and jointly differentiated from all other clusters. Such integration of experience develops a great deal of metaconscious knowledge, which gives rise to intuitions. In the case of the toddler, certain words that correspond to these clusters intuitively begin to make sense. As a result, associations between the relations of a toddler's life that are established by this differentiation and integration gradually constitute a metalanguage (a subject to which we shall return shortly).

The differentiation of the relations precedes any ability to carry out an explicit analysis of what the members of any cluster have in common and what distinguishes them from anything else. As yet, there is no possibility of analysing what the members of a cluster have in common and how this differs from what is shared by the members of adjacent clusters. Wittgenstein has convincingly argued that there is no fixed list of attributes shared by the members of a symbolic linguistic category and that what they share instead is a 'network' of criss-crossing and overlapping family resemblances.[11] For example, card games, board games, and field games are all games, but each subset shares different attributes, some of which may overlap with other subsets. Compared to all other daily-life activities, playing different games has more in common than the details that distinguish them. Within these differences, some may matter more to a person than may others, thus leading to a preference for certain kinds over others. However, such preferences do not challenge the differentiation of the playing of games from all other daily-life activities.

Living these distinctions in our lives is symbolically and metaconsciously sustained by the dialectical ordering of our experiences. The process of differentiation is metaconscious in the sense that we have no direct access to it and that it goes beyond individual experiences by relating them to all others in a systematic manner according to their being essentially alike or fundamentally different. Our simple model is not intended to make predictions of how this differentiation actually works in the brain-mind. It cannot be extended by linking the individual points to represent neural and synaptic connections that are strengthened or weakened by subsequent experiences.

The unique learning biases of human brain-minds distinguish us from non-symbolic species. It is not a question of the senses being connected to interpretive mental structures that warn about danger, inform about food, and guide mating behaviour and other life-functions. In children's brain-minds all this is immediately associated with everything else on the way to symbolically mediating all contacts between themselves and the world. In my detailed study *The Growth of Minds and Cultures* I suggested that the process of symbolization can be understood in two ways. One is that of the process of differentiation, which metaconsciously and symbolically assigns everything its unique place in a person's life according to a culture, and this 'place' reflects its meaning and value relative to everything else in that life. The second is the metaconscious and symbolic process of integration, which creates a great deal of metaconscious knowledge about groupings and patterns, thus permitting an intuitive grasp of the many implications of relating everything to everything else in a person's life. When a person acts on these intuitions, it is usually accompanied by further cumulative and non-cumulative differentiation. For example, there are many accounts in the literature of how highly creative people who had put a task aside when they were no longer making any headway had later intuitions about how to proceed. Their experiences of grappling with the task were metaconsciously and symbolically related to many others in their lives by means of a process of differentiation. Such incubation may generate additional metaconscious knowledge, which in turn may lead to intuitions that may break into conscious awareness when they are doing something entirely different. In this previous study I suggested that this interplay between differentiation and integration helps to explain the cumulative and non-cumulative aspects of the developmental stages of babies and children. It marks the period that is characterized by a playful attitude to the world. They remain open to what is new and surprising because the unknown has not yet been symbolized (a subject to which we shall return in a later section).

Our simple model may be used to illustrate several aspects in the development of babies and children related to memory and skill acquisition. Each experience in our lives is dialectically formed in relation to all other experiences, thus constituting it as a moment in our lives. This dialectical enfolding of each experience into our lives has significant implications for what we can and cannot remember. The lives lived by babies, who have little or no awareness of the distinction between their bodies, the bodies of others, and the world, are next to impossible to remember later on, when this awareness is highly developed. Our simple model illustrates why this is the case. The constellation of dots in relation to which the original experience was formed and lived has been completely transformed. If the experience had been represented by an isolated point, all the nearby clusters would have changed. If the experience had been represented by a point in the cluster, this cluster would have been transformed along with the neighbouring ones. In addition, since the point would have been enfolded into all the metaconscious knowledge that helped form the life at that time, it as such can no longer exist. In other words, remembering the corresponding experience by symbolically reliving it is impossible. In the constellations of dots representing forms of life based on a clear distinction between the body and the external world, there simply is no place for any experiences lived without this distinction. Any such experiences can no longer be lived or remembered. The same argument holds for the forms of life based on a clear distinction between one's social self, the social selves of others, and non-human life. This is also true for the forms of life that emerge as a result of the acquisition of a language and a culture. By this time, children have evolved the organizations of their brain-minds to the point that all the earlier experiences lived with little or no physical, social, and cultural selves can be remembered only in exceptionally rare cases, and they can certainly not be lived. Every human being becomes a person of a certain place, time, and culture, and this can never be entirely changed.

We do not need to assume that our episodic memories develop later. What we need to recognize is that symbolization involves the dialectical enfolding of our experiences into our lives, with the result that remembering forms of our lives that we have long left behind is no longer possible, either in general terms of what this was like or in specific terms of particular moments. The episodic memories of our lives in the very early forms are extremely limited, if not impossible. We can attempt to reconstruct them by means of our imagination, but this is not the same as remembering them. Those experiences were symbolized in relation to and enfolded into forms of our lives that we have left behind.

Our simple model also illustrates some aspects of our becoming members of a symbolic species as a result of culture acquisition. A culture is a design for skilfully making sense of, and living in, an ultimately unknowable reality. Each dot in the model represents the skills of making sense of, and coping with, a particular kind of daily-life activity. The more that babies and children become aware of the physical, social, and cultural dimensions of their lives in the world, the more the processes of differentiation and integration become consciously directed and the less they resemble a kind of autonomous information processing. Moreover, learning to live a life by means of a culture requires that the dialectical enfolding of the child's experiences into a life must be accompanied by the dialectical enfolding of every learned skill in order to reproduce and embody that culture in each new generation. All this is combined in the way we intentionally live the moments of our lives by means of a culture. Human lives and their cultures are dialectically enfolded into one another with a subjective and individual dimension as well as an 'objective' and collective dimension.

Consider the following example. A student sitting in a classroom may focus her attention on what the professor is explaining and relegate to the background the feeling of becoming hungry and of having sore muscles from the previous night's workout. Alternatively, the feeling of hunger may distract her to the point of creating a foreground with the thought of what to do for lunch. It is also possible that the student is having difficulty concentrating that day because she has received an email from a friend asking for advice on a difficult situation. If she is bored by the subject of the lecture, she may decide to study for a quiz she has later, dream about a date she has accepted for that evening, ponder the strange colour scheme of the professor's clothes, listen in on a whispered conversation behind her, follow up on an intuition regarding a problem she has been trying to solve, wonder about a strange odour in the air, and so on. Although the student has no control over her process of differentiation, she directs it as she lives moments of her life while relating to her surroundings.

Symbolizing a moment of our lives occurs on four interdependent levels of interpretation. These levels may be distinguished according to the wholes (in the sense of the whole being more than the sum of the parts) created by the process of symbolization on a particular level because they cannot be reduced to their constituent wholes created on the level below, nor can they be dissolved into the larger wholes created on the level above. The first level is that of sensations. This level deals with the metaconscious interpretation of sets of stimuli resulting in particular elements of experience in a dimension that corresponds to one of the five senses. For example, looking around while sitting on a bench in a park may yield visual sensations of trees, paths, children at play, or adults on a stroll. On the second level, depending on a person's focus of attention, all these visual sensations are metaconsciously interpreted as a foreground against a background, thereby constituting a visual gestalt. Similarly, all the aural sensations are jointly interpreted according to their relationships with the foreground and the background, to constitute an aural gestalt. The joint interpretation of all the olfactory sensations – freshly cut grass, the cigarette smoke of a bypasser, and the odour of a nearby garbage pail – also constitute a gestalt. So also are the sensations of the hard bench, a full stomach, and a sense of being overdressed for the weather. Some of these gestalts may only contribute to the background. On the third level, all these gestalts are jointly interpreted into an overall experience of a moment of a person's life as a still higher-level gestalt. Finally, this experience gestalt is interpreted in the context of all previous experiences as symbolized in the organization of the person's brain-mind. Each level of association and interpretation anticipates, prepares for, interacts with, and depends on all other levels according to the focus of attention and engagement during a moment of the person's life.

The process of symbolization becomes more complex when a girl sitting on the bench in the park accepts an invitation to join a nearby soccer game that she was watching. Becoming an active participant involves her physical embodiment into the game while her entire life continues to operate in the background. In other words, the process of symbolization as the relating of everything to everything else now includes externalizing her intentions in movements and actions. What sensations are for internalization, movements are for externalization. Examples include bending a knee, taking hold of something with a hand, or changing the field of vision by turning the eyes or the head. Similarly, what perceptual gestalts are for internalization, action gestalts are for externalization. Examples include walking, running, and talking. In terms of her life working in the background, her prior knowledge and experience of soccer now play an active role in supporting the way she reacts to or leads in the playing of the game.

The process of symbolizing a moment of the girl's life now involves the internalization of what is happening in her surroundings and the externalization of her intentions as to how to deal with the situation. In other words, the foreground of any experience gestalt may involve running in the direction where a player has just lost the ball, and an ongoing interpretation of what is happening in preparation for forming a more specific response. The experience gestalts now include action gestalts in the foreground. In turn, these experience gestalts are enfolded into the organization of her brain-mind by being differentiated from her prior experiences of watching and playing soccer. All this may be interrupted by a distant bell tower striking the hour. Although this aural sensation is initially in the background, it may remind her of a phone call she was supposed to have made some time ago.

The above classification may have little to do with how the processes of differentiation and integration are carried out by the brain-mind. As noted, iconic associations are based on a similarity between a sign and the reality to which it refers. Indexical associations are based on a physical or temporal connection between the two, while symbolic associations link the two by means of a culture. Babies and children must gradually learn to distinguish between similarity, correlation, causality, and tradition. For a toddler, a photograph of his mother may become iconic, the scent of her perfume indexical, and the term *mother* symbolic. All this is dependent on the level of development of the child's interpretative abilities, that is, of his ability to give something a meaningful place in his life. Generally speaking, a certain level of development of iconic associations is required for indexical associations, and a certain level of these for symbolic associations.[12] In this way, babies and children gradually learn to live a life instead of associating one moment to the next. Everything they know and have learned to relate to implies all the others, and hence their whole lives. Similarly, innate interpretations and responses are gradually extended by the processes of differentiation and integration to become individually unique and culturally typical, as we shall see.

It is now becoming apparent that the process of differentiation on each level of interpreting a situation and responding to it has four dimensions. The first differentiates the focus of attention from its surroundings in each dimension of experience (corresponding to the five senses) that is directed towards the external world and those dimensions that are directed towards the inner world. This focus of attention constitutes the foreground-background distinction. How babies and children learn to make this distinction varies widely from culture to culture, and historically within a culture. Every culture is unique in how it makes sense of human life in the world by differentiating the individual from the group, the human group from animal and plant groups, and all these groups from what we differentiate as ecosystems and the biosphere. For example, early traditional societies differentiated human groups from animal groups, but not from what we call nature. In Western paintings humanity did not become differentiated from nature until the Renaissance. At the same time, the primacy of the group over the individual was gradually weakened, and this process was greatly accelerated as a consequence of industrialization. In paintings this was portrayed by a sharper delineation between a person in the foreground and other people or nature in the background. These differences have a profound effect on the way babies and children are socialized. My earlier examples are uniquely contemporary and Western. As a result of scientism, Western cultures abstracted entities from their surroundings and treated them as relatively distinct foregrounds. In earlier traditional cultures (and today in some Far Eastern ones), processes of socialization placed a much greater emphasis on the relations between these entities, thus blurring a sharp foreground-background distinction and reflecting symbolic universes that were much more enfolded and indivisible. In these cultures, the foreground was not so much the entity itself but its relations with other entities, which it symbolized. In this way, the foreground-background distinction expresses a variety of metaconscious values as to the importance of how everything is related to everything else. As will become apparent in subsequent chapters, the foreground-background distinction sharpens with growing levels of desymbolization.

The second dimension of the process of differentiation refines the interpretation of the focus of attention by differentiating it from itself. Variations in this focus of attention can come from many sources, including our moving about as we interact with it, thus perceiving it from different vantage points; the focus of attention itself being mobile and moving around; changes in the surroundings resulting from weather conditions and light; and everything evolving in our lives and the world. Once again, this happens in all the dimensions of experience corresponding to the external world as well as those corresponding to the inner world related to bodily functions, feelings, emotions, intuitions, and thoughts (about real and imagined things). For example, a toddler playing with a dog never experiences it in quite the same way as they both move about. The engagements with the animal can be differentiated from one another. Similarly, a child learns to recognize an object (such as a dinner plate) from any vantage point. He will learn to recognize his best friend's voice even when the friend has a cold, shouts, or whispers, for example.

The third dimension of the process of differentiation creates diversity within a unity, while the fourth dimension creates unity within a diversity. Returning to our earlier example, a toddler may first experience four-footed animals as being essentially alike. As she gains more experience with them, their differences relative to one another are enhanced by the third dimension of the process of differentiation. This diversity grows with additional exposure to four-footed animals until it can no longer be contained within the unity of their appearing essentially alike. One or more sub-clusters may emerge of animals that are essentially alike but fundamentally different from the members of other sub-clusters and the original cluster. At this point, the fourth dimension of the process of differentiation creates a new unity within this diversity, based on the world of four-footed animals being populated by two or more different kinds. Continued exposure leads to further break-ups until the distinctions that the child has learned to make are compatible with those of the adults in her speech community. Even then, this diversity and unity can be refined when later the child takes an interest in animals and reads about those not encountered in her surroundings.

The third and fourth dimensions are particularly important in making sense of a living world in which nothing ever repeats itself in quite the same way. Even our most routine habits exhibit slight variations. Our nervous systems have evolved in relation to a living world, to the point that they do not cope very well with monotony and repetition. We are unable to indefinitely repeat a word or a phrase or cope with sensory monotony or deprivation. The latter can have serious long-term consequences.[13] Similarly, the living of a life cannot be split into two as, for example, on the assembly line, where people are included in terms of their hands but largely excluded in terms of their mental capacities. This hand-brain separation and the suppression of a great deal of a person's life can lead to nervous fatigue, which over time undermines peoples' physical and mental health.[14] A living world is very different from a machine-information world based on repetition and algorithms. The human nervous system did not evolve in relation to such a world.

The processes of differentiation and integration also sustain the symbolic development of a metaconscious self. Consider the example of a hyperactive child – a condition associated with his growing up in a lead-contaminated neighbourhood. His relationships with others are likely to involve a great deal of disapproval, if not rejection, as a result of the following kinds of situations. As soon as his parents come home from a long, tiresome day at the office, he storms into the living room, going a mile a minute telling them all the details of his day and showing them some drawings he made at school. The parents, having just settled down to relax, may urge him to slow down or to come back in a few minutes when they have finished their glass of wine. He may get the same kind of reaction from many other people when, for example, he bursts into the classroom in the morning, attempting to get the attention of a teacher who has thirty other children in her class. As the organization of his brain-mind sustains his living by symbolizing, interpolating, and extrapolating these experiences with others, the metaconscious knowledge may first be that many people do not share his enthusiasm. Differentiating these reactions from how these people deal with other children may imply that he is not a very likeable person. Eventually his differentiation and integration of these kinds of experiences may metaconsciously develop a low self-image, which can have long-lasting negative effects on his life. The point is well known: the way the processes of differentiation and integration help sustain our relationships with others includes in essence a metaconscious mirror in which we see ourselves.

As we have already noted, our earliest experiences enfold only a limited physical self, soon followed by a social and then a cultural self. As a result, there is no 'output' that the processes of differentiation and integration pass on to the self. A person is increasingly embodied in each and every experience. Just as the DNA is enfolded in each cell of the person's body, so the personal self becomes enfolded into each and every experience. Each experience becomes a unique moment of a person's entire life, as well as the expression of the whole person. The human consciousness of ourselves and the world develop together. Although embryonic during the early phases of the development of babies and toddlers, there is always a whole person living in a world. The differences are entirely related to the level of differentiation and integration. From the outset, there is a person living his or her life in the world.

Language and Order

The acquisition of language by toddlers and children begins to evolve their lives towards forms that gradually converge with those of the members of their language community. The earliest forms, at first characterized by a lack of awareness of being physical selves, followed by a lack of awareness of being social selves, are so different that their episodic memories cannot be relived through resymbolization later on in their lives. With the onset of language, however, the organizations of the brain-minds of toddlers increasingly work in the backgrounds of their lives in a way that remains individually unique but that now also becomes increasingly typical of their time, place, and culture. As a result, some episodic memories begin to take on forms that can be relived because the metaconscious knowledge into which they are embedded has been further differentiated and integrated by years of additional experience, but not to the extent that it is incomparably different.

The gradual development of a more enduring episodic memory is indicative of two developments. The organizations of the brain-minds of babies and toddlers have evolved sufficiently into forms that are able to constitute metalanguages for the language of their community. The other development is the gradual acquisition of entirely new and distinct processes of differentiation and integration to create new forms of metaconscious knowledge that manifest themselves as growing semantic memories. Toddlers discover a symbolic universe in which everything has a name and is endowed with meaning and value. This discovery was particularly dramatic in the life of Helen Keller, who suddenly intuited that language was a gateway to a new life in an entirely new world, even though she was blind and deaf.[15] She immediately began to explore this new world with a great deal of motivation, and it changed her life. As words begin to make sense as limited signs, the physical embodiment that gave rise to social embodiment begins to lead to a cultural embodiment in the child's world.

The onset of language is not a cumulative process of acquiring words and phrases that have taken on some sense in the context of the organizations of the brain-minds of toddlers acting as metalanguages. At first, the aural stimuli of the word *dog* spoken by a mother as she points one out to her child must become associated with this four-footed animal, much like the aural stimuli of barking. If the toddler's experience of the dog is sufficiently differentiated, the aural stimuli may take on a meaning and become aural sensations. At first, these sensations are merely differentiated from all the other aural sensations corresponding to the sounds that have taken on a meaning in the toddler's life. The aural sensation is simply another sound that has a meaning. However, the corresponding metaconscious developments gradually lead the toddler to intuit that the aural sensations of words and phrases spoken by others are very different from the sounds made by everything else in his surroundings. Gradually, the development of metaconscious knowledge gives rise to the intuition that sounds related to what he later recognizes as language have much more in common with each other than they do with all other aural sensations.

In terms of our simple model, the aural sensations related to language begin to form a cluster on the level of sensations within the aural dimension of experience. This development gradually paves the way for another on the next higher level, that of interpreting sentences as language gestalts. When this happens, a language gestalt within the aural dimension of experience can constitute the foreground on the next higher level, that of interpretation corresponding to experience gestalts. Gradually the experience gestalts characterized by a foreground of a language gestalt will be differentiated from one another, and jointly from all other kinds of experiences in terms of the model. This means that these language-related experiences begin to form a distinct cluster on this level of interpretation. When these developments become enfolded into the organization of the brain-mind on the next higher level of interpretation, the toddler is on the way to becoming a member of the community sharing that language and culture.

At this point in their development the brain-minds of toddlers begin to differentiate the language-related elements of experience from all others on the levels of sensations, gestalts, and experiences. Gradually, a metaconscious knowledge develops to the effect that what constitutes a language is what all other experiences are not. A distance is thus created between language and experience and between the corresponding forms of metaconscious knowledge. The metaconscious knowledge related to non-language experiences acts as a metalanguage in the context of which words and phrases take on initial meanings, but these are increasingly supplemented by the meanings derived from directly differentiating the language-related elements from one another. This differentiation entirely on the level of language builds a new kind of metaconscious knowledge, which forms a secondary context in relation to which these words and phrases take on a further meaning, and so on. Eventually, the meanings of words and sentences are dialectically built up from what all the others do not mean, and secondarily from the meanings derived from the metalanguage created by experiences not involving language. A double referencing system develops. What all the languages in the world share is their compatibility with the metalanguages developed in the brain-minds of babies and children, to the point that they can be readily learned.[16] Within a language, each word and sentence is dialectically differentiated from all the others, with the result that their meanings are what everything else does not mean. In this way, a language can be learned without reference to definitions or grammar rules. The compatibility between the organizations of human brain-minds and human languages has been assured by a long co-evolution. The variety of human languages would appear to indicate that this compatibility is very accommodating.

The new kind of metaconscious knowledge related to language implies an entirely new development within the organizations of the brain-minds of toddlers and children, which permits the emergence of a semantic memory, distinct from the episodic memory. As a result of brain damage, it is possible for people to lose their episodic memory while retaining their semantic memory.[17] The onset of a semantic memory thus signifies the beginning, for a child, of becoming a member of a symbolic species. In terms of our simple model, there is a growing cluster of experiences that now have a language gestalt in the foreground.

As toddlers and children begin to use words and phrases, there is a corresponding development of movements and action gestalts, since speaking a language on the levels of phonemes and words involves various patterns of muscle usage related to the mouth and vocal cords. Such motor skills correspond to the expressions of the self via the language of the community.

It should be noted that deaf children who are learning to sign a language develop corresponding clusters of language sensations and gestalts within the visual dimension of experience for the internalization of language, and clusters of movements and action gestalts for signing and possibly speaking it. Deaf-blind children who are learning a language face a much more difficult task in acquiring similar developments within the tactile dimension of experience for the internalization of language and hand movements, and action gestalts for the externalization of language by means of keyboarding or alternative signing systems. Learning to read a language leads to additional developments within the visual dimension of experience for children who are able to use all their senses and for deaf children, and within the tactile dimension for blind and deaf-blind children. It goes without saying that for deaf-blind children indexical and symbolic associations are acquired with much greater difficulty.

We have already suggested that babies and children learning to differentiate their experiences do so in terms of what makes sense in their lives at a certain point of their development. The processes of differentiation and integration have nothing in common with those of taxonomy or any formal classification. Gradually, individual differences become constrained by what makes sense in the lives of the adult members of their language community. In the same vein, 'grafting' a language onto a metalanguage is a process that can vary enormously from culture to culture because it expresses a unique style for doing so, which itself is an expression of the culture. The implication is that language acquisition permits the ordering of the experiences and lives of children by benefiting from the experiences of countless generations passed on in the form of a culture, which functions as a design for making sense of and living in an ultimately unknowable reality. The rare experiences of children growing up in isolation or being brought up by animals permit us to appreciate the importance of this cultural heritage.

As will become evident in later chapters, the roles of languages and cultures in human lives and societies have undergone a profound transformation as a consequence of industrialization, rationalization, and secularization. For example, contemporary mass societies provide a great deal of evidence that there is no direct correspondence between a metalanguage and the language that symbolizes it. In such societies, words and phrases are used as tools to create a desired effect as perfected by public relations, advertising, and the integration propaganda of the mass media.[18] It is no longer the meaning that words and phrases have within the language that is used by the mass media, but the feelings and emotions that they can produce by associating them with very different contexts than in the language experiences of the members of these communities. The wedge that is thus driven between the metalanguage and the language of the culture of a mass society leads to desymbolization.[19] The desymbolized context of the mass media creates an entirely new relationship between the metalanguage and the language of the members of a culture. In the same vein, we must be very cautious in extrapolating to the past the findings of current studies of human brains by means of magnetic resonance imaging (MRI). As we shall see, the extent of desymbolization is so great that the results are bound to be affected.

Languages and the metalanguages on which they are grafted are built up by processes of differentiation and integration. We have already noted that what this means is that no absolute knowledge can be had of any experience. What an experience shares with others that are essentially alike is what fundamentally different experiences are not. In other words, what something is has a dialectical relationship with what it is not. Instead of an absolute knowledge of an experience, there is a relational structure that symbolically maps everything in relation to everything else. An expression of this is found in the dictionary of any language. Each word and phrase can only be defined in terms of others. The entire structure of the language is relational in a circular manner and thus is without any meaning or value other than in the way it is affirmed and lived by the members of a language community. Each word, phrase, and sentence relates first and foremost to others, and jointly to the symbolic universe of the community. As such, they each symbolize something of the whole of that language and symbolic universe. However, this is all limited to what has until now been experienced and lived by the members of the community, passed on from generation to generation.

What all this implies is that we have no direct access to reality that would permit us to arrive at absolute facts or absolute definitions of anything. Any fact or definition is suspended in the dialectical structures of language and culture that mediate our relationships with reality. A very long time ago science had to give up on the possibility of creating a neutral observation language that would be objective and value free. It is worth quoting the observations of some renowned scientists on these matters.

The relationship between language and reality was expressed by Niels Bohr in this way: 'Ultimately, we human beings depend on our words. We are hanging in language.'[20] When someone objected that reality lies 'beneath' language, Bohr replied, 'We are suspended in language in such a way that we cannot say which is up and what is down.'[21] The possibility of the unknown impinging on the known as an anomaly was intuited by Albert Einstein during a period of crisis. He said, 'It was as if the ground had been pulled from under one, with no foundation to be seen anywhere upon which one could have built.'[22] Einstein also intuited that it was impossible to discover anything about the unknown without the 'lens' of what is already known: 'It is the theory which decides what we can observe.'[23]

The Self and Cultural Order

There is a widespread belief that a unitary self and a non-pluralistic culture are ideological creations of the past. There is no doubt that a highly decentred existence in pluralistic societies is a significant feature of any contemporary mass society under the influence of technique. While it may be difficult to imagine what human life was like before these developments, our premise should be that human life and societies have always been subject to centring and decentring influences and that, while the latter have the upper hand today, this is no guarantee of its being a universal feature of human history. Possibly, the most forceful argument for this position is based on the extraordinary centring effect that symbolization has on the lives of babies and children before they encounter decentring experiences.

Five interdependent developments are involved when babies and children learn to make sense of and live in an ultimately unknowable reality: the development of their person through a growing self-awareness; the expression of the self in a growing diversity of relations constituting a life; drawing the objects of these relations from the unknown into a cultural envelopment; the symbolization of the unknown as more of the known yet to be discovered and lived; and the changes in their life resulting from experiencing and internalizing all of this. From our analysis thus far, it is clear that a growing awareness of oneself leads to a growing diversity of expressions of the self. This greater diversity in turn results in a growing diversity of relations that can be established in one's life. These developments translate into a greater diversity of the objects of these relations being drawn into a cultural envelopment.

All these developments are linked in time by the following cycle. The person externalizes something of his or her self in behaviour that participates in his or her surroundings. The experiences of these surroundings are internalized and symbolized by neural and synaptic changes to the organization of the brain-mind. As a result, the self enfolded in this organization is modified and externalized in subsequent behaviour, and so on. As a result, there is an ongoing correspondence between the level of differentiation of the self, the life lived by the self, and the cultural envelopment within which that life is lived. At the same time, the corresponding levels of differentiation also ensure a relational centring of a self within a life and of that life within a cultural envelopment. The reasons are clear. The constituent elements that make up the differentiated self, as well as his or her life and cultural envelopment, take their symbolic place in relation to one another according to the meaning and the value they have within a person living a life in the world by participating in the way of life of his or her community. The process of symbolization thus ensures the joint centring of the five developments.

The same can be observed by reflecting on the experiences that are typical of a particular level in the same process of learning to make sense of and live in an ultimately unknowable reality. At a very low level of differentiation, experiences are little more than involuntary associations or simple stimulus-response behaviour. During the early phases of the development of babies and children, their experiences may be dominated by the objects experienced, but increasingly these experiences begin to flow from an active involvement in relationships being established with others and the world. The foregrounds of these experiences gradually shift from the objects of the relations to the relations of the self with these objects. Similarly, the backgrounds gradually evolve from a dialectical order of these objects into a person's world. The acquisition of words as signs helps to refine the differentiation of life in a dialectical order. The metalanguage prepares the way, ordering the lives of babies by dialectically separating its constituent elements. People are separated from animals, plants, and things, and vice versa. Within these dialectical separations finer ones are made to create a dialectically enfolded self, a life, a cultural envelopment, and a horizon of experience. What something is, in the lives of babies, is what everything else is not. It all constitutes a dialectically enfolded whole. This enfolding derives from the process of symbolization because the world of babies is not simply 'out there' but also symbolically mapped in the organization of their brain-minds. They are both internally and externally connected to the world into which they are thus enfolded. These relationships are the opposite of what we find in non-living systems, where each element exists in its own space and interacts with others via its external boundaries.

The acquisition of language accelerates the convergence between the cultural envelopment of children with the cultural order of their community. Children begin to participate in the cultural cycle that is operating in the lives of the members of their community. They externalize in their behaviour something of their selves and their lives. When they become full participating members of a community, some of this externalization may be objectivized and incorporated into its working culture and way of life to help the community respond to new situations. The internalization and symbolization of these changes by the members of the community is done in a way that is individually unique and culturally typical, and the resulting externalization now includes these changes in their behaviour. Hence, the centring of the self in a life, and the establishment of that life in a cultural order, becomes individually unique as well as culturally typical as children grow up to become adult members of their community.

The True and the Real

The considerable diversity of human cultures can be understood in part by the different ways in which the self, the expression of that self in relations, the integration of the objects of these relations in a cultural envelopment, and the unknown can be differentiated. For example, different cultures have ordered the 'world' of colours in different ways.[24] Some cultures have differentiated dark colours as shades of one primary colour, while other cultures have differentiated light colours as variations of another. One culture did not see the colour blue, because it did not differentiate blue from the other primary colours, seeing it instead as a shade of one of them. In the literature we find Greek and Roman authors who described the colours of the rainbow in different numbers and kinds of primary colours. In the same vein, we know that some early indigenous peoples did not differentiate humans from animals in the way that we do. To put it in modern terms, the human group was just another species in the biosphere. Some totemic societies also did not differentiate human beings from plants and natural objects but as beings all having a spirit.[25] In other cultures, the human group in the present was not differentiated from its ancestors. In general, traditional societies did not differentiate the individual from the group in the way that we do. Other cultures differentiated life in time very differently, as reflected in the tenses of their language. Since the metaconscious knowledge of the self is deeply affected by these kinds of differences, different cultures have produced very different kinds of human consciousness. All this is very difficult for us to understand because the cultural approach has been so weakened by desymbolization and so devalued relative to the scientific approach that we have a hard time taking it seriously. To clarify the situation, we shall introduce a distinction between what is true for the members of a cultural community and what is real for the members of a scientific discipline.

In drawing a distinction between what is *real* and what is *true*, I am not engaging in a philosophy of reality or truth. I am referring to the dialectical ordering of daily life by separating its elements as a consequence of living in the world and thus directing the processes of differentiation and integration in a culturally unique way. People have always been concerned with what is true, which includes but goes beyond what is real.[26] 'Do I have the best possible grip on the situation?' 'Do I have a good grasp on what this object is all about?' As Merleau-Ponty pointed out, we move around to find the 'right' vantage point for looking at a painting or manipulate an object to get the sense of it.[27] In other words, what is 'real' is subsumed under what is 'true' in terms of the relationship established with it. It is the meaning and value of everything relative to everything else, and hence to our lives and communities, that really matters. Language and culture create an order of what is true, not of what is real. The true must have the upper hand over the real in life if an ultimately unknowable reality is to make sense and if relativism and anomie are to be kept at bay. The distinction between what is real and what is true has little to do with our immediate experience of the world as detected by our senses and with its cultural interpretation resulting from our lives working in the background.

Suppose we attempted to go about our daily lives exclusively on the basis of the knowledge infrastructure provided by science. Let us begin with what we need to eat to live an active and healthy life. Determining a nutritious diet initially appears to be straightforward as we consult the most up-to-date literature. This knowledge is exact with respect to present theories and laboratory and field studies; it is not exact in relation to what was known a few years ago or what was known a few years before that, and so on. We encounter a kind of zigzag course as the health status of many foods is constantly being revised, sometimes in one direction and sometimes in another. Apparently, the scientific exactness of nutritional knowledge does not translate into reliability over time. This lack of reliability is a problem since our future health and well-being is significantly affected by what we eat today. It is with our food that we constantly repair and replace all the cells in our bodies, except for our brain cells, which are only repaired because these symbolize our lives. Our problems in deciding what to eat do not end here. There are important links to other disciplines which suggest that a nutritious diet is related to genetic make-up, lifestyle, and much else, as well as to where the food was grown and how it was processed, packaged, distributed, and stored. Many of these relationships are rarely studied comprehensively because they fall into the category of interdisciplinary studies, which is scientifically much less manageable.

Consider another daily-life decision faced by a couple, one of whom is a carrier of a possibly transmissible eye disease that leads to blindness. They decide to consult a specialist before starting a family. Once again, the tension between exactness and reliability appears: all the specialist can *scientifically* tell the couple is that 'today we think your chances of transmitting the disease are this, but five years ago we thought differently, and ten years ago we thought differently again.' The consequences of any decision last far longer than does the stability of the relevant knowledge base.

All this is confirmed by what we know from the history of Western science. As noted in the introduction, it gave up on studying human life in the world a very long time ago. The unmanageable complexity was somehow made manageable by dividing the world and the human life within it into different categories of phenomena and by parcelling out their study to autonomous disciplines. It divided an unmanageable complexity into 'intellectual monocultures' and thus opened the door to a measure of objectivity and control. The creation of any theoretical model involves decisions as to which details must be incorporated into it because of their importance and which details may be omitted because of their lack of importance. Such decisions cannot be made without reference to a set of values unless an unintelligible complexity is reduced to one category of phenomena. In other words, an intellectual division of labour based on parcelling out the study of one category of phenomena to one discipline was the key to an illusion of objectivity and control, in the sense that this implies a commitment to a belief that human life and the world can be understood one category of phenomena at a time. Western science thus separated itself from human experience and culture, which seek to relate everything to everything else in human life and the world.

Each group taking the responsibility for studying a category of phenomena sets out on a journey that, in retrospect, can be likened to a zigzag course. For example, Thomas Kuhn has shown that the Western search for physical and chemical reality exhibits such a course.[28] Aristotle's attempts to find physical reality pointed in one direction, Newton's in another, and Einstein's in still another. Beginning with Newton, investigators constructed a theoretical order of physical reality based on mathematics. In the course of elaborating and refining this logical order, a discovery was eventually made that was incommensurate with it. Its anomalous status with respect to the order of physical reality could only be changed by reconfiguring what was known into a different order. In other words, a cumulative period during which an order is elaborated and refined is sooner or later succeeded by what Kuhn called a scientific revolution. During such a non-cumulative period it is impossible to objectively measure the gains and losses of knowledge. Without any access to the physical reality that underlies all theories and experiments and from which future discoveries will continue to flow, it is impossible to say whether Aristotle's, Newton's, or Einstein's theoretical and experimental orientation came the closest. All we can say is that a particular fact, model, theory, or belief is exact in relation to the current experimental apparatus, experimental findings, and theoretical framework. In other words, exactness means commensurability with the current state of knowledge. A distinction must thus be made between the reality as known by the members of a discipline and a deeper reality from which future discoveries will flow.

For a theory to be exact it must predict the experimental evidence, but for a theory to be true there must be a guarantee that there never will be another scientific revolution in the field. Since such a guarantee cannot be provided scientifically, any claim of it being true implies an absolutization of a theory by means of a myth (in the sense of cultural anthropology) that ensures its truth. Exactness requires nothing else but commensurability between data, experimental design, and theory within the context of a discipline or specialty. We have confused knowing *more* with knowing *better*. The exponential growth of scientific knowledge made possible by trading breadth for depth is accompanied by the exponential growth of ignorance of how all scientific findings fit together into one known reality.

Fortunately for us, daily-life knowing depends on a culture that seeks to understand everything in relation to everything else by creating a cultural (that is, a dialectical) order. As a result, we know enough about the food we eat, the ground we walk on, and the nature that surrounds us to confidently eat, walk, and garden. We know full well that there is a great deal about these and other activities that we do not know and may never know, but we know that what we know is true. In other words, when we learned more there would simply be more details, but these would not threaten our daily-life activities.

The situation is somewhat analogous to an artist who, with a few pencil lines, is able to create a likeness of a person's face. The people who know this person immediately recognize the likeness, and no matter how many more details the artist may fill in to create an almost photographic image in due course, the image is true all along. Alternatively, this process may gradually reveal that the original proportions were not quite right. In the same vein, the very best photographers take pictures that are so characteristic of the person that they are a true likeness, as opposed to being merely real. For discipline-based knowledge to sustain daily-life activities, the unknown must be lived as an extrapolation of the known. Anything discovered in the future would simply add more details to what would still not be the order of what is true, restricted as it is to one category of phenomena. Moreover, this is the case only during the cumulative periods between scientific revolutions. As a result, scientific disciplines are unable to sustain true daily-life relationships.

Language, culture, and myth orient human life to what is true. What we need to know is whether things make sense and that this sense can sustain our lives. When things no longer make sense, we risk becoming anxious and depressed.

It is for this reason that I have insisted on the processes of differentiation and integration being of the order of the true, that is, a dialectical order established by any culture. It is a question of the meaning and value of everything relative to everything else, and thus for my life lived within a community. It has to do with the dialectical ordering of individual and collective human life within an unknowable reality. I know very well that there are many currents within the social sciences that reject references to meaning in language. They are correct to the extent that this language has been desymbolized under the pressure of technique (but this is the subject of a later chapter).

Two case studies may be illustrative. The first is that of a subject in an experiment who was asked to identify playing cards projected on a screen.[29] Some of these cards had been made anomalous by altering their usual colour or by putting a different border around the suit symbols. After having gained experience with this new world of playing cards, now populated by five suits, most subjects intuited that something was wrong, and looking at the images more carefully, they soon discovered what it was. They had originally learned to differentiate the world of playing cards into four suits, a concept that they now had to adjust. However, one of the subjects, despite intuiting that something was wrong, was unable to reorder the world of playing cards in order to make sense of it. The result was a crisis: this person was no longer sure of what he was seeing, as if his knowledge of playing cards was no longer true and, worse, his symbolic link with reality might no longer be true either. Such a crisis could not have occurred if it were merely a question of the order of what is real, that is, of mistaken identity.

Another illustration comes from one of the many fascinating accounts of Oliver Sacks.[30] A pianist who had lost his right arm was able to continue to show his students how to finger particularly difficult passages by using the parts of his cortex that he used to play the music. The organization of his brain-mind symbolized his many prior music-making experiences. As he continued to teach music after the loss of his arm, the corresponding portion of his cortex was not taken over by other activities, as is usually the case. Even though his mentally fingering the music was no longer real, it nevertheless remained true, and this permitted him to continue to teach his students.

We have taken a first look at the differences between what is true for human life and what is real for science and technology. There is a distance between what we have come to know by means of a way of life and culture, or by the practice of a discipline or specialty, and the reality that lies beyond, which manifests itself by an endless stream of new discoveries. The meaning and value of any new discovery can be established only in the context of what is already known. Cultures do this by a process of symbolization that dialectically relates everything to everything else in human life. It creates an order of what is true for that life. Within this order there can be no stand-alone facts or closed definitions but only elements with meanings and values that depend on those of all other elements, and vice versa. New discoveries and possibilities are added to this cultural order by the processes of differentiation and integration. The order of what is real is established, one scientific discipline or technical specialty at a time, by means of a unique intellectual and professional division of labour. Each builds and evolves a rational domain, the elements of which are exact with respect to all the others, giving it a logical consistency. There are no scientific or technical methods for integrating these domains into a single order encompassing every category of phenomena that is scientifically known or technically employed.

The access to an ultimately unknowable reality by means of the cultural approach differs from that of the scientific approach in the extent to which context is taken into account and how that context is organized. The cultural approach uncovers what is true within the context of the dialectically organized experience of a community, passed on from generation to generation by means of its cultural design for making sense of and living in an ultimately unknowable reality. As the community evolves in relating to that reality, it inevitably encounters a growing number of specific experiences that fit poorly into this design. Some of these may be interpreted as symptoms of the differences between what is known and lived and the reality that lies beyond, while others may be seen as symptoms of this evolution being non-cumulative. Both kinds of experiences may contribute to the embryonic beginnings of a new cultural design that gradually undermines and replaces the old one, thereby ushering in a new period in the history of the community. Such a non-cumulative transition usually involves many generations. If a new cultural design fails to emerge, the old one will be less and less able to sustain the community, thereby resulting in its disintegration and collapse.

The scientific approach uncovers what is logically or theoretically compatible with the current knowledge of a single category of phenomena, which the members of a discipline pass on from generation to generation as a scientific design for making sense of a tiny portion of an ultimately unknowable reality. Sooner or later, the cumulative evolution of this design in relation to reality uncovers an anomaly that is logically or theoretical incompatible with it. Such an anomaly is a symptom of the reality as it has become known by the community of practitioners being different from the reality beyond. Without having access to the latter, the scientific design must be reorganized in such a way that the anomaly, and as much of the previously acquired knowledge as possible, can be logically or theoretically integrated into a new discipline-based design. Such designs include a diversity of elements, such as exemplary solutions to problems, models regarding the 'nature' of certain processes, analogies for conceptualizing certain phenomena, and a variety of values to guide decision making during cumulative and non-cumulative periods. As a result, the scientific designs are sometimes likened to disciplinary or professional 'cultures.'

Neither the cultural approach nor the scientific approach has any direct access to the reality that lies beyond what it knows. In the case of cultural communities, their interactions with reality are metaconsciously interpolated and extrapolated by the organizations of people's brain-minds to constitute a metaconscious knowledge of what is known and lived by a community. This metaconscious knowledge may or may not be made explicit in world views or ideologies. For a cultural community, a fact is a detail of the dialectically organized experience of human life in the world. For a scientific discipline, a fact is a detail of a logically constructed domain, or of a theory, and can be backed by experimental investigations in a laboratory. Only scientism can transform a scientific fact into an aspect of reality, based on an unlimited faith in the omnipotence of the scientific method and a complete lack of a critical awareness of its limitations. The experiences of the members of a cultural community and those associated with the practice of a discipline are internalized and symbolized by the organizations of the brain-minds of the people involved. As we shall examine in the next chapter, scientific 'experiences' are separated from experience and culture and lead to different forms of metaconscious knowledge. However, metaconscious knowledge embedded in experience and culture and metaconscious knowledge separated from experience and culture result from the interpolation and extrapolation of specific encounters with reality.

Every cultural community creates its own cultural order in an ultimately unknowable reality. Their comparison reveals what each one takes for granted about this reality as a result of interpolating and extrapolating these encounters by means of the deepest forms of metaconscious knowledge, generally referred to as myths. Myths transform these cultural orders into orders of what is true, thereby enabling people to confidently get on with their lives in an ultimately unknowable reality by keeping relativism, nihilism, and anomie at bay. Similarly, all scientific disciplines create scientific orders regarding one category of phenomena in an ultimately unknowable reality. During cumulative periods, scientific practitioners behave as if these scientific orders were reality itself, minus some missing details yet to be discovered. We shall later show that this interpolation and extrapolation of their encounters with reality also involves myths.

Myths and the True

How can I possibly assert the existence of myths in a secular society? Is this not a projection of the past onto the present? I shall gradually develop the thesis that it is because of the desymbolization of human experience and culture in our contemporary societies under the pressure of technique that the very role of myths as well as their forms has changed. Searching for traditional myths in contemporary ways of life and finding none, we have jumped to the conclusion that we have been liberated from myths. However, myths constitute the deepest form of metaconscious knowledge developed in the organizations of people's brain-minds. Thus hidden away, they play a very important role in the symbolization of the unknown. This role is completely transformed when, during a time of social and historical transformation, myths can often be intuited to take on an explicit form commonly referred to as mythologies. The symbolization of the unknown by contemporary ways of life almost certainly involves secular myths whose roles have been greatly weakened by desymbolization.

I am well aware that any admission of even a limited role of myths in contemporary human life would undermine a great deal of contemporary social science. For this reason, my approach to myths will be sociological and historical, which has most in common with the approach of Jacques Ellul.[31] It is also influenced by the pioneering work of Eliade in the history of religion,[32] Ricoeur in hermeneutics,[33] Panikkar in theology,[34] and what I believe to be implicit in the historical analyses of Arnold Toynbee[35] and Thomas Kuhn.[36] As someone who has spent a great deal of his working life in both engineering and the social sciences, I have always been struck by the realization that our relationship with contemporary technology would be incomprehensible without the acceptance of the existence of myths.

A central component of culture acquisition is the gradual formation of metaconscious knowledge in the organization of the brain-minds of children and teenagers, which in some social sciences would be referred to as myths. It has to do with the symbolization of the unknown, implicit in the way the processes of differentiation and integration sustain life. Beyond the cultural envelopment babies and children have learned to make sense of and live in the unknown. This unknown is added to their cultural envelopment by a stream of discoveries of whatever else can take on a symbolic place in their lives. All this presents no particular problem until the order of their lives within their cultural envelopments begins to converge with that of the culture. How can this cultural order be relied on when so much remains unknown? We have already discussed how the organization of the brain-mind may be regarded as the symbolic interpolation and extrapolation of a person's experiences into a life and a world. For babies and children, the organization of their brain-minds is constantly undergoing changes. On the micro-level, the process of differentiation in all its four dimensions is constantly making adjustments. On the macro-level, the process of integration is working in the background, developing metaconscious knowledge that opens up new kinds of possibilities.

We have seen that at first babies do not differentiate their own bodies from the remainder of the world but that the progressive differentiation of their experiences contributes to the development of metaconscious knowledge in the background. When this metaconscious knowledge has grown sufficiently to support intuitions that they are physical selves, they can put these intuitions into practice and explore entirely new kinds of relationships with everything else 'out there.' In turn, their enhanced ability to physically express themselves paves the way for non-verbal communication, which in turn paves the way for the formation of metaconscious knowledge and intuitions regarding the role of language. Each time they begin to make use of a significant development in their metaconscious knowledge by means of intuitions, they change their selves, the expressions of these selves in new relationships, and the objects of these relationships within their cultural envelopment. So much is new, and so much constantly changes, that infants adopt a playful attitude. It is as if they and their world constantly changed before their very eyes. As a result, no symbolization of the unknown is as yet possible or existentially necessary.

Language begins to change all of this, bringing children into contact with the cultural order of their community. Metaconsciously, they discover that, when talking to others, they cannot move their eyes in an arbitrary manner. Eager to participate, to belong, and to fit in, they soon learn that they need to move their eyes according to the eye etiquette that is a part of this cultural order. The same is true for the conversation distance, their body language, their learning to live in space and time, the values inherent in all the relationships they establish, the social roles of others – in short, the way of life of their community as a design for living in an ultimately unknowable reality. In the face of this cultural order, their playful attitude works only up to a certain point, and then the adults begin to place limits on their playful orientation. In other words, gradually the organizations of their brain-minds symbolize the cultural order of their community as they participate in it. The organizations of their brain-minds converge as they learn to live the individual diversity and cultural unity of their community. They learn to internalize the cultural order to which they become increasingly committed, to the point that they can no longer live outside of it, any more than a fish can live outside of water. Only myths can fully accomplish this.

As the organizations of their brain-minds begin to symbolize the cultural order of their community as experienced through their own lives, the unknown can present them with two kinds of possibilities. The first is related to events, discoveries, and everything else a person may encounter in the future and which will be more or less lived as interpolations and extrapolations of earlier experiences, with the result that they can take some kind of symbolic place in his or her life. This component is the dominant one throughout a person's life.

The second possibility is constituted by what may be regarded as the 'silences' between the words and the experiences of a person's life. It must be remembered that language opens the door to the world of a cultural order. The 'silences' refer to the possibilities that are never spoken, thought, or in any way lived. These include, but are not limited to, how the members of distant cultures might deal with similar situations, but which for the person of a particular culture simply cannot come to mind and certainly cannot be lived, because these could not find a symbolic place in his or her life and world. These silences also include other possibilities that are entirely hypothetical in the sense that no specific culture has as yet dealt with them. These possibilities are as close as we can get to a scientific anomaly in a dialectical cultural order. It is worth noting that when people are confronted with situations that have a threatening potential to their lives and their world, they protect themselves. For example, when cultural anthropologists observe disturbing rituals in another culture or when people in a laboratory confront artificially crafted, disturbing situations, 'counter-transference reactions' have been documented by Devereux that make sense of these phenomena by reducing local dialectical tensions in the organizations of their brain-minds.[37]

The silences between words and experiences are a manifestation of metaconscious knowledge and are usually referred to as myths. The unknown is implicitly symbolized by the organization of the brain-mind as an interpolation and extrapolation of a person's experiences into the life and world ordered by their culture. The unknown is transformed into elements of the known order yet to be discovered or encountered. In other words, the unknown becomes the horizon of individual and collective experience, beyond which the members of a community can only encounter more of the same. It may be unexpected and surprising but never threatening. It is as if what a community knows and lives has been absolutized but not in the sense envisaged by Feuerbach and others in the sociology of religion.[38] It is a kind of absolutization, in the sense that nothing that is radically 'other' can be experienced and lived as long as the cultural order is healthy and vital. During such times, a culture, via the organization of the brain-minds of its members, acts in the background to sustain their lives by giving them meaning, direction, and purpose. Anomie is minimal, and relativism and nihilism are kept at bay.[39]

The kind of absolutization that has banished every possibility that is radically 'other' and thus threatening to a cultural order has a second dimension. To our simple model once again, every cluster of relationships in the organization of the brain-minds of the members of a culture implies metaconscious values. For example, the cluster of many differentiated face-to-face conversations distinguishes those occasions when someone made too much eye contact from those when the eye contact sustained the conversation, the occasions when someone stood too close (thus appearing pushy) from those when people stood at the normal conversation distance, the situations where the usual synchrony between body rhythms was broken (because someone was uncomfortable with a person who had a disability) from those where this synchronicity helped to sustain the conversation, and so on.

Similarly, eating experiences differentiate those occasions when people were annoyed because someone messed around with his or her food from those when everything proceeded normally, forming the metaconscious knowledge that eventually leads to the explicit value that food is not to be spilled. In other words, every cluster of differentiated relationships builds metaconscious knowledge about which of these are good or bad, useful or useless, polite or rude, and so on. Since these clusters of relationships are differentiated from each other, a metaconscious hierarchy of these values is also built up. Such a hierarchy thus metaconsciously identifies what in a culture is so important and essential that without it people could not imagine who they would be, how they would live, and what their world would be like. Such an entity has literally made the members of a community and their world who and what they are. To use religious language, it has created and sustains their cultural order. Owing to the absolutization of what is known and lived, this most important element in the life of a community becomes much more than simply the most important and hence the greatest good that is known to its members. Any radical alternative becomes unimaginable and unliveable; hence the greatest good that is known becomes the absolute good in all of reality. It becomes the central myth or sacred that during prehistory was related to what we call *nature*, during history was related to *society*, and more recently was related to the *life-milieu of technique*.[40]

A sacred and myths together constitute the deepest and most fundamental metaconscious knowledge of any culture because that knowledge includes each and every experience of a person's life lived by means of a culture. They symbolically create the lives of all of us by transforming our experiences into moments of our lives, and these moments into the lives themselves. They are much more than a kind of spiritual backbone, because they root and orient these lives in an ultimately unknowable reality and thus protect us from relativism, nihilism, and anomie. Their role is the equivalent of DNA in physical human life. Together with the hierarchy of metaconscious values that it anchors, this metaconscious knowledge works in the background to sustain human life in several ways. First, it creates a cultural order within the unknown that is capable of protecting itself from all foreign elements that cannot take their symbolic place within it. The unknown becomes the horizon of human experience, and the relationship between people and their cultural order becomes analogous to that between fish and water. Second, within this cultural order, situated within the horizon of human experience, individual and collective human life is given meaning, direction, and purpose by means of a hierarchy of metaconscious values anchored in the sacred or central myth. It is this second component that is most vulnerable to desymbolization, as will become evident in later chapters when we examine the emergence of mass societies. It is this desymbolization that postmodernism has mistakenly interpreted as a positive development.

Third, without the creation of a sacred and myths, a dialectical cultural order would not be an order of what is true, as opposed to what is real. When an order of what is true sustains individual and collective human life, the sacred and myths transform everything people know and live into what is true and reliable. Across the horizon of people's experiences will flow more of what is yet to be discovered and lived, in a manner that cannot disturb the cultural order. During a historical epoch of a society or a civilization, people can confidently get on with their lives because what they know and live is true, and what they do not yet know and will experience in the future will not alter this in the least. Once again, their culture sustains their lives by working in the background via the organization of their brain-minds. Of course, there is nothing absolute about a sacred and myths, and reality will gradually impinge on them to usher in a non-cumulative transition to another way of life. This marks the beginning of another historical epoch in the journey of a community, or if this cannot be achieved, a decline and possible collapse may follow.[41] Because a cultural order symbolizes what is true and reliable and not what is logically consistent, a non-cumulative development in the journey of a community has little in common with the appearance of an anomaly and the triggering of a scientific revolution in the history of a discipline or a specialty.

A few details of the above development are worth noting. Initially, the boundaries between the lives of babies and children and the unknown constitute the loci of playful discovery. This playfulness is safeguarded by two factors. The first is the protection provided by the processes of differentiation and integration, which prevent their lives from being swamped by the 'noise' of undifferentiated stimuli representing the unknown. What can be experienced and lived is thus restricted to what can take on a symbolic place relative to everything else in their lives. In addition, the dialectical ordering of the relationships that babies and toddlers can establish allows for no gaps. The potential 'silences' between the differentiated elements of the self, the expressions of that self in relationships, the objects of these relationships that form the cultural envelopment, and the unknown are not threatening, because they are unaware of these silences. Hence, the potential threat of these silences to the order of their lives recedes as ongoing differentiation unfolds their lives. This is accompanied by a diminishing of their playful attitude by which they sort out what is threatening and what is not. Second, the courage and commitment to this playful attitude is rooted in the loving support of others, ushering them into a life and a world that make sense. The situation is analogous to the one in which a child confidently strolls into a large store at the hand of a parent. As the parent stops to examine an article on a shelf, the child may feel secure enough to explore another part of the store. This confidence can be shaken when the parent is out of sight and no support, such as a watchful glance, is forthcoming. The unknown may now appear overwhelming, and the child bursts into tears!

It does not take long before undifferentiated stimuli largely disappear from the lives of children. The earlier example of a child being taken to the zoo and pointing to a giraffe as a funny horse is illustrative. This situation occurs when a child's prior experience is sufficiently differentiated, but something appears somewhat alike, with the result that he or she can relate to it in a tentative manner. The unknown gives rise to embryonic perceptions and relationships while it is increasingly experienced as more of the known with some surprises. There are myth-related dimensions to these developments as well. These may be illustrated by the simple model developed in an earlier section. As a consequence of the process of differentiation, what something is derives from what everything else is not. What something is therefore depends on the overall development of the child's life in the world. The more this life is differentiated, the more what something is (as a consequence of being differentiated from everything else) is dialectically compatible with the 'true' that the child has come to live. Any relationship with something new thus reflects the order of what is true in the lives of children and teenagers as a conjunction of their selves, their lives, and their world.

In parallel with these developments and as a consequence of the process of integration, what something is can be intuited as a result of the metaconscious knowledge that has been accumulated. Again, such knowledge depends on the level of dialectical ordering of children's selves, their lives, their cultural envelopment, and its relationship with the unknown. This ordering also reflects the embryonic development of the myths of their community.

The embryonic developments of these myths manifest themselves in daily life by the possibilities that are never spoken or acted on, thought, dreamed, hoped, or feared. Linguistically speaking, these are the silences between words and sentences. These possibilities, which can be explored by another culture but cannot be lived in this culture, are as important as those that *are* lived. As noted, this is evident from the diversity of human cultures that have dealt with comparable situations in incompatible ways or that have prevented such situations from occurring altogether, for the same reason – incompatibility with their way of life and culture. The embryonic development of myths increasingly removes from the domain of the possible everything that is radically other and thus dialectically incompatible with it. This role of myths in earlier societies is relatively well known and accepted.

The pursuit of what is true does not merely eliminate what is incompatible with a way of life and culture. It also orients human life, since everything is not equally meaningful or valuable. In other words, the reality as it is lived by a community is transformed into what is true and reliable for individual and collective human life.

In contemporary societies, all this is very different because of the high levels of desymbolization of their cultures, via which other cultures are perceived. We take on or discard 'deep' elements from other cultures, such as their moral, religious, or artistic ones, much like we do clothes or cars. Without these high levels of desymbolization and the accompanying decentring of our lives and cultures, all this would be entirely impossible. Our lives are centred primarily on a technical order and only secondarily on a greatly weakened cultural order, to which we can add or subtract anything as long as it does not interfere with the former.[42] This makes it very difficult to understand other places in the world where traditional, moral, artistic, or religious elements manifest their traditional myths. All this distracts us from our own situation of having to root our existence into an ultimately unknowable reality by ruling out everything that is radically other and therefore has the capability of reintroducing a threat from the unknown.

Today, under the desymbolizing pressures of science and technique, we have confused reality with truth, taking the former to be what can be measured and technically manipulated through the course of scientific research. Everything that is enfolded and dialectical and thus non-logical cannot be included. It is easy to understand why, during a cumulative period in the development of a scientific discipline when a particular 'intellectual map' is being filled in, we take it for granted that we are getting closer to reality. Then comes the inevitable scientific revolution, and we reconceptualize what is real, but we continue to confuse exactness with truth. In the same vein, what people in the Middle Ages (or any other period, for that matter) considered to be real in daily life is not real to us. Nevertheless, future generations will surely look back to our time and disagree with what we deemed to be real.

The same kinds of conclusions can be reached by recognizing that the symbolization of any life-milieu (such as the one we call nature) creates a sphere of freedom. For example, symbolizing a dead tree trunk floating down a river in terms of its potential meaning and value for human life is to go beyond immediate experience to create a sphere of possibilities, beginning with imagining and then building something equivalent to a canoe followed by new modes of fishing, travel, and warfare. Doing so for every constituent of a life-milieu greatly enlarges this sphere of possibilities and distances the group from that life-milieu. Such a sphere is enlarged still further by symbolizing all human activities and events. For example, the symbolization of certain human relationships opens the door to the institution of marriage. Similarly, going beyond the immediate experience of death by means of symbolization opens the door to ritual burial as an integral part of a religion.

However, within this newly created sphere of human possibilities some order has to be established if the symbolic relating of everything to everything else is to fend off relativism, nihilism, and anomie. Symbolization creates a distance between a human community and its life-milieu. This distance opens up a sphere of human freedom, but to make human life liveable within that sphere, the symbolic relating of everything to everything else that created this sphere of freedom must be anchored in myths. During the growth phase of the culture of a community, myths make the newly found freedom acquired by symbolization liveable by anchoring its meanings and values. As the culture matures, these myths prevent the community from effectively dealing with new discoveries and possibilities that reveal the unknown to be very different in some ways from what is known and lived.

It is worth noting that the Jewish and Christian traditions, which so deeply influenced Western civilization, demanded that human life remain open to the One who was entirely other, with the result that these traditions were understood by some as being iconoclastic with respect to the myth of the culture of one's community in general and with respect to its morality and religion in particular.[43] These traditions recognized that human life in an ultimately unknowable reality cannot escape the necessity of myths (symbolized as false gods). These teachings may be interpreted as an astonishing insight into the role of myths in human life. When communities turn for help to the greatest good they know, they are not aware of its limitations. Having symbolized the unknown as more of the known, the greatest good they know has become the greatest good in the universe, past and present. Myths thus make it all-powerful. When the members of a community are confronted with situations that reveal its limitations, they find it next to impossible to turn to radically other modes of thought and action, thereby making things even worse. This will gradually lead to a growing desacralization of this greatest good and a subsequent mutation in the culture or it will usher the community towards a descent into chaos and its eventual collapse. We shall soon discover how this describes our current predicament which, because of secular myths, prevents us from seeing that, despite the spectacular successes of the scientific approach to knowing and the technical approach to doing, these approaches are like any other human creations. It is our inability to recognize their limitations that prevents us from effectively confronting our economic, social, and environmental crises.

The True, the Other, and Myself

We have suggested that myths transform what we know and live into reality itself, minus some details yet to be discovered and lived. It is the result of the unknown being symbolized as more of the same. The possibility that human life in the world may be radically different has symbolically been eliminated. Now all that is left is our lives evolving in relation to our life-milieu. Prior to industrialization this placed a community in a symbolic universe that was true and utterly reliable. Today it places us within a universe in which desymbolized facts are not only possible but are the only reliable elements, thanks to our current secular myths.

It is difficult for us to imagine the roles that languages and cultures played before industrialization. We have likened the relations between a community and its life-milieu to those between fish and water. Until our own civilization, the personal word has always launched communities towards the order of what is true as a way to get an ultimate grip on everything.[44] Moreover, the personal word protects the freedom of the speaker and the listener, but it also opens up the possibility of ambiguous and paradoxical meanings and the violation of what is true by the lie. I am free to speak a word to another person or to remain silent. When I do speak to others, my words leave them free to do and think as they see fit. The situation would not be the same if my personal words were of the order of what is real. If my words were simply 'facts' and did not go beyond what is, others would not be left their freedom. Of course, there are many situations in which the word is transformed into an instrument of propaganda, advertising, or public relations for the purpose of influencing others as efficiently as possible to think or act in certain ways. In these cases, there is no personal word but only objectivized language in the service of technique. There are no speakers or listeners, and nothing meaningful is communicated.

The use of personal words places me before the mystery of others and thus before that of myself. Being face to face with others, I am not sure how they will interpret my words or what effect these may have on them. It confronts me with my own mystery of not fully understanding myself. Hence, I cannot be entirely sure of the meaning of my words and thus of what I say to others; they, in turn, cannot be certain that they have understood the meaning of my words. As a result, there is uncertainty, and I may have to start over as I seek to understand the other and myself. Such uncertainties also come from the profound ambiguous nature of our lives in the world despite the presence of myth, especially when cultures are young or old.

The personal word establishes dialectical relationships between the members of a community. We have lost track of the fact that the use of words has little in common with the transmission of information. Words have the potential of creating and sustaining relationships within the order of the true. If the other and myself are the same, there is nothing to talk about since we would both know the same things. On the other hand, if I have nothing in common with the other, no words can have any meaning because what they symbolize exists in one world and not the other. Hence, words depend on two people establishing a dialectical tension between being individually unique as well as culturally typical. In this way, the other can enrich my life by bringing in something new, without that new thing being dialectically incompatible with my life, and vice versa. At the same time, this implies the possibility of misunderstandings, conflicts, and lies since these also are of the order of what is true even though they violate it in some way. At best, I can reveal something of myself, the relationships of my life, the objects of those relationships in my cultural envelopment, and the unknown to the other. I do this in the knowledge that all this is different in the other, that I am aware of how uncertain, ambivalent and contradictory all this is in my life and the other, and that the words we speak to each other are the most fragile means that, once spoken, stir the air and disappear.

This kind of dialectical immediacy vanishes when the other is not present, and I can communicate only by transforming my words into some kind of object on paper or some other means of recording them. If the other is not there and I do not know him or her, then I can only speak or communicate with others who are reified as 'my average reader or listener' or a targeted group with a certain 'profile.' As a speaker, I am also reified because my words no longer reveal anything of myself but undermine the freedom of others by manipulating them on behalf of whatever they believe I represent. For those on the receiving end, there is no longer a speaker either but simply anonymous communication objects. Since the appearance of newspapers, followed by the mass media, these 'word-objects' have become so common that we tend to confuse them with what little personal word is left in the service of the order of what is true. Coupled to this is the scientific approach of reducing all this to what can be measured, quantified, and incorporated into a system, which we then treat as the reality of language and communication. Meanings are the first to disappear, followed by the speaker and the listener. Everything is reduced to what survives scientific abstraction: communication objects, functions, and structures. Anything else is subjective, arbitrary, and thus inadmissible to what is real. The link between language, culture, and the order of what is true is all but destroyed in our contemporary civilization.

Despite all these issues, talking to babies and children continues to play a decisive role in their lives. We have seen that people can enrich each other's lives by speaking together, provided that they are sufficiently different but similar enough to understand each other. This is certainly the case when infants and children talk to the adults around them. Such enrichment strengthens the bonds between family members and friends and constitutes the vitality of any social group. A group is viable when each member enriches the lives of the others by contributing new and interesting things. In the same vein, societies are vital when there exists a strong dialectical tension between individual diversity and cultural unity. Over time, these dialectical tensions tend to run down. People become more familiar with each other, with the result that it becomes more difficult to enrich each other's lives with something new. The commitment to a relationship or group may weaken unless collegiality, friendship, or love revitalizes and recreates it. In general, if in a society the cultural unity imposes itself too strongly on the individual diversity, there is little reason to belong to that society, other than the security found in conformity or the fear of the consequences of non-conformity.[45] On the other hand, if individual diversity undermines the cultural unity, people may find themselves living in different worlds with little in common, resulting in the danger of the community's disintegration.

All this is far from theoretical in the lives of babies, children, and teens. Learning to live with others in the world begins with dialectically separating elements of their lives in the order of what is true. By means of language, dialectical tensions are established with adults to create the kinds of social bonds that are essential for human well-being. The influence of science and technology on our awareness of ourselves and the world has contributed to a great deal of confusion as well as an oversimplification of the complexity of new generations learning to live in an ultimately unknowable reality. In the next chapter, we shall consider how desymbolizing experiences have made this process a great deal more complex, making growing up in a contemporary mass society extremely difficult.

2 Desymbolization: Losing Touch with Ourselves and the World

Undermining Symbolization

Babies and children growing up in our civilization soon encounter situations that are very different from the ones described in the previous chapter. Thus far, the limitations imposed on their development have been those of the organizations of their brain-minds, which excluded from their lives whatever could not be lived by taking on a symbolic place there. Such limitations play a positive role; they protect the children's lives from being swamped by disorder. The new kinds of situations to which we now turn are characterized by the limitations to symbolization resulting from these situations themselves, as opposed to arising from the levels of development of the organizations of their brain-minds. Making sense of and dealing with these situations now has a measure of imposed desymbolization. Anything communities create or do that impedes symbolization will diminish the abilities of babies and children to fully live their lives with others in the world by relating everything to everything else in order to maintain the integrity of their lives, the lives of others, and the world that sustains these lives. In other words, these situations impede symbolization to make it something less than what babies and children are capable of, given their level of development. Any such impediments create selective pressures on the evolution of the lives of individuals of any age, on the historical journeys of their communities, and possibly on all of humanity as a symbolic species. However, unimpeded symbolization is no guarantee of good and responsible human lives. In combination with other factors, it can help to create and sustain pathological cultures.

The consequences of any desymbolization may involve some or all of the following: limiting the growth and integrity of the self within a tissue of relations with others and the world; restricting the range of self-expression; limiting the relations that flow from this self-expression, and thus the ability to evolve this tissue of relations with others and the world; interfering with the integration of the objects of these relations into a liveable cultural envelopment for a community in an ultimately unknowable reality; and obscuring the horizon that encompasses the experiences of a community, by symbolizing the unknown as more of the known yet to be discovered and lived. Significant desymbolization may lead to a failure to make sense of things: being overwhelmed by circumstances, leading to a sense of helplessness and powerlessness to do anything about it; unsatisfactorily sustaining and being sustained by social ties with others; and a general inability to get on with life, as a result of being cast adrift by being uprooted from an ultimately unknowable reality. These conditions may lead to anxiety, depression, withdrawal, substance abuse, violence, mental illness, and, in the extreme, suicide. In this way the members of a community can become culturally ill. When this illness becomes widespread in a society or civilization, it may lead to relativism, nihilism, and anomie. Eventually a 'cultural death' may follow, when a society or civilization collapses and disappears because its cultural design for making sense of and living in the world is no longer viable.

As a symbolic species we are vulnerable to cultural illnesses and death. As discussed in the previous chapter, the organizations of our brain-minds metaconsciously interpolate and extrapolate all our experiences into lives lived within communities that are engaged within the world. Doing so creates the deepest forms of metaconscious knowledge, which correspond to the very fabric of our lives and our world, and thus to the myths of our cultures. Any undermining of these myths will tear at the fabric of our lives, leading to the above kinds of symptoms. In other words, our dependence on myths makes us, as a symbolic species, spiritually very vulnerable, except when these myths are safely out of reach of our conscious awareness. All this may have been dismissed by some pundits as our inability to let go of the past on our way to becoming a secular humanity. This explanation may once have been a hopeful idea, but it now fails to satisfy those of us who live in the industrially advanced world, judging by such indicators as the epidemic use of antidepressant drugs and our widespread impotence in creating and sustaining intimate relations with those we love.

Any attempt to face our cultural fragility must not be confused with an idealization of the past because of its apparent guidance by viable moralities and religions. It was precisely such moralities and religions that were attacked in the opening chapters of the Jewish and Christian Bibles because they were a threat to humanity's freedom and thus to our ability to commit ourselves to anyone or anything. Doing so becomes impossible as a consequence of being alienated, which, according to these traditions, involves our being possessed by someone or something in the way that a slave was possessed by a master in antiquity.[1] A slave may commit himself and his life to the woman whom he loves, but all of this amounts to nothing if his master decides to sell him the next day. Freedom is a prerequisite for love. We shall return to these issues because Western civilization cannot be understood without the influences of these two traditions, and the forms of desymbolization now being considered are unthinkable apart from them.

Desymbolization occurs when a situation restricts embodiment, participation, commitment, and freedom, and thus the ability of our lives to fully work in the background. We shall begin by considering the restriction of embodiment. It is difficult to exaggerate the influence that our embodiment in the world has on all our experiences. Our bodies constitute the vantage point from which we experience the world as a consequence of the kinds of senses we have and where and how these senses are located in our bodies, as well as how well we move these bodies around to arrive at more favourable vantage points for making sense of and being involved in each moment of our lives. Just imagine how different our experiences would be if our eyes were located on the tops of our feet, if we had only one ear located at the back of our head, or if only certain portions of our skin were sensitive to touch. Our bodies, as our physical vantage points, permeate all our experiences and hence our lives.

Moreover, we deliberately use this vantage point to improve our grip on a situation, as when, for example, we manipulate a new tool to get the feel of it, or examine a painting from different distances and angles. If we stand very close to a painting, we may get a good grip on how the artist's technique of brushing contributed to the creation of various details, but we shall fail to do so in terms of how these details contribute to the overall picture. To accomplish this, we need to back up in order to get a good grip on a cluster of details, including their interdependence, but we may still not get a good grip on the picture as a whole. If we back up a little more, we may still be close enough to appreciate the details, but they have now become local manifestations of the picture as a whole. We may now be at the point that if we back up any further, we may continue to get a sense of the overall picture, but we shall progressively lose the nuances of its local details. In order to further appreciate the painting, we may move sideways to explore how the painting appears when it reflects the light from different angles.

The influence of our physical embodiment goes well beyond our experiences. Recall that babies exploring their physical embodiment develop an awareness of their physical selves and that this awareness in turn helps to build an awareness of their social selves, and eventually their cultural selves. Our social and cultural embodiments in the world are thus permeated by our physical embodiment as well. Furthermore, the influences of our embodiment in the world on our experiences affect the organizations of our brain-minds and thus the way our lives work in the background of each experience. Such working in the background clearly has a physical, social, and cultural dimension. It should also be noted that the way our lives work in the background can be different for the focus of our attention and the remainder of that experience. This can occur, for example, when watching television.

Our embodiment in a situation is physical, social, and spiritual via a culture, or non-cultural via a scientific discipline or technical specialty. Hubert Dreyfus accurately predicted the failure of artificial intelligence to build daily-life knowledge and skills into computers, on the basis that these devices are not physically embodied in the world.[2] This ought to come as no surprise. Without a physical embodiment, it would be next to impossible for babies to make some sense of the blur of stimuli detected by their senses. They learn something about themselves relative to certain blurs of stimuli, and about these stimuli relative to themselves. It is the beginning of metaconsciously and dialectically separating everything from everything else to create an order of what is true for their lives. Nothing in this order can be separated from everything else by means of a definition so as to constitute the equivalent of a fact of their life. With the onset of language, their individual orders of the true begin to converge with those of their significant others and eventually with the cultural order of their community. In other words, while the order of their lives is initially centred in their physical selves, it eventually becomes centred in the cultural order of their community.

Our embodiment in a situation can be limited in many different ways. Full embodiment occurs when we talk to someone else about the immediate situation or when we work with our hands. When we talk about something that is not a part of the immediate surroundings, our minds are embodied in the foreground of the experience and our bodies in the background, although they frame the conversation with our eye etiquette, facial expressions, and gestures. Children experience this when they are read a story or learn to read a book. It probably matters little whether the foreground takes place in the human world or in a fictional non-human world because children can relate to them as extensions of their own world.

Our embodiment in a situation is often restricted when a relation is technically mediated by a device or a technique. Compared to a face-to-face conversation, a telephone filters out the eye etiquette, facial expressions, and gestures of the people talking, which is why some people prefer not to use it for important or sensitive conversations. In a bureaucracy, people deal with one another according to organizational techniques that distribute its mission over several departments, which in turn are broken down into functional units, job descriptions, and reporting relations. All this mediates between the people in the bureaucracy, and this necessitates the well-known adaptations in their behaviour. Similarly, hand tools mediate between people and the workpiece. Such tools permit some feedback, in contrast to computer-controlled machinery in a plant, which does not permit operators to develop a feel for the materials with which they work. The learning or application of mathematics in a scientific discipline or technical specialty can only occupy the foreground of experiences, because these bodies of knowledge constitute domains built up from separately defined entities that enter into logical relationships expressed by mathematical equations. Hence, the foregrounds of these experiences are incommensurate with their backgrounds, reflecting a dialectically enfolded world.

Desymbolization can also occur in situations that restrict our participation. Our involvement can range from being very passive to being highly active and from being unidirectional to being reciprocal. It is likely to be minimal if the situation has little meaning or value for us. Low involvement may also result when the consequences are not understood or largely beyond our control. One of the most common examples flows from the presence of the mass media in our lives. In the middle of the nineteenth century Kierkegaard became concerned with the rise of a disinterested curiosity that first occurred when it became common for people to discuss in coffeehouses what they had read in their newspapers.[3] Although it was seen by some as rational disinterested reflection and an essential feature of a democratic society,

Kierkegaard saw it as a source of levelling. People were talking about situations in which they were not involved, of which they had no direct experience, and yet about which they formed opinions without much thought or reflection. Holding these 'public' opinions did not commit them to anything, nor did they have to act on them, and hence there were no real consequences for their lives. People were not involved in any meaningful way. They applied 'principles' to judge events found in their newspapers. They could talk about these events and change their opinion as they read more or listened to others. All this could go on forever, and no one had to stand behind these opinions or be engaged by them.

Hubert Dreyfus points out that today we are engulfed by these kinds of situations when we live on the Web.[4] It contributes to the creation of what is called *the post-modern self*, open to all possibilities but committed to none, and hence incapable of suffering from disappointment. We shall shortly see that the use of computers and living on the Web create a similar existence. These kinds of analyses are complemented by those showing that, following the Second World War, entirely new kinds of societies emerged that were, and continue to be, characterized by highly desymbolized cultures, which must be reinforced by integration propaganda.[5] In the case of children, low involvement tends to occur when their playful attitude to others and the world is not adequately sustained by a sense of confidence and security that comes from being loved.

A passive involvement in a situation is characterized by our lives working in the background in such a superficial manner that we are largely excluded from a relation or a situation. Everyone in a mass society participates in the evolution of public opinion, which is very different from private opinions. The latter involve a great deal of active participation by gathering information, critically reading and evaluating it, and arriving at an opinion for which a person is willing to take some responsibility. There are many other situations that make it very difficult for our lives to work in the background in meaningful ways. For example, the mechanization and industrialization of human work required its reorganization in the image of the machine by means of the technical division of labour. It led to the separation of the hand from the brain. Simply put, those who organize the work do not use their hands, and those who do are not encouraged to use their brains. In either case, people's lives cannot fully work in the background. Those who work with their hands must suppress their education, experience, creativity, and motivation. Those who work with their brains frequently do not understand, and hence cannot benefit from, the experiences of the people who work with their hands. Similarly, computer-based work must be reorganized to take advantage of the strengths of machines, which are based on repetition and their context-limiting approach to relating to things by means of rules and algorithms. As a result, this kind of work is diametrically opposed to human work, which is based on a nervous system that was evolved to make sense of and deal with a living world in which nothing ever repeats itself in quite the same way. We shall examine all this in full detail in the following chapters. For now, we shall simply note that machine-based work does not permit participation on human terms.

In assessing our ability to participate in relationships, it is essential to recognize that many technical devices simultaneously function as means and as mediation. For example, telephones permit conversations with people who are far away, but at the same time they mediate these conversations in a non-neutral manner by filtering out eye etiquette, facial expressions, body rhythms, and subtle sounds that are not picked up by the microphone. Some devices do not allow for an immediate response or for any response at all. From this perspective, our participation in a conversation with someone else diminishes when it is mediated with a videophone, a cellphone, a regular phone, an email, a fax, or a Twitter message. Similarly, our participation in selecting a candidate for political office in a town hall meeting diminishes when it is mediated by talking to someone who attended it, by watching the event on television, by listening to it on the radio, or by reading about it in a newspaper. Our involvement in making something with our hands diminishes when it is mediated by hand tools, power tools, human-operated machines, or computer-controlled machines. In all these cases, our experiences of these relations are impoverished to some extent as a result of a technical mediation, which may not constitute a problem provided that we are aware of what is being filtered out.

Finally, participation in technically mediated relationships is often unidirectional. Such is the case for the mass media and almost all modern organizations. We can neither participate in nor learn from these kinds of relations because of a complete absence of any meaningful feedback. In all the above examples, the restrictions on our participation in situations reduce the extent to which they might have been symbolized by being fully related to everything else in our lives and the extent to which these diminished lives can work in the background. As a result, these desymbolizing effects are cumulative and self-reinforcing. The more desymbolizing situations we encounter, the more the integrality of our lives will be diminished, the more our symbolic capacities will be weakened, and the more all situations will be affected, even when they themselves do not constrain symbolization.

Desymbolization may also occur when situations restrict our commitment. The level of commitment becomes important when we move from situations that minimally involve people to those that engage them. Interpersonal relations cannot develop very far without mutual commitment. The more another person learns about you, the more vulnerable you become to him or her. You must trust them not to take advantage of you. As the relationship deepens, so must your commitment to one another. We all become vulnerable because of deep-seated anxieties about our future, uncertainties about our ability to handle some situations, sensitivities about past failures, worries about unresolved issues, shortcomings in meeting our own or other people's expectations, and generally the gap between who we are deep down and what we appear to be on the surface. Hence, we are only willing to take a relationship so far unless it is accompanied by a growing trust that the other will not take advantage of what he or she knows about us. At the same time, we must not take advantage of the other person. Both people need to commit themselves to the relationship, to the point where they become less and less tempted to gain personal advantage by revealing something of the other when circumstances may tempt them to do so. Learning to trust someone and committing to a relationship with him or her is made all the more difficult in the context of a modern mass society characterized by other-directed personalities and an orientation towards performance.

Hubert Dreyfus again points to the importance of our physical embodiment in relationships, noting that our bodies are effortlessly working in the background.[6] Momentary glances, shrugs, or gestures can provide us with a sense of the relationship and where it stands in terms of commitment and trust. I look at the other, and he or she responds in a way that tells me exactly where he or she stands. For example, when we are both listening to another person seeking to convince us of a new idea, the other may reassure me in a subtle way (that goes unnoticed by anyone else) that it is best to hear the speaker out despite the fact that none of it is convincing in any way. A different reaction may be communicated in the same subtle manner. People who know each other well can do so with a certain look, a shrug, a raised eyebrow, or a slight tilt of the head because their entire life together works in the background. As a result, the slightest indication can inform the other of everything they need to know about how he or she is ultimately reacting to what is going on. Much of this is so subtle that teleconferencing does not provide us with a sufficient form of presence for our bodies to work successfully in the background. In a face-to-face situation, people have a much greater freedom in directing their body to test the grip they have on the situation or to direct subtle hints, and they have greater access to other people doing the same thing. They can thus deliberately advance or restrain the synergy between these mutual efforts, based on a deepening sense of the relationship.

Interpersonal relations are not alone in requiring a deepening commitment if they are not to remain on the surface. Much of modern teaching remains little more than acting as a 'talking head' that is accompanying computer-generated images of lists of bullets, equations, tables, and graphs. This kind of behaviour has little in common with real teaching. A teacher may move around a classroom to obtain a better sense of the mood of the class and its level of interest and understanding. If we accept the five-stage human-skill-acquisition model developed by Stuart and Hubert Dreyfus,[7] the teaching of a particular skill will begin by the transmission of elementary rules that have the well-known format of 'if ..., then ...' As students advance to the second stage of skill acquisition, they begin to discover that the rules do not fit certain situations and that additional rules are required. At this point, learning a skill requires little involvement and commitment on the part of the students. They did not invent the rules and thus have little personal stake in how well they work. During the third stage they learn to differentiate enough situations that they get a sense that very few situations really fit the rules. As a result, they must be encouraged to engage in problem-solving, asking in which respects the situation fits this or that rule, and in which other respects it does not fit. They must reason this through, and as they begin to invest more of their personal understanding, they become more vulnerable. If they really misjudge a situation, they could be reprimanded by the teacher and snickered at by other students. During the fourth and fifth stages of human skill acquisition, these kinds of risks become even greater. The students must now rely on an ever more developed metaconscious knowledge that has been built up from diligent practice of the new skills. Any failures may thus be interpreted by the teacher or other students as a lack of practice, signifying a lack of commitment to mastering these skills. It is increasingly likely that, part of the time, the practice of these skills may involve clients under the watchful eye of the teacher. At this point, it is the students and the teacher who become vulnerable to the client. A poor performance of the skill may damage the reputations of students and teacher alike and could turn clients away, with damaging results. All this is even truer when the subject matter cannot be desymbolized to any great degree, as in the case of the humanities, for example.

In the same vein, business relations cannot advance very far without building trust and commitment. Imagine the situation of two chief executives intending to negotiate a deal that has far-reaching implications for their companies and thus for their personal reputations and careers. In such cases, there is a limit to what can be negotiated by telephone or teleconference. Their lives cannot fully work in the background because of a limited physical embodiment, and hence their full experience cannot be brought to bear on the situation. In order to overcome these limitations, chief executives will often spend some time together to build up a metaconscious knowledge of the personality of the other by sharing various activities such as a workout, a game of golf, drinks before dinner, and a leisurely meal. As they talk, their physical embodiment in the situation communicates much more than what they are explicitly saying. 'How committed is the other to this deal?' 'How confident are they about its success?' 'If things get difficult, can they be relied on to help work it through or are they likely to hang me out to dry?' 'If the other person does not appear to be too committed to the deal, why should I stick out my neck?' These kinds of deeper intuitions about the deal can be tested only in situations where the experiences and lives of the two executives can fully work in the background.

In contemporary mass societies permeated by technically mediated relationships, our lack of embodiment exposes us to greater risks because it is this embodiment that directs what we can experience by means of symbolization (differentiation and integration). 'Is this or that subtle detail in keeping with our sense of the situation or is it a possible clue that we do not have a good grip on it?' 'Is this particular verbal or physical expression in line with my sense of where the relationship is going or is it a sign that something is changing?' In other words, directing the process of differentiation and integration makes it possible to experience details that are incommensurate with our sense of a situation.

In the previously mentioned five-stage human-skill-acquisition model, a great deal of emphasis is placed on risk taking in the process of advancing to the higher stages. Commitment expresses itself in a determination to not only learn the skill but also excel in it. Each time a person makes a decision to tackle a situation in one way or another, he or she takes a risk. If you do not have a good grip on the situation, you may well engage in the wrong course of action. By responding to situations in ways on which expert teachers can improve, students run risks, but without taking such risks it would be impossible for them to perfect their skills. Without outright errors or imperfect responses it would be impossible to advance a skill beyond a certain level. Risk taking is indissociably linked to getting feedback that is essential for learning. The importance of all this is far from being recognized in discipline-based knowing and doing. Trapped in the triple abstraction, most of the consequences of the decisions made by students or practitioners fall beyond their domains of specialization, where they cannot 'see' them.[8] We should ask ourselves what the students are learning and how practitioners know they have done the best possible job. The answer is obvious: what they learn is to obtain the greatest desired results from the requisite inputs as measured by output-input ratios, which are criteria that are only 'internal' to a particular discipline or specialty. Nothing can be said regarding the context, that is, how it will fit into and contribute to human life, society, and the biosphere. The flurry of attempts to strengthen professional ethics is therefore doomed from the start. So are the efforts to build in a consideration of criteria for sustainable development, or the addition of the 'end-of-pipe' optional courses to compensate for what is excluded by the discipline-based approaches.

In the case of children, the playful exploration of new situations, even with a minimal ability to deal with the consequences, tests the grip they have on a situation. By moving around and trying it again, they may get a different sense of it, and all this is essential for the building of confidence and trust. If things go terribly wrong, they may have to rely on an adult to step in. Passive activities such as watching an automated toy contribute little to this discovery process.

Finally, desymbolization may occur when our freedom is restricted. On the level of individual situations, freedom is a prerequisite for committing ourselves to someone, for learning a skill to the best of our ability, for receiving teaching that sustains our professional and private lives, and for being the best we can be in our lives. The aspect of freedom and its opposite, alienation, is rooted in the reciprocal character of our relations with each other and the world: as we change our surroundings by our behaviour, our surroundings also change us. In a great many instances, such subtle influences can synergistically reinforce each other to produce lasting effects in the course of years and generations. For example, since many of our daily-life activities are mediated by one technical device or another, the joint effects of this mediation are far from trivial.[9] Since this mediation also has a desymbolizing influence, the overall effect on our experience and culture is extensive.

Owing to this reciprocal character of our relations with each other and the world, the issue of freedom and alienation arises. If the influence that our surroundings have on us exceeds the influence that we can have on our surroundings, the situation is alienating. On the other hand, if our influence on a situation is much greater than the influence it has on us, we have imposed a measure of freedom. The concept of freedom has no meaning other than in relation to what alienates us, and each situation contributes to both the freedom and the alienation of our lives.[10] Any talk about human freedom is entirely meaningless unless it is in relation to what alienates us.

The tensions between freedom and alienation in all daily-life situations are rooted in the process of symbolization. In relating everything to everything else, it necessarily unfolds our lives in relation to a particular time, place, and culture, which is symbolized as the order of what is true. Our lives become bound (enslaved) to that order, absolutized by metaconscious myths. This enslavement protects us against relativism, nihilism, and anomie. It protects us, individually and collectively, against human life deteriorating into an endless series of possibilities that are all equally good or bad, liberating or enslaving, sustaining life or undermining it, and so on. Traditional religions and morality bound people together in this enslavement and protection from anomie. It allowed people to journey together and to make history because of the inevitable process of sacralization and desacralization.

No traditional religion or morality began by considering what would alienate the community, in order to deliberately create a sphere of freedom in relation to its alienation. If this had been the case, the process of sacralization and desacralization would have been driven by a daily-life striving to become less enslaved and more free. Such a struggle for freedom is different from iconoclasm or demystification, which would simply weaken the roots of a community without building new ones. The historical journey of a community is the exact opposite. For a time, its current cultural order may represent the interpolation and extrapolation of almost all individual and collective experience. However, the cumulative adaptation of this order to ever new situations gradually runs into more and more difficulties, and eventually a mutation becomes necessary. This process is not a struggle for freedom, but a defence of what has protected the community from relativism, nihilism, and anomie. As soon as this protection becomes too weak, another must be found. Hence the process of sacralization and desacralization. The role of the sacred and its accompanying myths represent our spiritual commitment to what makes individual and collective life possible in a community sharing a culture. In Western civilization the messages of Judaism and Christianity proclaimed a liberation from this commitment. They depicted it as an enslavement to sin, but Christianity itself quickly became morality and religion, with the same disastrous consequences as all other moralities and religions.

Nevertheless, there is a tendency to emphasize the alienation created by these myths, as if we were free. It must be remembered that these myths come about by dialectically separating the elements of our selves, the expressions of these selves in relationships that weave the fabric of our lives, the objects that form our cultural envelopments, and the symbolization of the unknown as the horizon of the experiences of our lives. This dialectical separation (differentiation) and ordering (integration) is accomplished under the condition of alienation and not of freedom. Without any exception that I am aware of, what is missing is the establishment of a dialectical tension between this condition of alienation and freedom by means of a culture of a community.

Such a tension cannot be established by means of lofty principles. We know this all too well. Whenever a political regime claims to possess the truth as a consequence of communism or of democracy, the results, though very different, are disastrous. Besides, if we could ever achieve the truth, human life as we know it would come to an end. Everything in our lives and community would be perfect and thus incapable of evolving. Life would end in a death of endlessly repeating this truth. If life without myths is impossible, and if myths alienate our lives, then we need other commitments that can challenge this alienation in order to create some liberation from these bonds. A complete demystification is out of the question, however, because it would immediately translate into anomie.

To sum up: living a particular situation can vary enormously, and, along with it, the level of symbolization. If we are fully embodied in a situation and are entirely involved because of our commitments, we shall undoubtedly distort the situation and our lives by defending what makes these lives liveable but which is ultimately not true. We may seek to create a measure of freedom in our alienation so that the processes of differentiation and integration can reach their full potential and help us live as full a life as possible. We can hardly expect this to be the usual situation. As a result, a fully differentiated and integrated self, life, cultural envelopment, and horizon of experience are impossible. There can never be a unitary self living life to the fullest in a true cultural envelopment within a clear horizon. We are all divided selves living in alienated cultures within fuzzy horizons. However, it is clear that the conditions in a society are both the cause and the effect of the degree to which all this occurs. In the remainder of this chapter we shall focus on the most common desymbolizing experiences encountered by children and teenagers as they grow up.

Television as an Introduction to What Is Real

In the first chapter we showed that by means of the process of differentiation and integration, symbolization seeks to establish what is true for individual and collective human life, as opposed to what is real. Our civilization has all but turned this around. It primarily relies on an entirely new order necessitated by the process of industrialization. Although the creation and insertion into the cultural order of this new order of the real will be examined in more detail in the following two chapters, we shall see how babies and children encounter this order and are prepared for it by the watching of television, which gives rise to a set of desymbolizing experiences in their lives.

From the above perspective, it is clear that people's experiences of watching television involve little embodiment, participation, commitment, and freedom as compared to their other experiences. To begin with, the physical embodiment in what is seen is decided for the viewer by the camera persons and the producer. The skill of a camera person includes the selection of a vantage point, camera angle, close-ups, illumination, et cetera to obtain the best possible grasp of what is being seen. The skill of the producer includes selecting sequences of frames in order to decontextualize them from the rest and then to recontextualize them to jointly make a program that is likely to attract and hold the interest of the target audience. The social embodiment of these camera people and producers in the events being shot and recreated for a television audience is usually very limited. For example, to help make a news event, a television crew is dispatched to a community into which they are rarely socially embodied. Most likely, they will have little understanding of the developments that led up to the event and how it may affect the community in the future. As a result, the event essentially comes out of nowhere and may as well cease to exist once the crew packs up. The aspects of the event that are shown are those that 'work' on television, which requires at the very least that some of the event's prominent features be visible. However, showing someone who is distraught means little without a supporting sound bite explaining what cannot be seen. This sound bite may be followed by a comment from the journalist related to the significance of the event. Nevertheless, television is all about images and their primacy over the word. It is visual-audio and not audio-visual. Given the almost complete absence of social embodiment, what the television viewer may take away and what a person living in the community may directly experience are bound to be vastly different. It is also clear that when a crew is sent out to report events in other societies with different languages and cultures, these issues are even more significant, with the result that the event will be distorted by the culture of the television crew. In such cases there is no cultural embodiment whatsoever.

From the perspective of the viewers of a television program, the concept of embodiment is virtually meaningless. There is no daily-life equivalent of someone telling them a story or reporting on events. What they encounter on their screens are the results of an organization engaged in a process that is guided more by the necessity of making it all work for television than by getting a grip on things through symbolization. It makes little difference whether or not these results are presented by a spokesperson. The viewers will symbolize what appears on their screens as an 'image-story.' The foregrounds of these experiences are discontinuous with the backgrounds. The former present a highly decontextualized and recontextualized representation of human life in the world, while the latter represent the surroundings in which the television set is watched. The viewers' participation in making sense of these foregrounds is essentially passive and metaconscious. The resulting experiences participate in the formation of public opinions, as opposed to private opinions that are arrived at through active involvement in exploring the issue further by a variety of means to get a better grip. Nobody is committed to public opinions, which are as fickle as the wind. In a mass society, viewers have only a limited freedom over their watching of television if they want to stay in touch and have something to talk about with others.

How deeply the experiences of watching television are desymbolized can be seen from the following six consequences for human life and society. The first five apply mostly to adults, but the sixth will show how children are prepared for all of them. The more the members of a mass society rely on television programs to stay in touch with the world in which they live, the more they are exposed to the most discontinuous and fragmented view of things. Events come out of nowhere, appear on one or more broadcasts for at best a few minutes, and disappear into nowhere. The 'events' begin with the appearance of a television crew and essentially cease when the crew packs up. Making sense of this by means of symbolization is equally impossible for the crew, the producers, and the viewers.

To convince ourselves of how deep the problem goes, it is instructive to make an inventory of all the events appearing on a particular newscast for one week or more and compare this to how significant events in our own lives emerged and evolved. The differences represent a deep structural problem faced by any member of a contemporary mass society. We all need to stay abreast of developments around the globe because any one of them can have serious consequences for our lives. Nevertheless, it is impossible to rely on the organizations of our brain-minds to symbolize them in the way that we do daily-life experiences. The best we can do is to pick up the clues provided by the talking heads and sound bites coming from our televisions.

The clues generally relate these events to a kind of world view constituted by some of the following elements. Today science may not know everything, but there is nothing that, with further investigation, it cannot uncover in the near future, and although our technical capabilities may fall short of meeting this or that challenge, we shall soon find a means to do so. It is only by education that we have any chance of staying in the global race, and it is by economic growth and free trade that we are able to deal with unemployment and poverty. If the present government is inept, we shall soon be able to elect another and move ahead again. Medicine will soon deliver us from this or that serious disease. By buying lottery tickets, we have a chance at attaining a secular heaven on earth. These and other secular mythologies reassure us that, despite some difficulties, we are going in the right direction.

A further distortion of what is really happening to our lives in the world results from the fact that, unlike the visual experiences of daily life, an 'edge' is skilfully imposed around what we see each and every moment on our screens. Together, these edges constitute the horizon of our television experiences. There are no such edges in our daily-life seeing in the world because they can have no meaning for our lives. All we need to do is to shift our gaze or move our heads, and we experience no such edges. We have interpreted them out in the course of symbolization, with the result that the foreground can never be separated from the background. What we see on television appears to be interpreted in the same way. The 'visual silences' that lie beyond the edge of the screen are lived as more of the same. The one plane that crashes, the one train that derails, or the one person who is assaulted thus becomes symptomatic of what lies beyond what we see: a world that appears to be much more dangerous than it really is. Although this is a very obvious example, the same kind of extrapolation turns every story and every event into a kind of myth-image as a result of our metaconscious interpolation and extrapolation of what we see. Our symbolization of human life in the world is thus further distorted by television.

If everything is as highly desymbolized as it appears, why do we become so readily absorbed in our television viewing? Low embodiment, participation, commitment, and freedom surely ought to turn us away from this pastime. Part of the explanation is the artificial enrichment of any program by regular sensory 'jolts.' These include scene changes, camera changes, zooming in and out, flashbacks or flashforwards, sound effects, music, and much more. In daily life we occasionally encounter situations that startle us because we did not anticipate them. Finding ourselves unprepared, our nervous systems respond by heightening our level of arousal in anticipation of and in preparation for a reflex action. Television producers use this characteristic of our nervous systems by administering a regular dose of jolts to our senses, to improve the likelihood that our attention will stay focused on the program. It is the parallel of lacing junk food with excessive amounts of salt, sugar, or caffeine to jolt our taste buds. If we grow accustomed to these jolts, regular food may appear dull in comparison. Similarly, the unenriched daily-life world may appear dull and uninteresting compared to the enriched world of television. No producer will allow a person to talk for very long without administering some jolts, while in daily-life conversations such jolts would be annoying and distracting. All this leads to a third distortion of the symbolization of our life in the world.

When assessing the effects of something on a particular environment, it is important to recognize that the results will be a function of both the effects and this environment. When it comes to the effect that technology has on human life and society, this is often overlooked. We all know that when one of two identical grenades is exploded in a crowded bar and the other in a farmer's cornfield, the results will be very different. The same is the case for television. In a mass society characterized by highly desymbolized cultures what we see on television performs a similar role in our lives as did custom and tradition in pre-industrial societies. We are engulfed by a bath of images showing us what we must eat and drink, what we must do with our appearance, what we must wear, what we must drive, what we must own, what health issues we must discuss with our doctors, what we must invest in, what home improvements we should make, and a great deal more. This bath of images supplements our highly desymbolized cultures by providing us with meaning and direction for our lives.[11] The members of a mass society tend to have other-directed personalities instead of tradition-directed ones, as a result of using the organizations of their brain-minds as a kind of radar, scanning what everyone else is doing in order to belong and move with the flow.[12] The same processes result in public opinions and in a statistical morality that transforms what most people do into what is normative. As a result, the previously discussed consequences of television for human life and society penetrate very deeply by supplying what highly desymbolized cultures can no longer provide. The implication for human life is that, although everyone participates in the evolution of a culture, people now become addicted to their new 'cultural supplement.'

Much of what we see in our news shows is even more deliberately engineered by public relations firms on behalf of their clients. A great deal also comes from so-called research institutes, whose experts are funded by industry or other deep pockets to prepare reports that are almost irresistible to producers because they are usually of high quality and they are free. All this further distorts how television portrays human life in the world. It may be objected that these research institutes serve the common good by presenting the findings of scientific studies done by qualified experts. To believe this is reassuring, but it does not stand up very well to critical scrutiny. As we shall show later, expert knowledge is so highly specialized that it can be recontextualized into human life and society in many different ways. Moreover, any issue addressed by research institutes is so complex that its scope spans a great many disciplines and specialties. There is no scientific way of integrating the findings of all these disciplines and specialties; hence those who do attempt to do so rely extensively on knowledge coming from beyond their domains of competence that is not scientific in character. They tend to rely on patterns of explanations that fit their values and beliefs. No human being is free from this. As a result, these research institutes will employ scientists whose patterns of explanation beyond their domains of expertise match our own goals. This problem has become particularly latent in some right-wing think tanks. I am suggesting not that left-wing think tanks are free from this problem but that their funding is usually so limited that their voices are rarely heard on the mass media.

Before examining the sixth and most fundamental consequence of television for human life and society, a few general remarks are in order. The production of television programs must amplify the above consequences as much as possible. Since this production is rather expensive, television programs must be capable of attracting and holding large audiences in order to expose them to commercial messages without which the bills could not be paid. Public broadcasters cannot escape this constraint, because their government subsidies can be justified only if the numbers show that the public is watching. However, these financial constraints mask a more fundamental one: the need for integration propaganda in mass societies whose cultures can no longer provide adequate guidance because they are so highly desymbolized. Without integration propaganda, mass societies would lose their social cohesion and begin their decline towards disintegration and collapse. Contemporary mass societies cannot survive this desymbolization without a technically engineered social cohesion. We shall examine this further in the next two chapters.

All this raises another fundamental issue. The cultures of traditional societies that have very low levels of desymbolization depend on the metaconscious knowledge implicit in the organizations of the brain-minds of their people, as do the highly desymbolized cultures of contemporary mass societies. Nevertheless, there is an important difference. In cultures with low levels of desymbolization the metaconscious knowledge implied in the organizations of people's brain-minds acts as a commons through which everyone participates in the transmission, evolution, and adaptation of the culture. There is a close symbolic correspondence between the metaconscious knowledge supporting episodic remembering and that supporting semantic remembering. In the case of the highly desymbolized cultures of contemporary mass societies the metaconscious knowledge supporting semantic remembering is deeply possessed by public opinions, statistical morality, and myth-images created by means of integration propaganda. It is now a cultural commons dominated by those who can afford the technical means to participate in its evolution. These participants are constantly exploring ways in which to distance the metaconscious knowledge associated with semantic remembering from the metaconscious knowledge associated with episodic remembering, in order to evolve the former with as little constraint from the latter as possible. This distancing is the art of public relations, which desymbolizes language by using words as tools to create desired effects, as opposed to using them in accordance with their meanings in the dialectical order of language.

Many studies have had a sense of the desymbolization created by television, but few have adequately assessed the consequences in terms of the sociocultural contexts in which this desymbolization occurs. Jerry Mander has argued for the elimination of television.[13] Neil Postman has argued that television is 'amusing ourselves to death.'[14] Marshall McLuhan has argued that it helped to transform human consciousness in a new 'global village.'[15] Guy Debord has suggested that we have become spectators to our own lives.[16] We need to go deeper by examining the sixth consequence.

We have previously argued that the organization of the human brain-mind is such that the early experiences of babies and children contribute to the creation of a great deal of metaconscious knowledge, which acts as a metalanguage. Babies and children are thus launched towards the symbolic mediation of their relations with others and the world by means of a culture. The visual dimension of experience is fully integrated in these developments, by which they create an order of what is true in their lives and evolve this order towards the cultural order of a community. This cultural order cannot be entered into other than by means of language, and it is language acquisition that launches babies and children towards becoming full members of a symbolic species. Such a development implies that the visual dimension of experience is integrated into the aural dimension of experience on the latter's terms. However, this integration becomes fragile under the pressure of desymbolization.

The fragility of the relation between hearing and seeing is rooted in the very different ways that we apprehend our lives in the world by means of the corresponding senses. When we open our eyes, our surroundings can be seen through a sequence of glances. If we close our eyes and open them again, everything we saw is still there. There is an immediacy and permanence that the apprehending of our life in the world by means of language and culture does not have. The sounds made by a speaker are briefly present and then disappear forever.

The listener cannot bring them back following a momentary distraction. To understand what a speaker is saying, a listener must wait until a temporal sequence of sounds has been completed. We are well aware of the need to interpret what the speaker is saying because of the common occurrence of misunderstandings. In contrast, although the interpretation of our visual experiences is highly complex, we have little awareness of it, and it all appears to be rather direct, immediate, and certain. If our hearing and seeing were not integrated into the symbolization of our experiences, we might well conclude that our eyes and ears bring us two very different worlds. In the world of seeing, we do not depend on other people; the world is simply there for us, and we all appear to see it the same way. In the world of language and culture, we depend on others as we alternately act as speakers and listeners. We share something of our lives in the world in a manner that is temporal, fragile, and ephemeral. It leaves us free because as speakers we cannot impose an interpretation of what we are saying, and as listeners we can never be entirely sure that we have correctly interpreted the intentions of the speakers. Of course, neither the speakers nor the listeners are free from their cultural embeddedness. However, I shall argue that this embeddedness is very different from most postmodernists' interpretations.

Babies and children gradually learn to speak and listen in the context of discovering the cultural order of those who nurture and love them. Of course, babies and children have no idea how fundamental are language and culture for their lives. It is not until much later that they may encounter the stories of children who are growing up deaf-blind, being brought up by animals, or spending much time in isolation; however, they will most certainly have a sense that the permanence of what they see is not shared with what they hear and that their spatial world is different from their temporal, cultural world. Language brings out a very different kind of world. It comes from symbolization and not from the dream, as held by Lewis Mumford,[17] nor from the imagination, as believed by Cornelius Castoriadis,[18] nor from myths, as argued by Roger Caillois.[19] Symbolization creates a distance between a symbolic species and its life-milieu, which is ordered by means of a culture.[20]

The characteristics of the world we see are very different from the world we access by language. First, everything we see exists in its own space separate from everything else we see. When we focus on a particular point in our field of vision, it can be only one thing. It cannot simultaneously be something else. Consequently, this world implies the principle of non-contradiction. In contrast, the world of language and culture is dialectically ordered.

Second, because we see everything in its own space, distinct and separate from everything else we see, the world of seeing implies the principle of separability. No enfolding occurs in this world. No such separability exists in the world of language and culture, where the meaning and value of everything is dialectically enfolded into the meanings and values of everything else.

Third, since everything in the world accessed by seeing can be separated from everything else, it is possible to define everything on its own terms. Definitions can be closed. In the world accessed by language and culture, everything is dialectically related to everything else, with the result that no such closed definitions are possible. Many of them are essentially contested.[21]

Fourth, since the world of seeing implies non-contradiction, separability, and closed definitions, everything within it can be measured and quantified. Doing so opens the door to mathematical representation and applied logic. Such mathematical representations will be internally logically consistent, and externally consistent with what they represent. Hence, such mathematical representations are of an applied (as opposed to a pure) character. None of this is possible in the world of language and culture, which is a world of enfolded wholes whose meanings and values overlap with those into which they are enfolded and of which they are constituted.[22] Consequently, these meanings and values have multiple dimensions that may be ambiguous and even contradictory.

Fifth, because the world apprehended by seeing implies the principles of non-contradiction and separability, its representation can be built up one constituent at a time. Each additional constituent will not affect any of the previously added ones. Complexity is built up one constituent at a time. In contrast, the complexity of the world accessed by language and culture grows by progressive differentiation. The world of babies and children emerges out of undifferentiated stimuli. What emerges is first differentiated from what has not yet been differentiated, and then from everything that has previously been differentiated. It is a relational complexity that cannot be assembled one discovery at a time. The concept of a fact is unthinkable. Since we live in an ultimately unknowable reality, we are unable to intellectually grasp a chunk of it and call it a fact. All human knowing and doing based on symbolization is ultimately relative and thus must be grounded in myths to protect it from relativism, nihilism, and anomie.

These five differences further illustrate how far seeing is removed from hearing, and the word from the image. Generally speaking, traditional cultures attempted to keep this potential duality at bay by subjecting what was seen to what was heard. The Jewish tradition associates the word with human freedom, and the image with the subjection of this freedom to moral and religious enslavement.[23] Making these associations is not a moral issue but a structural one rooted in reversing the hierarchy between the word and the image.[24]

Our civilization has made a very different choice. It regards as *real* whatever can be separated, defined, measured, quantified, and preferably expressed in mathematical models. Whatever cannot be grasped in this way is considered less real and objective or downright subjective and unreliable. In the following two chapters we shall show how our civilization has built an order of the real within the order of the true and how this has changed everything. The order of the real is the order of desymbolization. It is accessed by means of the image instead of the word. Its complexity is rational and built up with the principles of non-contradiction, separability, formal definitions, measurement, quantification, and mathematical representation. This order of the real places all of us before a dilemma: our contemporary ways of life are focused on developing the order of the real, even though this cannot be done without undermining everything that has made us a symbolic species. The resulting desymbolization has undermined the ability of contemporary ways of life to sustain us and, in turn, to be sustained by the biosphere.

As language acquisition launches children towards becoming full members of a symbolic species, their encounters with television launch them towards the order of the real. It is as if they constantly have to choose between two worlds: the one they share with others and the one on television. Whenever they wish to withdraw from the former, they can enter the latter, and vice versa. It is not long before the differentiation of their experiences will imply a metaconscious knowledge of how these worlds are different. Initially, they may prefer the artificially enriched world of television, but as their personalities develop, they may become more torn between the two. Simply put, for shy and introverted children the television may represent a liberation from the world with which they are less comfortable. For more sociable and extroverted children the world of the television robs them of the social support that they derive from sharing their lives with others. The attitudes of the adults towards television in general, and towards the programs aimed at them in particular, may play a moderating role. However, the children discover soon enough that the world of television is more real and important than their daily-life world and that many programs give them advice as to how to behave in the daily-life world.

What all this means for babies and children is that the differentiation of the experiences of watching television from all the other experiences creates two different relationships between seeing and hearing. The acquisition of language gradually leads to the primacy of hearing, while watching television is characterized by the primacy of seeing. These differences in turn lead to others involving embodiment, participation, and commitment because of the different ways in which their experiences are lived. In watching television, there is no relationship established between the speaker and the listener by means of the word supplemented by images. They cannot embody themselves in these experiences, and they soon learn that the program does not respond to them. There are no consequences, and no commitment to anything is required. Children are thus immersed in the kind of situation described by Kierkegaard and elaborated by Dreyfus. In addition, children encounter first and foremost a world of images in which the word plays a supporting role, and these impressions are powerfully reinforced by the attitude of adults. In other words, their television experiences gradually introduce them to the conflict between the order of what is true and the order of what is real that reigns over our society, as we shall see more clearly in later chapters.

Initially, all this may amount to little more than the differentiation of the experiences of watching television from all other experiences. The foreground of these experiences related to the artificially constructed world of television is discontinuous with the background into which they are physically, culturally, and socially embodied. Within this foreground, there is another foreground-background distinction as determined by the program they are watching. The world of television is very different from their daily-life world, as may first be noted by the hypnotic rapidity with which perceptual jolts occur, which are almost irresistible to their reflex actions. The attitudes of the adults may encourage the children to integrate the world of television into their lives as more real and true than is their daily-life world. For example, people who are on television are much more important than those who are not. A great deal more may be intuited as children's levels of symbolic development grow. However, the kinds of developments described in the previous chapter are clearly undermined by television watching.

The experiences of watching television and their differentiation from all others gradually establishes two worlds in the lives of children: one comparable to the order of what is true and the other to the order of what is real. The differences are evident in terms of their embodiment, participation, commitment, and freedom. Human life in the world as 'experienced' on television is, according to the adults with whom the children interact, much more objective and true than is their daily-life cultural envelopment.

The following story is illustrative. I once agreed to help a television producer create a series of spots sponsored by a transnational corporation, which were designed to help the public understand the difficulties associated with the use of technology. One of these covered the influence of cellphones on our lives. I wanted to make the point that the influence of cellphones was much greater than people realized. I decided to illustrate this point by standing on a beach to explain that the erosion by the incoming waves could not be understood in terms of particular waves and that this was equally true for the overall effect of individual technologies on our daily lives.

As the television crew was setting up, I was asked to say 'anything at all' so that the sound levels could be adjusted. Not wanting to repeat myself, I mentioned obvious effects such as the blurring of our working lives and our private lives. When I later saw the spot on television, I was dismayed. What the television audience saw was an 'expert' standing on a beach with the waves coming in behind him, talking about something that everyone already knew while his credentials were flashed underneath. The producer had selected the off-the-cuff remark that I had made while the sound levels were being tested, despite the fact that it had nothing to do with the background and the fact that you certainly did not need credentials to observe what was obvious to anyone. I telephoned the producer to inform him that I would no longer participate in the project. His response was interesting. He asked me to defer my decision until I had talked to a number of friends and acquaintances about what they had seen and heard. The response was exactly as the producer had predicted: the 'meaning' and 'value' of the event was that I had been on television, and not a single person remembered what I had actually said, never mind the fact that it was out of keeping with the background of the waves and my qualifications. The 'significance' was that I had been included in the order of what is real, and this was the message. In addition, what most people already knew from their experiences with cellphones had now been made real by an expert.

The above argument is not likely to appear very convincing, because all of us interpret it in terms of a highly desymbolized culture. It has so diminished the order of the true in relation to the order of the real that we hardly notice that government has become a question of management, justice a question of efficiency, the economy a question of growth, communication a matter of exchanging information, and love a matter of sex. Everything takes on its 'meaning' and 'value' in relation to the order of the real, which has little or nothing to do with what all this means for individual and collective human life. Moreover, what comes on our television screens has very little to do with what *is*. It is a reconstruction of what *is* into a non-reality in which language can participate only on the terms dictated by the images. It has little or nothing in common with the word, making the concept of audio-visual meaningless.[25] Television establishes a false relationship with a false reality. It immerses us in a bath of images that socially control our lives and alienate us from ourselves, others, and the world. Jacques Ellul concludes that television provides us with a non-language of images that refer neither to what *is* nor to what is true.

Joel Bakan provides us with a powerful illustration of advertising aimed at children.[26] The 'nag factor' approach is the solution to marketing products to children, who have no money of their own with which to buy them. The solution resides in directing children to nag their parents to buy these products for them. Advertisers and child psychologists have studied the ways in which children spontaneously nag their parents, the success or failure of their tactics, and the factors that influence this (such as age and socio-economic background). In this way, these studies identify the most effective nagging behaviour and how children can best be induced to adopt these tactics for a particular target group. Single parents who may feel guilty for not having enough time with their children, couples in the middle of a divorce, and professionals who are often away are all particularly vulnerable. As an example, take eating junk food. After the film *Super Size Me*, one might have expected a drop in parents' taking their children to the McFoods of this world. After all, do we not test drugs by exposing animals to high dosages for a short period of time? In this case, a human being was exposed to high doses of junk food, and the results were so alarming that, had it been a drug, it would have been withdrawn from the market. The well-funded public relations machines began performing the same role for the fast-food industry that they had performed for the tobacco industry earlier on. It has all been turned into a question of parental responsibility; hence the current epidemic of obesity and diabetes cannot be blamed on the corporations. The public relations campaign, supported by the nagging behaviour of children, has completely neutralized the message of the film.

The research of one corporation shows how effective all this is: 20 to 40 per cent of purchases would not have occurred without this nagging. Children have an enormous influence over the products that their parents buy, and it is far from restricted to food and toys. It even reaches to a large variety of adult products, including cars. Of course, the advertising agencies and their clients know very well that young children cannot understand what these ads mean for their lives, nor can they understand the difference between the ads and the programs. However, the people involved completely compartmentalize their work from their homes, and particularly their relationships with their children. In this way, most of them avoid confronting the psychopathic nature of technique and its most important institution, the corporation. Advertising completely undermines the experiences that come from children learning to make sense of the food they eat and the significance it has for their health and their lives. It helps to turn upside down the relationship between seeing and hearing: if a picture is worth a thousand words, and American children see an estimated thirty thousand ads per year, imagine how much talking parents need to do. In addition, kids encounter ads in their schools, on their streets, and on their television screens. The order of what is true is under heavy siege, and so are children's chances of learning to live meaningful lives with others in the world. And this is only the beginning. The World Trade Organization is slowly but surely, directly and indirectly (as governments learn to censor anything that impedes the rights of corporations), making it impossible for communities to democratically impose restrictions on these extreme excesses. Coupled to this is the desymbolization of contemporary cultures as the traditional moral and religious force capable of resisting what will undermine a community. The result is a growing strengthening of the order of what is real in children's lives and an ongoing weakening of the order of what is true in their daily-life experiences. All this helps us understand a little better what is going on in our high schools.

Marshall McLuhan has suggested that the mass media have extended the range of our senses of seeing and hearing to include the entire world.[27] As a result, we have become members of a global village with a different awareness of ourselves, others, and our world. Such an analysis has two pitfalls. First, why stop at the way technology is extending our senses, when it is simultaneously augmenting a great many other human abilities? Second, if this is the case, can the overall influence on our lives be understood by limiting our attention to the mass media? Despite these shortcomings, this approach to understanding the mass media has the merit of making us aware of how deeply television watching affects the lives of babies and children.

Computers as the Playground of What Is Real

The tensions between the order of what is true and the order of what is real, which are introduced into the lives of children as a consequence of television watching, are greatly reinforced by their contacts with computers and, via them, with the Internet. What they observe on their computer screens is a simulation of a world generated by a program based on rules and algorithms. Consequently, whatever world is simulated on their screens has all the qualities of the order of what is real. No matter how realistic the images may be, anything living must be reconstructed in the image of the machine. This process implies the principles of non-contradiction, separability, closed definitions, measurement, quantification, and mathematical representation. The complexity of the image has all the characteristics of a non-living machine: distinct parts and clear causal relations that operate without reference to the whole. Even when the differences between the visual dimension of experience of the simulation and the daily-life world are subtle, the differences between the corresponding overall experiences are very great indeed. For example, it is possible to win in the simulated worlds of computer games (or other games, for that matter), but it is impossible to clearly win in daily-life situations. In a dialectically enfolded world, any action we take is related to everything else as a result of symbolization. Everything has a great many consequences, with some of them being positive, others negative, and still others neutral. Moreover, all these consequences are related to, and inseparable from, one another. In contrast, the simulated world of a computer program has none of these qualities. It is a defined domain completely separated from the order of what is true with distinct and separate elements interacting in causally limiting ways, thus eliminating any possible meanings and values other than performance and the end result of winning or losing. After all, performance is a measure of things on their own terms and separate from all the other things in the domain. In turn, the domain is completely separate from the order of what is true that children establish in their lives by means of symbolization.

There is no possibility of children (or adults) embodying themselves in the domains displayed on the screens of computers. These domains cannot be understood in terms of language and culture. They must be accessed by means of the displayed images, and language can at best play a supporting role. When children begin to interact with computers, the tensions between the image and the word are reinforced, as well as those between the order of what is real and the order of what is true. It leads to desymbolizing experiences that have similar effects on children's lives to those of watching television.

The involvement of children with their computers is more active than is their engagement in television watching. Nevertheless, any involvement whatsoever occurs on the terms of the program as designed by the organization that sells it. It is the program that embodies children in the domain, and their responses are limited to a specified kind at a specified time. Moreover, any interaction takes place via the keyboard, mouse, joystick, or touch screen. Physical embodiment is thus limited in the extreme, to repetitive and reflex-like movements of the hands. Initially children may treat this much like their interactions with their daily-life world, until they learn that nothing of this passes to the computer. There is no social or cultural embodiment. The fact that the program environment (a domain separate from life in the world) with which children interact frequently simulates or reminds them of living entities in a living world amounts to a technological bluff: that what is alive can ultimately be reconstructed from dead entities. We may enjoy the humorous portrayal of these kinds of issues in television programs like *Star Trek* when, for example, the robot Data perfects the rules simulating dating behaviour without ever being able to pull it off in a human fashion. It is much less amusing in the case of Western philosophy and of artificial intelligence seeking to capture human experience by means of algorithms and logic.

Many of us can become very absorbed in winning a computer game or in successfully completing a programming task. Yet it is difficult to speak of being committed to such a game or task in the usual sense of that word. There is nothing to commit to, other than a logical domain that operates on the basis of a set of rules. There is no question of a genuine vulnerability, and there are no risks for our lives with others in the world. These kinds of situations should not be any different from being totally absorbed in a chess game to the point of losing all track of time and the other things we should be doing. Any game is a domain that is completely separated from our lives in the world. We get involved in it, amuse ourselves with it, and then we need to get on with our lives. The problems with commitment cannot occur except when people live to play games or to program. They then live in and through these domains by keeping everything else at bay as much as possible without ever being able to succeed. Everything other than playing games or perfecting programs loses its meaning and value. At this point, it is a question not of commitment but only of addiction. Such people have lost the freedom to play games or do programming because they thoroughly enjoy it. They have sacrificed their lives in the world and hence all possibility of freedom.

Children's experiences with computers place them before a dilemma. If they treat the programmed environment as more real and interesting than their daily-life world and begin to deal with the rest of their lives in the same way, they will succeed in the one at the expense of the other, their ability to live in a human world. However, if they intuit the human living world as primary, they will encounter difficulties relating to programmed environments that adults take as a key to surviving in the future. How children will respond to these possibilities depends on everything else that is happening in their lives as well as their own resources and preferences.

Some will be attracted to the programmed environments because they appear to be more predictable and reliable. Besides, with diligent application, it is possible to win, and even when problems occur, there is a definite and final solution. In comparison, the human social world is uncertain, ambiguous, and risky. There is no escape from being vulnerable to others in ways that do not occur in programmed environments. There are no possibilities of unambiguously winning, nor are there any final solutions to anything. However, its dialectical and enfolded character that is ever new provides a subtle richness of experience that cannot be matched in programmed environments. It is also the world of children's social support. When upset, disappointed, or in need of help, they can turn to others as others turn to them. Although this social support comes with risks, life would appear to these children as being unbearable without it. Watching television and playing with computers may bring children to a subtle and gradual leaning towards one or the other possible development and turn them off of or on to computers. In the extreme, it may lead to a child's becoming either a computer hacker or someone with a computer phobia.

In a programmed environment or a program-mediated interaction with others the immediate consequences for a person's life and for the lives of others are often almost imperceptible and are easily presumed to be non-existent. The long-term consequences of becoming poorly adapted to a living social world appear more like theoretical possibilities than what is actually and potentially happening to a person's life. All this is particularly true for children growing up in a world in which parents and teachers themselves have difficulties getting a grip on their daily-life technologies. It brings us back to the earlier story of likening this to the effects that waves have on a beach: it is the ever-changing wave patterns that must be understood, which is no easy matter. For children, parents, and teachers there is little to go on other than some deep-seated intuitions based on the evolving metaconscious knowledge implicit in the organization of their brain-minds.

Incidentally, one of the waves that are eroding the beach is undergoing far-reaching changes. Cellphones are being transformed into mobile computers capable of a whole new range of functions such as text messaging and accessing the Internet. Children and teenagers are becoming perpetually connected to their devices. They are inventing a new language of signs as an art of text messaging. These developments are intensifying the kinds of issues we are examining.

As the immediate effects of technologies on people's daily lives are subtle and almost imperceptible, and the long-term consequences are far away, it is easier to fall back on the order of what is real. For example, children may be unaware of the far-reaching implications of placing personal details on the Web or of bullying classmates in this way. The former may well be the equivalent of their having been on television, something of their daily lives joining what is 'real,' and having the social status associated with this. It is also what makes bullying on the Web that much more serious. It is next to impossible for children not to transfer their commitment away from learning to make sense of and live good lives with others in the world towards embracing an artificial reality that is dead by virtue of its logical character and yet simulates what it is to be alive. The extent of this transfer of commitment and its consequences become evident when children play war games and take satisfaction in their ability to destroy lives and cities. It illustrates a high level of alienation.

Hence, it may be expected that the interactions children have with their computers have a significant influence on their perceptions of who they are and what it is to be alive, and how this is differentiated from what it is to be a dead object. In turn, their perceptions of themselves deeply influence how they express themselves in the relationships they establish and how these relationships are woven into their lives. All this in turn influences the grip they have on the objects of these relationships, as well as the constitution of their cultural envelopments. These developments help shape the horizon of their lives and their participation in evolving the myths that make their lives and their community true.

Some of this has been illuminated by the pioneering research of Sherry Turkle.[28] The manifestation of this in adult life has been researched by Craig Brod.[29] It is essential to take cognizance of the time in which most of this research was conducted. Sherry Turkle's *The Second Self* and Craig Brod's *Technostress* both appeared in 1984, which reflects a time when the personal computer was making inroads into family life. As a result, the adults still remembered very well what life was like before the computer, and the ambivalence of their computer-related experiences undoubtedly had a profound effect on the children around them. In other words, their computer-related experiences could still be differentiated from those of human life before the computer. From this perspective, the research reported in these two books provides a particularly interesting snapshot of what has since evolved into a situation in which people's recollections of life before the computer are dimmer or non-existent and where the effects on their lives are accepted at face value.

Sherry Turkle's research shows that children between the ages of four and eight are interested in the meaning and significance of the computer as a curious object in their lives. At ages nine and ten, their metaphysical preoccupations are overshadowed by their actively seeking out what they can do with these machines. As adolescents, all this gets caught up in the development of their identities in relation to maturing bodies, peer pressure, and a desymbolized culture that provides little guidance, leaving them vulnerable to the pressures exerted by the mass media and the Internet. I shall integrate her findings into my broader analysis of children's learning to make sense of and live in the world.

Young children face the difficulties of differentiating the computer from everything else in their lives. Doing so involves everything else because what the computer is, everything else is not, and vice versa. At first, the computer may enter the dialectical order of what is true in their lives by being directly differentiated from all other objects and, via them, from everything else. As they gain more experience with computers, this positioning of these objects in their lives will be less and less satisfactory. They may be experienced as less and less like other objects and a little more like human beings. Much of the behaviour of children appears to confirm this. Their intuitions, based on metaconscious knowledge implied in the organization of their brain-minds, lead them to question whether the computer can think or feel. Over the course of several years, the processes of differentiation and integration helps them get a grip on this unique object and come to some kind of understanding of its meaning and significance for their own lives and for their families.

Sherry Turkle reports many examples of children discussing whether their computer can cheat in a game, whether it knows when it wins or loses, and whether there are feelings associated with this. Their computers appear to be complex, smart, and somewhat alive. As a result, the dialectical order of their lives in the world is deeply affected. Their intuitions strongly motivate their play and explorations. 'Do I kill the computer when it crashes?' 'Do I bring it back to life when I reboot it?' 'If my mom or dad stop loving me, could they cause me to crash?' 'Could they reboot me if they miss me?' 'Will the computer be offended if I reprogram it?' 'Why is the computer smart enough to beat me but not smart enough to be proud of it?'

Gradually, this play and exploration results in more nuanced questions and convictions. This computer game always wins, which is impossible unless it cheats. Computers have feelings, but they never show them. In other words, they are like us in some respects and like things in others. Children may value in their computers what they cannot find within themselves, and they may value themselves for what the computer lacks. As a result, the computer may become most directly differentiated from human beings, and as such it will eventually affect children's perceptions of what it is to be human and what it is to be a thing. Television and computer screens act as a window into what may gradually emerge: the order of what is real becoming distinct from, and more important than, the order of what is true.

These developments may also lead to a dissociation between thinking and feeling because the latter is unique to human beings, while the former is shared with computers. As a result, thinking may eventually become associated with the expression of a cold, detached and logical part of the self, and feelings with the expression of a more primitive and non-logical part of the self. The two may no longer be seen as expressions of one and the same person. A kind of divided self may develop in some people that is part machine-like and part emotional. Such a self may give rise to different modes of expression, some corresponding to relationships that are more mechanistic with little human control, and others permeated by emotions and all that these entail. There may thus be different kinds of involvement, commitment, freedom, and alienation. As has been pointed out by many observers, as computers become a little more like people, people become a lot more like computers.

When the differentiation of the computer from everything else in children's lives begins to stabilize, they have learned a great deal about this object that is unlike all others and in some ways like themselves. The process of differentiation has built up enough metaconscious knowledge of computers to satisfy their metaphysical questions. Some children may show an interest in exploring exactly what these machines can do for them. The resulting experiences will further confirm and evolve the position the computer holds in the order of what is true that they have established in their lives. This shift from a preoccupation with the meaning and value of computers for their lives to pushing the performance of these machines will change how their lives evolve in relation to them. These machines will gradually be differentiated from their order of what is true. In other words, the process of differentiation will slowly create a metaconscious knowledge supporting intuitions that the world of computers is very different from their daily-life world. This world of computers is differentiated from the world of television, yet seems to have more in common than either one does with daily life. It is another step towards inserting the order of what is real into the order of what is true in their lives.

When children begin to play computer games and possibly try their hand at some programming, mastering what the computer can do and coping with the situations flashed on the screen become increasingly important. They are no longer content to be on the receiving end of this 'environment.' As they learn to understand it better, they want to assert themselves more and more. These kinds of assertions are very different from the ones directed to others and the world. The processes of differentiation and integration lead to intuitions that the program environment on the screen is not only a world in itself but one that is very different. It is less and less a question of what events on the screen mean by relating them to everything else in their lives but more a question of mastering them on their own terms as opposed to those of their culture. In the case of computer games, it is a matter of an all-out struggle with the machine, staying on top of what it throws at them and beating it in the end. In the case of programming, control is everything if the machine is to carry out the functions as desired. On some deeper level, children begin to realize that the programmed environment is internally differentiated into a different kind of order than the one associated with their daily lives. Children recognize that this kind of mastery and control simply does not occur in their relationships with others, nor is there any possibility of definitively winning or permanently beating a situation. On a still deeper level, they may have a vague sense that within a programmed environment there is no point in asking metaphysical questions, because its entities have no meaning and value in relation to one another, nor can they be related to the sense that their non-simulated equivalents have in the order of what is true. The entities of a programmed environment are better dealt with on their own terms, as distinct entities with distinct relationships to the other entities.

This dissociation is most evident when children play war games. A child will get excited when he is winning because he has destroyed ten cities, and it is only this number that matters. The cultural orientation of relating everything to everything else by means of symbolization is in some way distinguished from the technical orientation towards power, performance, and success. The destruction of the ten cities appears to have nothing to do with his urban habitat and his life and those of others within it. His many computer-related experiences further strengthen seeing over hearing, the image over the word, and the emerging order of what is real over the order of what is true. All this socializes children for the world that awaits them: a world that is unthinkable without technique and its orientation towards power, performance, and success.

With this shift towards performance, computers become a lot more addictive. It is no longer a matter of a child's requiring a high density of jolts administered by the program to keep his or her attention, but a determined commitment to master and control these jolts on the screen. Along with this comes the implicit recognition of the difference between what comes on a television screen and what comes on a computer screen. It is precisely because the latter is rule-based and algorithmic that it can be controlled, and even when control escapes the child who exercises it, it can be regained. All this is possible only in relationships dominated by images. These images do not leave the kind of freedom that speakers and listeners experience in face-to-face conversations. In these, there is no question of controlling one another by means of the word without destroying what that word is. If two people sincerely seek to understand one another and to share something of their lives, they are establishing as true a relationship as possible. Any element of control, manipulation, or force would negate this possibility.

Learning to use a program and exploring what it can do has little in common with learning to play a game like chess with others, using a traditional chessboard. The chess game has no autonomy from the players. The relationship between a programmed game and the players is very different, especially when each player manipulates a particular entity created by the program. As Sherry Turkle has observed, players tend to identify with the programmed entity by means of which they participate in the game, and this awareness merges with the actions they take. Unlike the actions in daily life, the game has no built-in tolerance for error. It relentlessly pushes the players to the very limits of their abilities to act as the program demands. The players know that if they make one false move, they are 'dead.' It is a situation of high demand and low control, which in the case of human work is known to lead to the kinds of stresses that are unhealthy and that cause physical and mental illness in the long term.[30] To the extent that the players identify with the entities they manipulate, winning or losing becomes a question of life and death in the game environment. They cannot stop but must push ahead, which results in the game having power over the players. Some people experience this as relaxing because it blocks out the pressures of life. The game demands total surrender and leaves no room for daily life's intrusions. It may be argued that there is a liberation of sorts, but there is also a new kind of enslavement. Players themselves speak of a fusion between their minds and bodies as they learn to 'think with their fingers,' as it were. It has even been suggested that computer games should also have a place in the Olympics.

There is another aspect to the quest for mastery, control, and victory, which contributes to the enormous holding power of computer games. Each new screen and each new round demand a greater performance, and this is being pushed to new limits by the military.[31] There is a kind of infinite that appears to give some players a religious feeling of sorts. There is also the possibility of infinite perfection: no matter how good a player is, the drive for perfection never diminishes. This evokes a certain anxiety about the player's perfectibility, as a kind of endurance race with the game. It makes it difficult to walk away from the game; you know you can do better next time, and you can become perfect in the end. It is the ultimate contest between the player and the game, which amounts to a contest with yourself. In this ultimate contest, no clutter associated with one's life beyond the game can be tolerated. The game must be pure, decontaminated from everything else in life. This leads to a total devotion in many players. These kinds of developments in the lives of children withdraw them from the order of what is true and contribute to the preparation of a relatively distinct order of what is real. The resulting experiences are desymbolizing in character, and they join the many others associated with living in a contemporary mass society.

It must be remembered that computers are but one of the many technologies on which children depend in their daily lives. In many respects, the computer is the culmination of what is more embryonically present when they play with battery-driven toys, watch television, build things that have no significance other than their ties with television programs or movies, and participate in the manipulation of many devices used by the adults around them. Since the early 1980s the growing use of computer chips has greatly intensified children's direct and indirect exposure to programmed environments. These environments simulate or carry out any function by technically dividing it into a sequence of independent and endlessly repeated steps that can only be changed by other steps. Whether any step is carried out by a device or a human being makes no difference, because a person will have to use the device on its terms. No machine, other than automatons built for amusement, have ever imitated the way human beings carry out a particular activity. The reasons are simple: machines thrive on repetition, but human beings are destroyed by it because our nervous systems have evolved to cope with a living world in which nothing ever repeats itself in quite the same way. The computer is not a tool that can be used as we see fit. All corporations have now learned this lesson. In order to make effective an economic use of computers and information technology, the corporation had to be re-engineered, which involved its reorganization in the image of the computer.[32]

For the majority of children who are unwilling to surrender their lives to computer games and programs, a curiosity about and a fascination with performance will soon be engulfed by other developments. As they gain further experience with computers, the differentiation and integration of these experiences may lead to a growing intuition that performance is an end in itself and thus incompatible with their order of what is true. They cannot escape computer games and programs, any more than they can escape television, but they are likely to incorporate them into what adolescents do. They must develop their social and cultural selves, not by differentiating what it is to be human relative to everything else or by pursuing the performance possibilities of one kind of object. They must develop who they are relative to the other people in their lives as their order of what is true converges towards the cultural order of their community, thus leading them to find other sources of their individual selves.

The implications of full or partial participation in any activity can be imagined by turning to our simple model of how the organization of the brain-mind symbolizes, differentiates, and integrates our experiences. There will be clusters of experiences in which children are able to fully participate in the relationships that they establish, and others in which participation is greatly constrained. In the extreme, these constraints are dictated by programmed devices and environments. As a result, the clusters of experience will range enormously in terms of the kinds of expression of the self, the relationships established by these expressions, the objects of these relationships and their participation in a cultural envelopment, and the contribution of all this to a person's horizon of experience. It also means that the contribution these clusters make to the metaconscious images of a person's social self will vary enormously. In a great many cases, who a person is and what it is to be alive have to be accommodated to a technical division of labour, a machine, or a programmed environment. As a result, growing up in a society that directly and indirectly depends on a myriad of technologies and whose habitat is a technical artefact leads to divided selves on a scale that was unprecedented in earlier societies.

How children learn to deal with the many daily-life technologies that they encounter reveals something of who they are. 'Am I the kind of person who is more than a little apprehensive about all these technological devices, seeing them as unavoidable intrusions? Or am I the kind of person who has a knack for these gadgets because they appear full of interesting possibilities?' There is little doubt that the computer helps to crystallize these kinds of questions and feelings. Sooner or later, children are confronted with aspects of their humanity that in other societies would have remained undeveloped.

For example, computers may help shy people to deal with their loneliness. They can always retreat to their computers, which are always there for them, and thus escape the awkwardness of having to deal with others. According to the studies, it appears that being alone with the computer is somehow experienced as different from being alone. Shy children are able to build a safe environment in which they can express themselves in ways that do not lead to the kinds of discomforts they experience with others. Instead of talking things over with others, these children would rather simulate different things on their computer so that they can think about the result and form opinions that way. They may develop a sense that people are too unpredictable and emotional to fit into their programmed environment, and they may come to regard it as more dependable, controllable, and predictable. In other words, it is these aspects of their humanity that they come to value at the expense of everything else. What they do not realize is that this preference for certainty and greater control leads to a new kind of alienation masked by the 'freedom' that is strictly delimited by the programmed environment. Even though today their computers may be networked, these children may continue to minimally use them to communicate, or even to communicate anonymously in a chat room. From what we have discovered thus far, the implications of shy children 'escaping' to their computers are very different from those of 'escaping' into a book.

For many other children, the competence they have gained in programmed environments opens the door to new forms of self-expression and new intuitions regarding their evolving identities. They also experience a tension between a programmed environment, where they have a great deal of control, and their social world, where they are vulnerable. This tension establishes a different relationship with programmed environments. They do not constitute a refuge but an impoverishment of their lives and their relationships with others. They use computers, but this does not tear them away from their lives. One thing is certain: psychoanalysis has focused much attention on the sexuality of our humanity, but the computer is refocusing this on the computational and informational aspects. Along with this has come a shift in the meaning of many words. As noted, we use the words *memory*, *information*, *expertise*, and *communication* in relation to human beings as well as machines. Apparently, we accept the fact that much of our humanity is informational. We have entered the age of *homo informaticus*.

Regardless of the preferences developed by children and carried over into adult life, our sense of ourselves and what it is to be human has been profoundly affected. Moreover, these effects have contributed to highly divided selves with a greater range of expressions, relationships, cultural envelopments, and horizons of experience. The result is a spectrum of experiences ranging from those that are symbolizing in the way they contribute to the organization of the brain-mind, to others that are desymbolizing and have a decentring effect on the organization of the brain-mind.

All these developments greatly contribute to the difficulties that teenagers experience. As adolescents, they must come to terms with a maturing body and a host of pressures from their peer group and the mass media, while receiving little guidance from their highly desymbolized culture. In their extremely vulnerable state, the security of a programmed environment may well become more appealing than their cultural environment, but any withdrawal will aggravate their difficulties with others.

The choices that children and teenagers make also have significant implications for their cognitive and intellectual development. Benson Snyder distinguished between numeracy and literacy as two distinct cognitive styles.[33] Numeracy is more appropriate for dealing with the subject matter of mathematics and the physical sciences, while literacy is more appropriate for that of the social sciences and humanities. In the latter, we are trying to understand a reality that is dialectically enfolded. Hence, these subjects cannot be examined one entity at a time, with little attention being paid to the larger whole to which it contributes. Ideally, everything must be understood in relation to everything else, with the result that the understanding of anything depends on everything that has already been learned and will continue to depend on what is learned in the future. There can be no definitive understanding or a complete resolution of any issue. The process of enquiry remains open ended. The rise of discipline-based teaching and research has greatly simplified these interdependencies by limiting them in relatively arbitrary ways. Nevertheless, a person's cognitive and intellectual styles remain context-sensitive within the confines of a discipline.

The cognitive and intellectual styles associated with the physical sciences deal with a world that, for a very long time, has been regarded as being essentially mechanistic. These styles are appropriate for examining a reality that is essentially mechanistic in nature. In other words, a subject can be learned one entity at a time, and little attention needs to be paid to the whole.

Some people are more comfortable with the numeracy mode, and others with the literacy mode. However, Snyder found that people who were strong in the numeracy mode and did not develop their literacy mode tended to encounter difficulties in their personal lives, frequently with unhappy outcomes.[34] The development of these cognitive styles in children dovetails with the ones we have already described, resulting in the reinforcement of some tendencies and the weakening of others.

The computer is highly catalytic in how children learn to think about themselves. To the extent that they regard themselves as part machine and part human, it constitutes a reifying development, a person partially becoming a 'thing.' This will have a significant effect on how children regard others and their social relationships with them. To what extent will a reifying tendency contribute to a non-self that expresses itself in a non-life in a non-world? Throughout human history there have been dehumanizing tendencies. In our own society, children cannot escape the computer and the Internet. However, during the closing decades of the twentieth century some eminent people have tried to limit their influence. One of the most distinguished computer scientists at the Massachusetts Institute of Technology spent a great deal of time attempting to convince school boards to get computers out of the classroom until children were older. A former chair of the computer science department at my university would not have computers in his home until his children were in their late teens. One of the most respected American materials scientists strongly discouraged people who were working in his labs from having computers on their desks because it tended to detract from their creativity. They and others recognize that our relations with computers are highly complex. It is once again a question of attempting to assert some freedom in the face of their reifying effects that are indissociable from those of the life-milieu of technique.[35]

When babies and children grow up with computers and the Internet, it should come as no surprise that many of them adapt very well to this programmed environment. It is increasingly a matter of social and intellectual survival. Some studies claim that this evolution in the organizations of the brain-minds of this 'digital' generation is mostly positive.[36] We have noted the obstacles that these studies face. The influence of computers and the Internet cannot be disassociated from the influences of many other technologies. In addition, the overall influence of all the technologies we encounter in our societies is a function of the characteristics of these technologies and of the social and cultural setting into which they are introduced.[37] It is rather embarrassing that this latter problem is almost entirely overlooked in the literature, because it is such an obvious one.

The children and teenagers referred to in this chapter grow up in mass societies characterized by highly desymbolized cultures. As a result, the members of these societies do not receive the cultural guidance that people did in the past in the forms of traditions, customs, morality, and religion. They do the best they can by employing the organizations of their brain-minds as a kind of radar instead of a mental map, in the sense that they scan what everyone else is saying and doing to ensure that they more or less 'fit in' and 'go with the flow.' It gives rise to the other-directed personality, public opinion, and statistical morality.[38] For children and teenagers this lack of cultural guidance creates a need for strong peer groups that serve as surrogate communities and for technologies that are helpful in this regard. The widespread participation in Facebook may almost certainly be interpreted as an attempt to participate in the order of what is real. It is the equivalent of what, in the past, was achieved by being on television. Many people now use the mass media as one of many resources available on the Internet. It offers active participation in the order of what is real, thus democratizing it to some extent. Similarly, the widespread use of Twitter and the like enhances people's abilities to extend their contacts with others, thereby empowering their other-directed personalities. In this way they can 'surf' the waves of every social and intellectual fashion in the making and keep abreast of, and participate in, public opinion and statistical morality. However, it creates more and more desymbolized experiences and hence a further desymbolization of a culture.

High School Science as the Imagination of What Is Real

Teenagers are introduced to Newtonian physics by means of free-body diagrams. These depict a few entities of this world. One of the first depicts a mass sliding along a frictionless plane as a result of a force having been applied to it. None of the constituents have anything in common with those of the daily-life world of teens, or anyone else, for that matter. There is a 'pure' force (that is not embodied in anything) pushing a mass (that occupies no space other than a single point), and this mass slides along an infinite plane (that is also non-material because it produces no friction). The surroundings of the force and the mass are also non-material, since they also produce no drag of any kind. Some time later, a free-body diagram is shown of a force causing a mass to fall along a vertical plane. Still later, these teenagers are shown a free-body diagram of a falling mass pulling another mass up an inclined plane by means of a string run over a pulley. Although the falling mass may be referred to as a weight, it is not embodied, nor is the pulley and the string. They have learned to see this situation in the world of physics as a composite of the previous two. The mass sliding up the incline as a result of a force being applied can be decomposed into a mass sliding along a horizontal and a vertical plane. In this way, students are guided down a path that leads towards the frontier of physics one step at a time. This path is designed to make the greatest possible use of what they have learned thus far by exploiting the potential of previously solved problems to act as paradigms or exemplars.[39] Each new physical phenomenon encountered in the mathematical domain represents a small creative leap forward in the application of Newton's laws of motion used in the previous situation, and itself becomes the jump-off point for the creative leap to the next situation.

The same approach is used for the study of physical phenomena within solid materials, liquids, and gases. Free-body diagrams are drawn to represent all the forces exerted on an infinitesimally small cube (called *a finite element*), located within a mathematical abstraction (called *a continuum* or *a fluid*). A continuum is a mathematical model of a solid material, in which the properties of this material are uniformly distributed throughout space. A fluid is a mathematical model of a liquid or a gas, in which its properties are uniformly distributed throughout space. Once again, Newton's laws of motion can be applied to derive equations governing the distribution of physical phenomena within these mathematical domains as a first step in developing the physics of solid materials, liquids, and gases.

In all these situations students encounter physical phenomena completely unrelated to those of their daily lives. The developments of babies and children described in the previous chapter are permeated by physical phenomena that they learn to master: crawling, walking, eating and drinking, playing with toys, riding tricycles and then bikes, playing ball, climbing trees, sledding, skating, and much more. There is hardly a daily-life activity that does not involve a certain understanding of, and mastery over, physical phenomena. However, these physical phenomena are encountered as being mingled with many other phenomena, and it is precisely this mingling that makes each of these activities unique. As they learn to differentiate these activities from all the others in their lives, a great deal of metaconscious knowledge is built up of the physical phenomena involved in these activities, as is evident when they skilfully ride their bikes or manipulate a soccer ball. However, the knowledge of the behaviour of physical phenomena is embedded in, and inseparable from, the activities in which these phenomena occur. It is impossible to detach this implicit knowledge in order to constitute an explicit body of knowledge of the experiences that babies and children have of the physical phenomena in their lives and the world. Physics classes do not begin by attempting a kind of Socratic dialogue to uncover and build on their metaconscious knowledge of physics. The situation is analogous to the knowledge that babies and children acquire of the grammar of the language of their community. Before they go to school, they exhibit an implicit knowledge of this grammar in the mastery they have achieved over their language. Once again, no teacher will attempt to build on it when beginning to teach them grammar explicitly.

The kinds of physical phenomena that students encounter in their high school physics classes are entirely different from those mastered in their daily-life activities. Physical forces are now disembodied, as opposed to resulting from situations such as throwing a ball or riding a bike. Few surfaces are smooth, and none are frictionless. Everything has inertia, and nothing is weightless. Free-body diagrams remove all these details because the physical phenomena are examined in a mathematical domain. Moreover, this domain can only be imagined and visually represented. Access to the domain of physics is by the visual dimension dominating the aural dimension of experience. Nothing can enter this domain without having been mathematically defined, and having once entered, any interactions with all the previously introduced entities must be specified by mathematical equations expressing Newton's laws of motion in some form.

As noted in the introduction, it may be objected that this is true only for the early stages of the students learning physics. After all, it is not long before they learn to account for friction, inertia, air resistance, and much more. The more the students advance in their mastery of physics, the greater the convergence between the mathematical domain and the real world because the physical phenomena in the domain of physics increasingly predict the behaviour of physical phenomena in the real world with greater accuracy. Were this not the case, the mathematical model of the physical phenomena would be a poor proxy for the phenomena occurring in the world in which the students are embodied. After all, this is what modern science does. However, an explanation to this effect is as common as it is false.

For those students who eventually reach the frontier of physics, the mathematical domain of their subject has nothing in common with the world in which they live. The reason is simple. At any time in their studies and in their practice of physics the mathematical domain of their subject is populated only by physical phenomena. All other phenomena are externalized and delegated to other disciplines. As a result, the free-body diagrams and the mathematical domain of physics always remain completely separated from the experience and culture of the students. It is impossible for these students (or for physicists, for that matter) to symbolize in a normal fashion what is represented in these free-body diagrams or in the mathematical domain of their subject. The experiences of learning and practising physics cannot be directly differentiated from all the others. The processes of differentiation and integration can only differentiate these experiences from each other and jointly from everything else that goes on in human life. It may be assumed, therefore, that the experiences associated with the learning and practice of physics result in developments in the organizations of the brain-minds of the people involved that are only weakly related to the rest of the world.

It is misleading to claim that physics examines one dimension of our lives and the world, because no reference is made to all the other dimensions delegated to other disciplines, how these may affect the phenomenon being examined, and what loss of understanding may have occurred as a result of abstracting one category of phenomena. Physics examines physical phenomena in intellectual and physical contexts as if no other kinds of phenomena existed, since these have no influence on either theory or experiment. It so happens that this is a good approximation of the kinds of situations examined by physicists, bringing it enormous success and making it the model for all scientific disciplines.[40]

Despite this success, some physicists have raised significant methodological issues. Different kinds of physical phenomena currently inhabit different mathematical domains within physics. A search is now on for a unified theory that is able to bring together the different branches of physics.[41] The lack of success in finding such a theory could perhaps be interpreted as a symptom that the physical nature of our universe and of matter and energy cannot be successfully modelled in terms of principles of non-contradiction and separability. This appears to be implicit in the work of physicists such as David Bohm[42] and Bernard D'Espagnat.[43] Similarly, if the observer influences what is observed, is it possible to separate the discipline of physics from other disciplines if the observer must be included in the experiment? Even though scientific disciplines such as physics are separated from experience and culture, their desymbolization cannot be absolute, given the organization of the brain-minds of their practitioners. Physicists appear to be unable to detach themselves completely from the vantage point of their culture represented by its myths.[44]

From the perspective of symbolization, the experiences of teenagers learning physics have much more in common with those of watching television and playing with computers than they have with the daily-life experiences discussed in the first chapter. The experiences of learning physics differ from the latter in three ways. First, the foregrounds of the experiences discussed in this chapter are discontinuous with respect to their backgrounds. The television screen shows situations and events that have been decontextualized from the world in order to be recontextualized in a program. The computer screen shows simulations of the world obtained by the rules and algorithms of computer programs. The blackboard in physics classes and the diagrams in textbooks visually represent the mathematical domain separated from experience and culture. In all three categories of experiences, the backgrounds are those of the dialectically enfolded world in which the people involved are embodied. The second important difference is that the students cannot be embodied in the foregrounds other than by being disembodied spectators. When using computers or learning physics, they depend on the use of a mathematical imagination and visual representations. The third difference is related to the different kinds of metaconscious knowledge generated in the organizations of the brain-minds of the students. There is a great deal of evidence to suggest that the experiences of learning physics are directly differentiated from each other and jointly differentiated from all the other experiences, with the result that the metaconscious knowledge built up in the development of the organizations of the brain-minds of the students is discontinuous from the metaconscious knowledge built up from all the other experiences of their lives.

This last difference manifests itself in several ways. Learning physics results in two parallel modes of knowing and doing related to physical phenomena. One has been referred to as *intuitive physics*, and the other as *school physics*.[45] As noted above, many daily-life activities involve a knowledge of, and mastery over, physical phenomena. Even physicists do not rely on their school physics when they ride their bikes or play squash. We would not dream of teaching physics to athletes in the hope of improving their performance. Our daily-life skills manifest the development of a considerable metaconscious knowledge of the physical phenomena enfolded into these activities. This metaconscious knowledge includes, but is far from limited to, what is referred to in the literature as intuitive physics. Furthermore, this intuitive physics is clearly not replaced by school physics.

The existence of parallel forms of metaconscious knowledge related to physical phenomena can be induced from the results of experiments in which subjects were presented with different kinds of situations involving physical phenomena.[46] If these situations were described in daily-life terms, students tended to give answers based on their intuitive physics. If, on the other hand, these situations were described in terms associated with school physics, they tended to give answers conforming to what they had learned about school physics. The answers were often different because, from the perspective of intuitive physics, school physics appears to be counter-intuitive.

Further light can be shed on the different kinds of metaconscious knowledge involved in intuitive physics and school physics by another experiment. It provides evidence of how the intellectual worlds of the physical sciences are qualitatively different from those of the humanities, with the implication that this reflects corresponding differences in the phenomena they study. The experiment explored barriers to the full and equal participation of women in the sciences.[47] Diaries were kept by a group of graduate students and faculty members from the social sciences and humanities who were taking a course in university physics or chemistry and by a group of graduate students and faculty members from the physical sciences who were taking a course in English literature. The members of the group learning physics or chemistry found that the lack of an open view of the subject matter presented at the beginning of the course with ongoing references to it constituted a real barrier to learning the subject. They had learned to use such overviews for interpreting particular topics, for situating these topics in relation to one another, and for understanding how each topic contributed to the whole. In other words, they were accustomed to lower levels of desymbolization than they had encountered in their physics or chemistry course. They tended to find the relatively cumulative development of these courses, in which the bigger picture tends to emerge only towards the end, difficult to accept. Despite assurances from the professor that they did not need to know the bigger picture in order to learn the subject, some of the people taking the course strongly felt that it inhibited their ability to put what they were learning into their own words and hence to fully engage themselves in the subject matter. Others had a sense of being lost, with no idea of where they were in the course and where they were going, which made it difficult for them to participate. Still others felt they were being treated as children, being spoon-fed small amounts at a time and having to trust that everything would work out all right. Some people complained that they had difficulties doing the problem sets because they had no idea of the reason for them or of their significance within the larger scope of the subject matter.

Their professors, on the other hand, had difficulty understanding what they were talking about because they themselves came from disciplines with much higher levels of desymbolization. For them, not to expose the students to the full complexity of the subject matter at the beginning of the course but to gradually introduce it to them represented the best and possibly the only pedagogical strategy. They could not understand why their students were so uncomfortable with not being able to make their own contributions to the problems and being essentially limited to imitating the examples given by the professors in class and by the textbooks. They could not understand why the students felt that they were merely asked to master certain approaches for dealing with different kinds of problems, as opposed to struggling with the fundamental concepts and structure of their subject matter, and that they were getting feedback only on how they did on their problem sets and not in their mastery of the complexity of the subject. Hence, the professors had difficulty taking the questions of their students seriously, and this further aggravated the situation. The 'hidden curriculum' (what is implicitly transmitted in the classroom) appears to suggest that the intellectual context and the history of the subject do not matter. For the students, it appeared that physics or chemistry had no structure and no context since only a few relations were connected in each chapter and no larger connections appeared to be necessary. In terms of personal involvement in the subject, this appeared to be limited to mastering problem-solving techniques. The contribution of personal opinions and intellectually pushing the limits of the subject matter were not welcomed or deemed necessary.

In a mirror-image part of the experiment, graduate students and faculty members with a background in the natural science disciplines took a course in English literature. Much the opposite kinds of situations resulted. This subject matter is much more contextual and cannot be broken up into relatively self-contained building blocks. Such a subject requires a higher tolerance for ambiguity, and it certainly cannot be reduced to mathematical equations without extreme reductionisms. Some members of this group complained that there were just too many words. The professor found this impossible to understand, because it was with words that an understanding of the subject was built up. However, the students thought that the course meandered in the absence of a clear rational progression. They also had difficulty dealing with concepts that were not stripped of their cultural and historical 'baggage.'

These two phases of the experiment confirmed the presence of two very different intellectual cultures. The differences can in part be explained in terms of their levels of desymbolization. The intellectual cultures of the physical sciences are separated from experience and culture, while those of the humanities are desymbolized to a much lesser degree, depending on the school of thought. Consequently, the students' experiences of being socialized into these very different intellectual cultures were very different in terms of the relationship between their foregrounds and backgrounds, the level of desymbolization of their foregrounds, and the differences in the kinds of metaconscious knowledge built up with these experiences. The intuitions of the students taking courses in these different intellectual cultures were very different, and the way in which students and professors articulated them also varied considerably from person to person. Nevertheless, it seems reasonable to assume that the symbolization of free-body diagrams and mathematical domains in the disciplines of the physical sciences is very different from the symbolization in the humanities.

The differences in the levels of desymbolization in the disciplines of the natural sciences and the humanities suggested by the above experiment have substantial implications for the levels of participation and commitment of the students. In their experiences of learning physics or chemistry, it is impossible for students to be as involved as when they are fully embodied in a situation. Possibly the closest that students can get to daily-life experiences is in doing puzzles or playing games. Temporarily, they may be totally absorbed in a puzzle or a game, but they know full well that sooner or later they have to get back to their lives with others in the world. Some students may come to enjoy the challenging mathematics, physics, or chemistry problems in the way that others enjoy puzzles or chess games, provided that a scientific ideology does not get in the way. However, the way that we differentiate these kinds of activities from all the others affects the way that we see ourselves in terms of our involvement. We do not expect to be involved in puzzles or games in the same way that we are involved when we get on with our daily lives.

In terms of our participation, there are significant differences between the disciplines of the natural sciences, social sciences, and humanities. The level of desymbolization is highest in the natural sciences, which has been built up as much as possible in mathematical domains. The concept of a detached observer can have real meaning more easily in the context of the way that students live these experiences than in the social sciences or humanities. For example, physical phenomena are examined in intellectual and physical contexts that are very different from the context in which people lived fully embodied lives. As a result of a build-up of metaconscious knowledge, students will gradually develop intuitions to this effect, on which notions of a detached and objective observer may appear as a plausible explanation. In the social sciences and humanities this is much more difficult.

Before the discipline-based intellectual division of labour developed within the social sciences, it was much easier to think of scientific activities as contributing to a better understanding of human life and the world. Success in this endeavour involved making a highly unintelligible complexity more intelligible by deciding that certain aspects were very important for this understanding, others less important, and still others only marginally relevant. However, what constitutes the basis for deciding such a hierarchy of relevance? Clearly, this cannot be done without an implicit understanding of what is important for human life and the world, and this is why the myths of a society inevitably leave their mark on science. For example, despite their differences, for Karl Marx, Max Weber, and Jacques Ellul, what was important for human life was our freedom. Being enslaved to someone or something resulted in unacceptable forms of human life. According to Marx, if capitalism enslaved both rich and poor, a society found itself in a position in which the system it had created to serve it had turned into something that enslaved it. Hence, our humanity demanded that we commit to transforming this system in order to make it serve society. According to Weber, the phenomenon of rationality was beginning to enslave societies by shutting human life into an iron cage. It drove Weber into a deep depression. Most recently, the theory of technique created by Jacques Ellul commits to the value of freedom as the most important feature of human life in the world.[48] By paying very little attention to the phenomenon of technology, the social sciences have had an easier time of claiming objectivity and detachment from a world so deeply permeated by technology. Generally speaking, it can be argued that for the last two hundred years our relation with technology has had two primary components: people changing technology, and technology changing people.[49] If the former is more decisive during a particular epoch in the history of a society than is the latter, human freedom with respect to technology is maintained; the reverse situation results in our being enslaved to our own creation. By ignoring how technology changes people (a subject that in any case falls beyond the scope of any social science discipline), the straw men of technological determinism and technological autonomy could be dismissed out of hand. This is a classical manifestation of the kinds of counter-transference reactions against which Devereux has warned us.[50]

Once the social sciences had come into the grip of the discipline-based intellectual division of labour, the concept of an objective and detached observer transforming an unintelligible complexity into a more intelligible one without requiring any human values became intuitively plausible. After all, reducing human life and society to a single category of phenomena is very different from living all these intermingled phenomena in daily life, even when these phenomena have been 'unfolded' to a considerable degree as a result of industrialization. This also makes it much easier to convince ourselves that quantitative approaches using survey methodology and statistical analysis of the resulting data are much more scientific than are traditional approaches, even though the questions on the survey are not informed by a comprehensive scientific understanding of human life in the world. Such an understanding would require a scientific approach for integrating the findings of the many relevant disciplines, but we have never succeeded in unifying the discipline-based sciences.

What we have endeavoured to show is that what the teenagers are introduced to by means of free-body diagrams is a domain constituted of mathematically defined entities that interact according to mathematical relations expressed by means of equations. This domain is gradually made more and more complex as new entities and relationships are added. The discipline of physics has built several of these mathematical domains based on the principles of separability, non-contradiction, internal logical consistency, and a limited external consistency with physical phenomena created in laboratory environments. These mathematical domains are not consistent with one another, and a unified theory has thus far eluded physicists.

The mathematical domains of physics have been stripped of all chemical, biological, social, and cultural relations. Each one does not describe even one dimension of the real world, because no reference to the other dimensions is made, nor is there a perceived need for giving an account of its dependence on all the other dimensions. Physics examines physical phenomena in domains where everything else might as well not exist, because it has no effect on the analysis or the experimentation. Of course, there is a practical value in all of this because industrial processes, machines, and engineering systems frequently manipulate one set of phenomena at a time. These developments have necessitated the reorganization of much of human life and society in the image of the machine; that is, they are treated as distinct and separate domains within which only a single category of phenomena at a time contributes something to the other domains. However, such applications are no justification in the search for a scientific understanding of our lives and the world.

The learning of school physics by teenagers is undoubtedly complicated by the scientific ideology voiced by many of their teachers. Students are not encouraged to use their imaginations to enter the abstract mathematical domain of Newtonian mechanics, which clearly exists nowhere. They are not told that this is an abstract model that does not approximate their daily-life world in any way, other than with respect to what at best may be one of its dimensions. Nor are they taught anything about the limitations of discipline-based knowing and doing and those of daily-life-based knowing and doing, and certainly not about how these parallel modes have unique and opposed strengths and weaknesses. Instead they are told that science is a controlled and objective way of gaining knowledge of the real world and that their experiences are merely subjective, with a very limited reliability. It is this ideology that helps them to develop the metaconscious myth of science.

The Emerging Order of What Is Real

In the previous chapter it became apparent that toddlers and children enter into the cultural order of their community by means of language. As long as we have been a symbolic species, language has been the exclusive gateway to first establish an order of what is true in the lives of children, followed by a growing convergence between this order and the cultural order of their communities. As they grow up, their symbolization of the experiences of living their lives in the world becomes increasingly typical of their time, place, and culture. Their growing up as members of a symbolic species requires the primacy of hearing over seeing and the word over the image.

In the present chapter we have shown that toddlers, children, and teenagers growing up in contemporary mass societies engage in activities such as watching television, playing with computers, and learning scientific disciplines in high school, all of which are highly desymbolizing. Owing to a limited embodiment, participation, commitment, and freedom, these activities impose severe limitations on how everything can be related to everything else in their lives, thus undermining first the order of the true and later the cultural order. They learn that the members of their communities regard being on television as more real and important than daily life, that computer skills are almost more important than any daily-life skills, and that scientific disciplines introduce them to 'real' knowledge that is more reliable and objective than daily-life knowledge. All this (and more) implicitly tells them that beyond the cultural order they are entering lies an even more important order. The real world is to be found on a screen, and their participation in it requires scientific and technical disciplines and specialties.

Growing up in a contemporary mass society thus confronts children and teenagers with a difficult dilemma. They encounter a community that sends a strong message that the activities in which they are least embodied, in which they can least participate, where little commitment is required, and the least freedom is enjoyed, are the most important ones in their lives. In contrast, the daily-life activities with much higher levels of embodiment, participation, commitment, and freedom, thus permitting a much fuller grip on their lives and their worlds, are devalued as being personal and subjective. It is more important to live on the Internet and to be a specialist in one small domain of a discipline or specialty.

The acquisition of a culture has always been inseparable from a growing reliance on myth and thus on being alienated. It was the price that had to be paid for protection against relativism, nihilism, and anomie. What will happen to us as a symbolic species when a new order begins to dominate the cultural order? To answer this question we need to shift our focus from the individual to the community of which he or she is a member, and on how these communities have established an order of what is real within the order of what is true during the last two centuries.

3 Colliding Orders and the Triumph of the Real

Re-engineering a Symbolic Species

Our civilization has once again transformed the world along with who we are as a symbolic species. The two are indissociably linked. First, the world had to be made ready before it could be populated with technologies of all kinds. Once introduced, these technologies rapidly spread throughout societies. Human experience became permeated by these technologies and all the changes that came with them. In the course of several generations the organizations of people's brain-minds were deeply affected, as was manifested by significant cultural mutations.

The potential for significant changes in human consciousness and cultures is rooted in our being a symbolic species. As we change our surroundings, these surroundings inevitably change us. We have already seen that the transformation of our surroundings begins with the symbolization of each and every constituent in relation to all the others and our lives. In this way, human groups and societies have always distanced themselves from these surroundings by interposing a sphere of possibilities implied in the symbolization of their surroundings in relation to themselves, and themselves in relation to these surroundings. It characterizes every step of our historical journey. From the very beginning there was a scientific dimension to this symbolization of people's surroundings, which led to indigenous peoples' acquiring a great deal of knowledge about the significance for human life of the constituents of their natural surroundings. Before any technological activities could emerge, whatever natural materials and processes they employed first had to be symbolized in human terms. In addition to this kind of symbolization, their economic activities required rituals for permission to appropriate plants or animals. All this was accompanied by the symbolization of human activities and lives in relation to an emerging cultural order. Further symbolization was required in the unfolding of social, legal, moral, and religious dimensions. In this way, human groups gradually transformed their surroundings into a life-milieu in relation to which their lives evolved. This process took place in a dialectical tension between a sphere of freedom established in relation to the constraints of that life-milieu, and the alienation that resulted from the creation of myths to protect that freedom from relativism, nihilism, and anomie. All this was internalized by each new generation as its members acquired a language and a culture.

There was another important aspect to the ways in which individual and collective human life evolved in relation to a life-milieu and how it evolved in relation to that life. The deepest metaconscious knowledge resulting from the organizations of people's brain-minds interpolating and extrapolating all their experiences implied that their life-milieu was the most valuable element in their lives. By providing everything necessary for life, the life-milieu had the power over life and death and over good and evil. Intuitions to this effect led to its sacralization as an ultimate anchor in an unknowable reality, on which morality could be founded. Unfolding the potential of the sphere of possibilities created by symbolizing human life in a life-milieu led to their joint evolution.

The first life-milieu that humanity created for itself was what we call *nature* but what the people at the time symbolized very differently. It opened up a sphere of possibilities for satisfying their needs and aspirations and for dealing with the complexities and fears of life and death. Nomadic groups thus developed a great deal of knowledge about their life-milieu, which expanded and transformed the sphere of possibilities through which they distanced themselves from the constraints imposed by the life-milieu. They thrived, and as their numbers grew, the food supply available by gathering and hunting became less adequate. They dealt with this challenge by making a more creative use of their vast knowledge of plants and animals in order to adapt their ways of life and transform their life-milieu by means of initiatives that shared some features with what we call agriculture.

As the competition for land increased, some groups may have been driven into the deltas of the great rivers, where the first civilizations were born. These were based on very different ways of life that required the cooperation of many groups to drain land, close it off with dykes, and control its irrigation with dams. Societies thus emerged, and these became the primary life-milieu for the group. It is via this life-milieu that the group distanced itself further from the life-milieu of nature, permitting it to be adapted to human life on a scale that was unthinkable until then and, in any case, unachievable. The new possibilities created by the life-milieu of society included a much greater social division of labour, which unfolded many more spheres of activities as well as new institutions. At the same time, these developments trapped the majority of people into hard physical work, leading to a significant reduction in the quality of their lives.

Thousands of years later, Western civilization unfolded the sphere of possibilities created by society (the primary life-milieu) in relation to nature (the secondary life-milieu) by means of the three 'perfections': Greek philosophy, Roman law, and Christian revelation. With hindsight it turned out that these perfections were contradictory, thus fuelling an unusual dynamism. Eventually the unfolding of all these possibilities led to the creation of a third life-milieu, mainly by exploring the possibilities of industrialization and urbanization.

Industrialization and everything that came with it had another profound influence on human consciousness, cultures, and the primary life-milieu of society for several reasons. First, it radically transformed the material and energy constraints of societies. Seen from the perspective of 'people changing technology,' it offered a greater mastery over these constraints. Seen from the perspective of 'technology changing people,' these constraints were qualitatively transformed. Before industrialization the material and energy constraints were symbolized, and thus dominated, by the cultural orders that had been created and had evolved in relation to a life-milieu. They were subjected to their human meanings and values, with the result that traditional societies struggled not to live by bread alone. Although constantly threatened by crises, this symbolic mastery ensured that such challenges were dealt with in terms of these cultural orders and not as arising from 'natural' economic phenomena. Out of necessity, industrialization caused the material and energy constraints on human life to take on a life of their own by escaping the cultural orders, owing to the failure of their domination by symbolic means. Eventually, entirely new approaches to knowing and doing that were separated from experience and culture had to be developed. All this led to a great deal of desymbolization.

Second, the introduction of machines into human life, and the adjustments this required, in essence created an order of what is real within the cultural order of what is true. The result was a collision between the two orders. It began with, and spread out from, the inability of human beings to repeat any activity in quite the same way. This impossibility is also rooted in our being a symbolic species. When attempting to repeat a cycle of motions, we escape by thinking of other things, our moods change, our fatigue levels increase, our need for food and drink gradually rises, and so on. As a result, these experiences will vary, and their internalization and symbolization will change us in small and subtle ways. Each time we attempt another cycle of motions, we shall have changed. Our nervous systems have evolved to cope with our lives constantly evolving in a life-milieu in which nothing ever repeats itself in quite the same way. Our inability to deal with sensory deprivation, monotony, and repetition without serious negative consequences for our lives is well known. We are back to the differences between an order of what is real as represented by machines and an order of what is true as represented by our living as a symbolic species.

No matter how much our lives and cultures may become desymbolized, we continue to be a symbolic species. Hence, technology and industrialization are inseparable from all of human life and society, and vice versa. To put it simply, as we change technology, technology simultaneously changes us. Industrialization and urbanization are as much a transformation of our world as a transformation of who we are. They constitute a vast experiment on what it is to be a symbolic species and thus on what has made us human until now. Seeking to understand this, one discipline at a time, is to behave as though we have already ceased to be a symbolic species. It makes a genuine intellectual life within the modern university next to impossible.

In the present chapter we shall attempt to show how human life, in the primary life-milieu of traditional societies and the secondary life-milieu of nature, was transformed by industrialization as a consequence of the material and energy constraints escaping from being symbolically dominated by a cultural order, only to become dominated by a technological and economic one. In the next chapter we shall attempt to show how these developments caused human knowing and doing to follow suit and become separated from experience and culture. Both developments opened the floodgates to desymbolizing experiences and our being compelled to live primarily in the order of what is real (and only secondarily in the cultural order), with vast consequences for our being a symbolic species. All this should have come as no surprise. We know all too well how, when human beings began to live in the life-milieu of societies, human consciousness and cultures (including their moralities and religions) were entirely transformed. By symbolizing this as the dawn of human history, as if no history preceded it, we have diminished the significance of the greatest human mutation after that of our becoming a symbolic species. It has also made us less aware of how we are again involved in a gigantic experiment: transforming ourselves as a consequence of the desymbolization of our cultures and thus of what has made us human until now. A few people may celebrate this as a liberation from religion and myths, but for the overwhelming majority of humanity, life has become extraordinarily difficult.

The Emerging Technological Order

A fundamental incompatibility exists between machines and the cultural order that human communities establish by means of symbolization. They represent diametrically opposite approaches to ordering human life in the world. Many traditional cultures appear to have had some sense of the incompatibility of these two orders, or at least of the potential threat posed by a mechanistic order to a living order. In their ways of life they limited the use of machines in one way or another. Some cultures built them simply to prove a point, after which they were destroyed; others restricted their use to specific areas of life; and almost all excluded them from production.[1] Some of this can be explained by a lack of adequate materials, processes, or knowledge, but this is not always the case. Some well-known examples include the ancient Chinese restricting the use of gunpowder to entertainment, the Japanese banishing the use of guns during a certain period, Alexandria applying technology to many domains except for production, the Greeks refusing to contribute engineering manuals to this city's great library, and Western European societies building sophisticated automatons without anyone applying these skills to production.[2] The reservations of many cultures regarding the use of machines may have been mere prejudices, but there remains a core of phenomena that cannot be so easily dismissed.

In this respect, the transformation of Western civilization towards the end of the Middle Ages is perhaps unique. A new cultural orientation was emerging, which philosophers expressed in part by a growing emphasis on the visual dimension of experience at the expense of the aural dimension.[3] This reversal is rather remarkable since access to the cultural order is via language: first listening to it and then learning to speak it. However, this most complex and essential human order worked in the background and remained largely invisible. As a consequence of a growing emphasis on seeing over hearing, the machine became widely accepted as a symbol of the most perfect known order. Descartes expressed this by declaring that the universe was a gigantic clockwork created and put into motion by God to function in accordance with his 'laws.' Christianity had all but lost its Jewish roots, including its caution about paying too much attention to what can be seen instead of to the word. These philosophical intuitions only very gradually permeated the mainstream of Western societies until industrialization compelled almost everyone to pay much more attention to what they could see as opposed to what they could hear. What could be heard of the new order of machines was noise, while the cultural order remained accessible mostly by speaking and hearing.

Industrialization produces a collision between the cultural order of a society and the order that emerges as a result of the employment of machines. Before any human work can be mechanized, it must be internally and externally separated from the cultural order of a society by means of the technical division of labour, which reorganizes that work to take advantage of what machines excel at: the precise, rapid, and endless repetition of distinct and separate production steps. It establishes an order based on repetition, which is diametrically opposed to any living order based on constant adaptation.

When human beings are organized to make pins in this way (to use Adam Smith's example), one person endlessly repeats the cutting of wire into fixed lengths, another worker makes the points, another the heads, and still another hardens the points by means of heat treatment. Together, they will be able to make far more pins in a day than if each person were to make entire pins, but it comes at a high cost to their lives and society. They can no longer fully live these work-related activities because they cannot be sustained in the background by the organization of their brain-minds, which symbolizes their lives (including their education, past experience, intuitions, and personalities). Work-related experiences are no longer an individually unique manifestation of the cultural order but an imitation of an externally imposed mechanistic order. As a result, a portion of the organizations of their brain-minds is no longer their own.

In addition, work-related activities are no longer externally connected to the cultural order of a person's society. People do not need to connect the production step in which they are engaged to the production process, the product being made, the needs and desires of the users of these products, and the contribution that they and the products make to the way of life of their society. Therefore, it is next to impossible to symbolize these activities. These production steps are a measure of themselves, the 'one best way' of transforming the required inputs into a desired output without any consideration being given to the broader context. Alternative forms of converting these inputs into the output are compared in terms of their output-input ratios, including the efficiency of the throughput of matter or energy, the productivity of labour, and the profitability of capital. In other words, it is a way of perceiving, organizing, and evaluating the world that is diametrically opposite to doing so on the basis of symbolization. No reference is made to the lives of the people directly and indirectly involved or how, via these lives, communities have changed and ecosystems have been affected. Nor is any reference made to the physical context such as the product, the production facility, and all the physical processes associated with them.

A useful distinction has been made between the 'horizontal' and the 'vertical' dimensions of the technical division of labour. The former divides the production process into cycles of operations, which become its building blocks. These are assigned either to machines or to people. However, the people involved do not participate in the planning, implementation, and operation of the work process. These matters are taken care of by the vertical dimension. Just as the horizontal dimension divides the 'hand,' the vertical dimension divides the 'brain.' It begins with the people on the shop floor who supervise those who carry out particular production steps. These supervisors report to the next higher level of coordination and management, and the people working on that level in turn report to the next higher level, and so on. A pyramidal organizational structure is thus created, with senior management at the top.

Today the people that make up this 'collective brain' are almost all university-educated practitioners of discipline-based knowledge. Their knowledge domains are built up rationally and often mathematically as opposed to dialectically and symbolically. They can therefore participate either informally using their daily-life 'intuitive' skills or formally by employing their specialties. As is the case for the people carrying out a particular production step, their informal participation is sustained in the background by the organization of their lives. Their formal participation is based on a portion of the organizations of their brain-minds that could not be symbolized in the usual way and which, therefore, excludes their non-professional lives. To use the simple model developed earlier, the organizations of their brain-minds can sustain their formal participation only on the basis of the differentiated clusters of experiences separated from their lives and culture, which derive from and are associated with the acquisition and practice of their specialties. Consequently, their formal participation is limited to determining the one best way according to their specialty, without any consideration being given to the context that lies beyond it. As noted, this one best way is 'best' in terms of output-input ratios, as opposed to being best for human life and society based on symbolization and cultural values. Just as the people executing particular production steps have no idea of the implications, for the world beyond, of paying special attention to this or that detail, so also do the specialists have no idea of the implications of their interventions, which mostly fall outside of their domains of specialization.

We do not sufficiently appreciate the problems associated with limiting the context of human work to a particular production step for people involved in the horizontal dimension of the technical division of labour or to a domain of specialized knowledge for the people involved in the vertical dimension. Chief executive officers may think that the buck stops with them, but this is limited to the constraints imposed on their technological role. Such has been the case ever since the beginning of industrialization. Suppose an entrepreneur-owner of a factory had a change of heart when he fully recognized the impact that factory work was having on the lives of his workers. He would have rapidly discovered that he really had no options in a cultural sense. If he used his personal wealth to supplement the wages he paid to his workers in an attempt to improve their lives, he would lose his factory when his fortune ran out. Alternatively, if he funded these efforts by raising prices, he would soon be competed out of the markets for his products and go bankrupt. Holding on to any (cultural) values of any kind as a factory owner became impossible unless the role of markets could be delimited by high labour and health standards based on these values. Even if these had been imposed at the same time, such standards would not have got to the root of the problem and would have merely limited the damage, as is the case for all end-of-pipe approaches. Hence, an employer could not symbolize his relationships with his employees via the work organization. The same was true for anyone else involved in the vertical dimension of labour. Their authority and power was constrained by their technological role, which excluded whatever values and beliefs they practised in their personal lives. Their work-related experiences could not be symbolized any more than could those of their workers. Human work and its organization had to be withdrawn from the cultural order by means of a corresponding desymbolization.

In both its horizontal and vertical dimensions, the technical division of labour reorganized human work into a sequence of cyclical hand or brain operations assessed on their own terms and connected to all the others exclusively by means of the physical or intellectual inputs obtained from the previous step and the output delivered to the next one. A technically divided work process was planned in exactly the same way as a machine was designed. Each production step was a separate and distinct 'part.' After each production step was optimized without any reference to its context, the steps were 'assembled' into larger units, which in their turn were optimized without any reference to context. These units were then assembled into the overall production plan, which was also optimized without any reference to context. In the order established by this production plan, everything was endlessly repeated. The operations involved in each production step produced the same results as much as possible, which simplified subsequent steps by creating interchangeable parts, sub-assemblies, and products. In this kind of order, everything was brought into 'existence' by first the making of distinct and separate parts and then their assemblage into larger units. Within these larger units, everything maintained its own identity as opposed to expressing an aspect integral to that of the larger entity into which it had been assembled. The relationship between the parts and the whole of a production plan, including its implementation and operation, was unique to a mechanistic order.

I am not suggesting that all this happened overnight. In the first generation of industrial societies it took more than two centuries. It is well known that Henry Ford was the first to perfect the horizontal dimension, and Frederic Taylor the vertical one, and their synergy resulted in the Fordist-Taylorist production system. Still later, information machines required further perfecting of the system. These machines, much like their conventional counterparts, are based on a technical division of intellectual labour. As noted, work organizations initially incorporated information machines as tools in the service of conventional bureaucracies. A fundamental incompatibility soon revealed itself, compelling the work organization to be reorganized in the image of these information machines. This was referred to as business process re-engineering, the management of technology, enterprise integration, and the re-engineering of the corporation.[4]

Within the overall production plan, any horizontal or vertical step can be assigned today either to a machine or to a human being. If it is assigned to a person, he or she will have no choice but to work in the image of a conventional or an information machine. There is simply no other possibility. As a result, the new order, which had gradually been established by first technically dividing human work in preparation for mechanizing, automating, and computerizing it, began to form enclaves in the cultural order of human life and society. Human beings had to participate within these enclaves on the terms dictated by the mechanistic order, but outside of them on the terms of a living biological and cultural order. It is reasonable, therefore, to expect tensions at the interface between these two orders.

Since technically divided work cannot be supported by our lives working in the background, the resulting desymbolization involves the suppression of the resources that we normally bring to our work. Moreover, working according to the one (technically) best way involves doing so with the least variation possible. Suppressing ourselves and our lives in this way requires a great deal of mental energy, with the result that nervous fatigue characterizes this work. During the beginning of mechanization and industrialization, people everywhere had to put up with a rigorous, brutal, and inhumane factory discipline for the first time in their lives.[5]

Ever since the beginning of industrialization the problems resulting from the technical division of labour have been exacerbated by its politicization, which pitted those on the political left against those on the political right. Work in the image of the machine and the order this imposes on any work organization cannot be symbolized and hence cannot be truly politicized, detached as it is from the cultural order of human life and society. Working in sweat shops and on assembly lines, and the supervision and management of the corresponding work organizations, has alienated people everywhere regardless of the political regime, and neither the left nor the right has ever come up with any solutions. Organizing work in the image of the machine, and thus withdrawing it from any and all cultural orders, excludes a culture-based solution involving human values. For well over a century there have been suggestions that our social ills can be cured by a return to the values and religious commitments of our civilization. I shall endeavour to show that this manifests a complete misunderstanding of what is happening to human life and society. The symbolization of human knowing and doing, to be taken up in the final chapter, would, if implemented, chart a very different course of action.

Industrialization thus began to transform human work as if it were a part in a mechanistic order. It was no longer an expression of human lives and of a cultural order, which caused it to be treated as a commodity belonging to another order. An economic order had to be created to extend the order of the machine. This would take care of anything that could no longer be regulated by the cultural order.

The impossibility of incorporating human work organized in the image of the machine into any traditional culture or of transforming such a culture to enable it to symbolize such work allowed ideologies of all kinds to step into this vacuum. There was certainly no shortage of such ideologies in the industrializing societies of the nineteenth century. None of them dealt with the root problem, and none of them facilitated or brought within reach a culture-based solution that might have made this kind of work liveable by means of symbolization. Humanity had always lived by creating cultural orders, and the impossibility of incorporating this technically divided work and its organization marked the beginning of a cultural crisis from which we have never recovered.

The growing acceptance of a new human nature, *homo economicus*, which maximizes the utility derived from a wage or the profit from allocating capital, ultimately implied that, deep down, human life runs on the optimization of algorithms dominated by money. It resolved the problem by claiming that human beings are essentially mechanisms that can be incorporated into those of markets. The extension of the order of the machine was thus legitimated. It also established the role of money by, as much as possible, substituting prices for cultural meanings and values. It paved the way for political economy to be replaced by the relatively autonomous discipline of economics. Nevertheless, the notion that human life is all about money and that everything can be expressed in terms of it is a development that would have been incompatible with any traditional culture unless a devastating desymbolization had been tolerated.

Without an ideology of one kind or another, all this has become a rather painful and embarrassing chapter in our history. Such feelings are usually avoided by arguing that there are simply no alternatives. Given the current global population, it is clearly impossible to go back to craft work. Nor are we likely to give up the benefits of interchangeable parts and mass production. We must face the fact that our present cultural context is the result of our never having taken these problems seriously. We have allowed, and continue to allow, human work to be swallowed up into the order of the machine. We treat this as some kind of inevitability and dismiss any protests as being naive. I have in mind the women at the gates of the Fordist-Taylorist factories protesting the destruction of the humanity of their husbands. I am also thinking of how in Japan, the land of lean production, women have complained that their husbands upon retirement are little more than empty husks, thanks to the endless drive to perfect the mechanism of production. It is surely no coincidence that, in the society that invented lean production out of necessity, death by overwork is a recognized disease. That all this was not inevitable is shown by alternate approaches that go to the root of the problem of work in the image of the machine. Had these been implemented from the beginning as an expression of the values of the cultural orders of traditional societies, things would have evolved very differently.[6]

A Growing Dependence on Matter and Energy

Another tension quickly developed on the interface between the enclaves of production ordered in the image of the machine and the surrounding cultural order of an industrializing society. We have already noted that a group of people making pins on the basis of a technical division of labour could produce a significantly higher output, even without mechanization, than if each member individually produced entire pins on the basis of a craft process. Mechanization increased this production output even more. These changes occurred everywhere that production was reordered by the technical division of labour and mechanization, thus increasing the throughput of materials and energy in all the mechanistically ordered enclaves. Prior to industrialization, craft production had always been embedded in the cultural order of a society, which supplied the raw materials and put the finished products to use. The capacity of the cultural order to do so became strained as the throughput of materials and energy increased. Additional activities had to be reorganized in a way that required the severing of their internal and external connectedness to the cultural order so as to be linked together into a separate economy that could no longer be organized on the basis of a culture, as was the remainder of society. This economy organized the exchanges of materials and energy between the mechanistically ordered enclaves of production independently from the cultural order. All this can be brought into focus when examining the process of industrialization through the lens of thermodynamics.

Human life in a cultural order depends on a throughput of matter and energy: matter for the ongoing repair and replacement of the cells, tissues, and organs of the human body, and energy for powering our physical and mental processes. As a result, even a simple activity such as blinking our eyes connects us to the biosphere and beyond through an intake of food, drink, and air and a discharge of what our bodies can no longer utilize. These connections to the biosphere are direct if we eat from our own garden and drink from our own well and indirect if we rely on others for growing, transporting, processing, packaging, distributing, and retailing our food and drink. Similarly, they depend on direct exchanges with the biosphere if we compost our own wastes and on indirect exchanges if we rely on waste and sewage collection systems. The same argument can be made for all daily-life activities. Clothing ourselves involves direct exchanges if we make all of our own clothes from what we have directly procured from the biosphere or indirect exchanges if we purchase them. It immediately follows that almost all daily-life activities, including making cellphone calls, using our computers, relaxing by watching a movie, and driving around in our cars, involve complex and diverse exchanges with the biosphere, usually via long chains of activities.

Industrialization moved more and more people off the land and turned them into wage earners who worked long hours and who could not procure most of their necessities directly from the biosphere, as was common for many peasants. As a result, more and more activities of an industrializing way of life became connected to the biosphere via long chains of other activities that intertwined to form a network of flows of matter and a network of flows of energy. It is via these two networks that individual activities exchange matter and energy with the biosphere. The network of flows of matter has to procure its inputs from a corresponding larger network representing these flows in the biosphere. The smaller network is suspended within the larger one since matter can be neither created nor destroyed at the interface between these two networks. The same argument holds for the network of flows of energy and its suspension within the larger network representing these flows within the biosphere. In its turn, the biosphere cannot create or destroy the required matter and energy, any more than can a way of life or an individual activity. Since there is a negligibly small input of matter from outer space, the larger network of flows of matter must 'produce' all flows by sequences of transformations in which the output of one transformation becomes the input into the next until the loop is closed. Any human usage of matter must tap into these cycles. In contrast, the larger network of flows of energy receives a constant input of solar energy by day and radiates low-temperature heat back into space by night. Every human activity is constrained by these connections, which can be studied through the lens of thermodynamics.[7]

Prior to industrialization, the networks of flows of matter and energy were enfolded into and evolved by the cultural order of a society because any exchanges of matter and energy between the activities constituting a way of life generally followed the patterns of kinship and other social ties that made up the fabric of close-knit local communities. The exceptions were the centres of trade and commerce, where different cultures encountered each other and intertwined their cultural orders. In addition, such centres extended the ecological footprints of the communities and societies beyond their boundaries. Since the flows of matter and energy were embodied in flows of goods and services, and since these flows generally involved a technology of some kind, the networks of flows of matter and energy may be regarded as the technology-based connectedness of human life and society. It was enfolded into, and an expression of, a cultural order.

The process of industrialization shattered all cultural orders by splitting off a considerable portion that was now primarily technologically or economically ordered and secondarily culturally ordered. This development paved the way for the making of a clear distinction between an economy and the society to which it belonged, as well as for regarding the former as the principal force behind the development of the latter. This interpretation was extended to all of human history by both the political left and the political right, be it in somewhat different ways. As we shall see shortly, this reflected the myths that ruled over the cultural orders of the industrializing societies of the nineteenth century. What remained of any cultural order would never be the same again as a consequence of a profound desymbolization. The process of industrialization unfolded what had been the enfolded dialectical orders into ever-more distinct scientific, technological, economic, social, political, legal, moral, religious, and aesthetic spheres. We shall see that as industrialization advances, it becomes increasingly plausible to parcel out the study of what were once deeply enfolded and dialectical cultural orders to the emerging disciplines of the social sciences. Nevertheless, doing so fails to recognize that, even with high levels of desymbolization, we remain a symbolic species.

The process of industrialization directly affected the technology-based connectedness of human life and society and, via it, everything else. Consider what would have happened to this technology-based connectedness when a machine was used to mechanize a human activity. These machines did not imitate this activity as earlier automatons did. We have already noted that an activity had to be technically divided before a machine could be used. Having thus translated this activity in the image of the machine, any production step could be carried out by a machine or by a human being working in the image of that machine. It became immediately evident that even without machines groups of people organized by means of a technical division of labour could produce a greater output than could each member of the group carrying out the entire original activity. The addition of machines further increased this output. It is for this reason that in 1776 Adam Smith predicted a new 'wealth of nations' resulting from the technical division of labour in productive activities.[8]

We are now able to build a general model of what happened to the technology-based connectedness of the way of life of any society when productive activities that exchanged matter and energy with it became technically divided and mechanized. The reconstituted activity was now capable of more rapidly transforming the inputs received from the technology-based connectedness into the outputs returned to it. Bottlenecks were thus created; the activities yielding the inputs to the reconstituted activity could not keep pace, and those receiving the outputs from the reconstituted activity could not transform them at the high rate at which they were received. The reconstituted activity could not operate within the local dynamic equilibrium of the technology-based connectedness that had existed prior to its reorganization. It was starved for inputs and became clogged with outputs that it could not exchange.

The local dynamic equilibrium could only be restored by also technically dividing the activities with which the originally divided activity had exchanged matter and energy or their composites. This in turn strained the connections between the reconstituted activities and their adjacent non-reconstituted ones with which matter and energy were exchanged, thus requiring further adjustments and so on. The entire dynamic equilibrium of the technology-based connectedness had to be adjusted in order to accommodate a greater throughput of matter and energy. The constraints it posed on the way that these adjustments were made could be expressed in the laws of thermodynamics: matter and energy can be neither created nor destroyed but only transformed, and all transformations are irreversible. These constraints are symbolized in terms of the futility of producing anything that serves no further purpose, other than producing something else or selling it. This prevents flows leaking from the technology-based connectedness back into the biosphere.

Additional adjustments to the technology-based connectedness were required because each reconstituted activity also required indirect inputs that did not end up in the desired outputs it produced. For example, in the case of factory production, indirect inputs included production equipment, lubricating oils, spare parts and tools for maintenance, buildings, facilities, and transportation systems. Any significant changes in throughput involved the modification or replacement of any or all of these. All such indirect inputs corresponded to the outputs of the activities that produced them, which in their turn required a variety of inputs, continuing in long chains that reached all the way back to the society-biosphere boundary. It was the locus of transfers of flows of matter and energy between the technology-based connectedness of a way of life of a society and the two networks of flows of matter and energy corresponding to these flows in the ecosystem and the biosphere.

The adjustments that resulted from the technical division of labour and mechanization thus triggered another tier of adjustments in the technology-based connectedness of the way of life of a society, corresponding to the changes in the indirect inputs. In addition to the desired outputs, there were also undesired ones, among which the most common were the pollutants released directly back into the biosphere or indirectly via a chain of treatment and landfill processes. The only difference between these particular undesired outputs and the desired outputs was that the former were products that could not be sold and eventually had to be disposed of at increasing costs.

All this reveals a fundamental characteristic of the process of industrialization: technically dividing and mechanizing productive activities is an all-or-nothing proposition for a society exchanging matter and energy exclusively with the biosphere. If it also exchanges matter and energy with other societies, the activities connected to these exchanges, either directly or indirectly, may have to be reconstituted as well. The assumption is that the members of each society are engaged in activities essential for evolving its way of life, and hence these people cannot be easily reassigned in order to increase the number of people in non-reconstituted activities and help them keep pace with the reconstituted ones. There is therefore little choice but to technically divide the non-reconstituted activities as well. For example, mechanizing the weaving of textiles makes little sense unless the production of yarn and the finishing of the woven cloth are also mechanized, and this only makes sense if other adjacent activities are mechanized as well. Similarly, increasing the output of textiles in a society makes no sense unless people want and can afford more textiles or if they can be exchanged with other societies for other necessities.

It is for this reason that the industrialization of a society shares some characteristics with a controlled chain-reaction process. Wherever a machine is introduced, a human activity is shattered and its fragments must be controlled and organized. Also, the local dynamic equilibrium of the technology-based connectedness must be restored. Doing so requires further mechanization, and once again a human activity is fragmented and additional disturbances in the technology-based connectedness occur, and so on until the dynamic equilibrium in this technology-based connectedness is restored. The consequences of all of this are far-reaching. Technically dividing and mechanizing a human activity into a network of relatively distinct fragments, endlessly repeated by machines or by human beings, does not alter the fact that the fragmented activity continues to participate in the technology-based connectedness as well as the cultural order of human life and society.

As a growing gap opens up between this technology-based connectedness and the cultural order as a consequence of industrialization, the fragments of human activities must be primarily an expression of either the one or the other, but they cannot be both. Either they express the necessities imposed by the local dynamic equilibrium of the technology-based connectedness or they remain internally and externally connected to the cultural ordering of human life and society. The results show unambiguously how industrialization leads to a growing imposition of the technology-based connectedness on the cultural order. The dependence of the way of life of industrializing societies on an ever-larger throughput of matter and energy makes it impossible to proceed on the basis of prevailing cultural beliefs and values. The chain-reaction-like process simply reflects the victory of the necessities imposed by the dependence on matter and energy. It is experienced as a kind of collective roller-coaster ride; human activities are shattered, and the fragments are doubly constrained, but more so by the technology-based connectedness than by the cultural order. What emerges is a new kind of technology within a different kind of society, still jointly sustained by the biosphere but in a way that gradually undermines it.

In nineteenth century Western Europe, despite all ideologies and reassuring explanations, a new kind of obedience to matter and energy imposed itself. For example, entrepreneurs encountered difficulties that were piling up at the boundary between the portion of technology-based connectedness under their own influence and the adjacent areas that were not. The result was a relative slowdown in the active area until the adjacent ones could be influenced or until the resulting bottlenecks attracted sufficient creative attention by other entrepreneurs or politicians, at which point the pace of change in the original dynamic area would pick up once more. The transformation of the technology-based connectedness of the way of life of a society in order to achieve a new dynamic equilibrium based on a higher throughput was generally very turbulent, but regardless of how the details played out in a particular country, region, or industry, the final result has been and continues to be a new dynamic equilibrium based on a higher throughput of matter and energy.

Here we see the double movement as the mechanization of productive activities unfolded the cultural order. As some human activities became primarily technologically or economically ordered, they remained activities within people's lives and hence within a cultural order, despite high levels of desymbolization. A non-dialectical tension was thus created between what was constrained primarily by technological or economic ordering, and secondarily by cultural ordering, and what was constrained primarily by cultural ordering and secondarily by technological or economic ordering. Since what was primarily technologically or economically ordered was next to impossible to symbolize in terms of its meaning and value for human life and society, ideologies of all kinds had to fill the void.

The symbolization of what was primarily culturally ordered was also affected because of the many experiences in which people's lives could no longer fully work in the background. People were compelled to live as if everything were no longer related to everything else in their lives. It was not a question of a separate and distinct technology but only of a sphere of technological activities that was so highly desymbolized that it became more and more universal with respect to the cultural order from which it had been unfolded. Similarly, it was not a question of a distinct and separate economy but only of a sphere of highly desymbolized economic activities that were increasingly ordered by the first universal institution invented by humanity, namely the Market. These developments in turn affected the moorings of the cultural order of an industrializing society in the biosphere. Its ecological footprint had to be expanded, and this soon required the appropriation of the ecological resources of other peoples by conquest, colonization, and war.

The Emerging Economic Order

Within the enclaves of production ruled a technological order that gradually ensured efficient processes, productive labour, and the lowest production costs. However, this order could not regulate how the goods and services that were produced in these enclaves, largely independently from the cultural order, were to be distributed. The cultural order had to surrender this task to an economic order at its own expense. Until after the Second World War, the new economic order ruled over the technological order.

The economic order became established in the following manner. As noted, when peasants (voluntarily or involuntarily) left the countryside to find work in the new industrial centres, they were compelled to work long hours. As a result, they no longer had the time or the means to provide themselves with most of the necessities for life. They now had to purchase them with their wages. Such wages were used to translate their needs and wants into demands for specific goods and services in various markets until they were spent. Entrepreneurs responded to this opportunity by allocating their investment capital to supply these demands as profitably as possible. The Market mediated between the efforts of wage earners to meet their needs best by allocating their income to products and services, and the profit-maximizing activities of the entrepreneurs in the many markets for these goods and services. The concept of a Market force represents the organizing force jointly exerted on society by all the markets for individual products and services. The Market took over the role that culture used to play in organizing and regulating this aspect of human life. It 'ruled' on the basis of prices established in markets by means of the mechanism of supply and demand, while cultures had done this on the basis of symbolization for the establishment of values. The difference between these two modes had immediate and far-reaching consequences. Symbolization seeks to understand and deal with everything in the context of everything else. As a result, the appropriateness of the technology of a society and the sustainability of its way of life could mostly be taken for granted until now, albeit with some exceptions. Out of necessity, the Market functioned with a much more restricted context in the establishment of prices by means of the mechanisms of supply and demand operating in individual markets.

The fundamental differences between the new order and the cultural order within which the former emerged in industrializing societies were in large measure attributable to the contexts taken into account.

Before anything could participate in the new order, it had to be treated as a commodity so that it could be traded in markets where its price would be determined. Anything that needed to be commoditized had to be desymbolized first so that it was no longer internally and externally connected to the natural and cultural orders. For example, a part of a living ecosystem that was helping to sustain local life had to be desymbolized, to the point of its becoming a piece of terrain that could be described on a deed, before it could be bought and sold in a real estate market. The portion of a person's life spent in a factory had to be desymbolized and then treated as labour before it could be bought and sold in a labour market. These kinds of examples could be endlessly multiplied, but before anything can be ordered by the Market it must be desymbolized and treated as if it were no longer an integral part of human life or the world. Treating anything in this way introduces non-dialectical tensions into human life and the world between what has been internalized into the economic order and what remains externalized because of desymbolization.

Next, the prices of any of these commodities had to be established in markets by means of a mechanism of supply and demand. The problem was that these prices were a very limited indicator of the values of these commodities for human life and society. Whatever could not be taken into account as a consequence of commoditization was not reflected in the prices established by subsequent market transactions. We have never been able to deal with this difficult issue. Any attempts to internalize the social costs of a market transaction have always involved so many arbitrary assumptions that total costing has escaped us. The scope of market externalities quickly became so vast that the ability of the Market to order the goods and services in a particular way of life was exceedingly limited as compared to doing so by means of symbolization and culture. Unfortunately no industrial society has ever invented a culture-based alternative.

Before going into detail, it is important to note that the backbone of the new emerging economic order was the technology-based connectedness that joined all technically divided and mechanized activities to each other and the biosphere by means of flows of matter and energy, mostly embodied in flows of materials, parts, sub-assemblies, and products. To the indirect flows we must now add those of commoditized labour and capital. When prices began to be substituted for cultural values in the establishment of the new order, capital became commoditized value.

This expanded technology-based connectedness was built up in a way that has more in common with the workings of a mechanistic order than with a living order. All resources extracted from the biosphere were concentrated and refined and thus 'decontaminated' from what belongs together in nature, in order to become relatively 'pure' finished materials and fuels. These commodities became the building blocks of all intermediate and final products. When entrepreneurs sought to meet the market demands of wage earners, they modified the corresponding areas of the technology-based connectedness, which were 'assembled' into the desired products and services. To accomplish this, entrepreneurs could limit their attention to that technology-based connectedness by organizing the required inputs into a particular production step, optimizing this production step to produce the greatest desired intermediary output, and transforming this output into an input into the next production step by creating suitable links.

The indirect inputs also had to be organized, but for a very long time the undesired outputs were ignored. In other words, industrializing societies behaved as if their technology-based connectedness were a vast mechanism built up from separate parts. All related thoughts and actions implied a desymbolization of whatever was temporarily borrowed from individual lives, societies, or the biosphere in order to constitute the necessary commodities. Despite such desymbolization, what was borrowed remained integral to the natural and cultural orders. Although externalized from the processes of symbolization, these connections did not disappear; they could only be strained. Technologically or economically ordering a sphere of human activities thus amounted to a distortion of the corresponding areas of the natural and cultural orders. Gradually this led to the reification of human life, the emergence of mass societies, and the environmental crisis. Industrialization is thus inseparable from reorganizing ever-larger areas of living orders in the image of the machine in order to create a mechanistic order by impairing the host orders.

A significant gap thus opened up between the way a culture would have ordered the flows of goods and services essential to a way of life on the basis of values and the way the Market did so on the basis of prices. The former was reduced to the latter by the commoditization of what was integral to human life, society, and the biosphere; the production of goods and services by transforming and assembling these commodities in the context of the local technology-based connectedness; the optimization of all production steps and operations on their own terms by means of output-input ratios; the pricing of all commodities in terms of their supply and demand in markets; and a disregard for all the consequences to human life, society, and the biosphere in the present as well as in the future. Market forces were now acting on the evolution of human life, society, and the biosphere. We shall briefly examine some of the consequences.

The environmental crisis constitutes a first example of a Market force acting on the biosphere. Beginning in the nineteenth century, the networks of flows of matter and energy corresponding to human economies were assumed to make such small demands on the corresponding networks in the biosphere that they could not diminish its capacity to regenerate these flows and thus sustain these economies. As a result, the borrowing of matter and energy from the biosphere did not have to be priced. Mainstream economics has never revised this assumption. It thus helped to produce a second Market force acting on society, which created underemployment and unemployment. Since natural resources were free (except for the costs of retrieval) and labour was priced, entrepreneurs sought to economize labour rather than both labour and natural resources. As a result, the economies they helped to build over-consumed nature and under-consumed labour. This trend was further reinforced by taxation policies that taxed what we increasingly had too little of (good jobs) instead of what we consumed too much of (nature). It should therefore come as no surprise that, for the greater part of the history of our civilization, underemployment and unemployment have been the rule.

Another Market force on human life and society resulted from the commoditization of labour. Adam Smith already predicted the dreadful consequences of repetitive work that would suppress and atrophy the resources and lives of workers, thus causing tremendous stress. Social epidemiology has identified this kind of work as a major source of physical and mental illness in human populations.[9] The dehumanizing and disease-producing effects are also known to spill over into workers' families and communities. Another Market force on human life and society relates to the basic needs of poor people, who cannot back these needs with money in order to translate them into market demands. This Market force has condemned a large portion of humanity to a condition of misery. From these examples, it is clear that much of what is essential to human life, society, and the biosphere received no attention in the new economic order. For these and other reasons, the growth of this order had to be accompanied by a dramatic increase in the role and size of the state. The reasons are clear: as the new order pushed back the cultural order, the many issues with which the former could not adequately deal had to be attended to by a second organizational locus. The growth of the state became a necessity, even though it went against every prevailing political, moral, and religious tradition.

There is one well-known positive Market force. What Adam Smith called the great 'invisible hand' of the Market would produce the best possible world for most people.[10] He rarely mentions it, and this should perhaps make us pause to ask whether we have not blown its benefits out of all proportion. Have we forgotten that the history of our civilization has been continually plagued by Market instabilities and crises

Today's transnational corporations constitute the largest and most powerful sector in the economy. In order to employ the latest highly specialized and technological knowledge, they had to resort to planning that included minimizing their dependence on markets.[11] As a result, this planning took over much of the Market's role. When celebrating the marvels of the great 'invisible hand' of the Market, we are engaging in wishful thinking about an economy that has not existed for at least half a century.[12] Attached to the great 'invisible hand' is an 'arm,' which, in the form of negative Market forces, knocks over almost everything that is essential to human life, society, and the biosphere. It would appear, therefore, that the new order, guided by performance measures and prices and regulated by the Market and the state as its organizational loci, turned out to be a very poor substitute for the traditional cultural orders it displaced. Doing so by central planning was even worse.

The great tragedy of these developments is that, all across the political spectrum, we have refused to treat the Market like any other human invention: very useful for certain things, irrelevant for others, and harmful to still others. It is an unavoidable situation that stems from the interrelatedness of human life, society, and the biosphere. Those who believe that we can entrust almost everything to the Market, especially by making it global, are like contractors who believe that the only tool they need to fix everything is a hammer. What is remarkable is not the fact that some people treat the Market as a new kind of secular god but that anyone listens to these people. Take the labour market as a case in point. Governments everywhere quickly learned that this market could not satisfactorily regulate the work of children. The remedy was not central planning but a simple restriction of the role of the labour market to adults. This is a typical example of the reason that governments everywhere had to step in: to become a compensatory locus for organizing human life and society, without which the Market would have made things much worse. Nevertheless, even in those historical situations in which the Market and the state functioned synergistically, their joined efforts were still far from what could have been accomplished on the basis of symbolization and culture.

The emergence of the technological and economic orders within the cultural orders pulled the fabric of human lives and societies in different directions. The technological order was subservient to the economic one, because no entrepreneur could make a profit unless the cost prices determined by the former were lower than the market prices established by the latter. The human activities that made up the economic order were only weakly connected internally and externally to the cultural order. The difference between these and all other activities was their high level of desymbolization, since very little of people's lives could work in the background. Constantly being on the edge of starvation, factory workers had to carefully allocate their wages to their most pressing needs and forgo everything else. As a result, their behaviour could be modelled as *homo economicus*, which may be compared to an algorithm that maximized the utility that could be derived from these wages. Under these conditions, their behaviour was so highly desymbolized that their lives were reduced to a few basic needs that had to be satisfied as best as possible in order for them to survive. Functioning in this way revealed a profound alienation by the economic order. People were compelled by necessity alone and were prevented from having normal relations with their spouses, children, and neighbours as a result of doing little more than working, eating, and sleeping. Almost everything of human life other than basic survival was externalized. The same was true for the factory owners. They had to allocate their capital in a manner that maximized their profits. As noted, if their behaviour did not conform to the model of *homo economicus* in their economic order, they would soon cease to play the role of factory owner. All this was in sharp contrast to the culturally ordered activities in which people's lives could more fully work in the background. Many more aspects of human life could then be metaconsciously taken into account, including their social relations and cultural values. *Homo economicus* is thus a sign of alienation, manifesting itself in terms of high levels of desymbolization. After all, symbolization permits us to live full lives by relating everything to everything else. Extreme poverty or wealth accumulation prevents people from doing so.

People had to somehow make sense of this broken situation in order to make it liveable. What was desymbolized now had to be provided with meaning and value by means of ideologies, and there was no shortage of them. As Polanyi has noted, human life had to be forced into the mechanism of the Market and into the activities organized in the image of the machine.[13] This new 'government' of technological and economic affairs resembled a mindless mechanism, and out of necessity humanity convinced itself that this mechanism had been there all along underneath everything else. Whether this was explained in terms of a mechanism of human history, when economic bases for human life succeeded one another as they became possible through technological innovation, or whether the 'growing up' of humanity would make it economically rational, socially progressive, politically liberal, and religiously secular, alters nothing fundamental. Either explanation was regarded as perfectly natural by the adherents of these ideologies.

As Polanyi also showed, from a historical perspective *homo economicus* was not natural at all. Just because the behaviour of wage earners and entrepreneurs within the economy could now be approximated in terms of an optimization process ensuring the greatest desired output from a monetary input does not make it natural. In all earlier traditional societies, *homo economicus* would have been regarded as deviant and destructive to the community and, in the extreme, as psychopathic. After all, behaving in a manner that can be modelled by an algorithm is to ignore all consequences for others. The values that guided the equivalent of economic activities in the past arose metaconsciously, like all others, as an integral part of symbolizing the associated experiences by modifying the organizations of the brain-minds that sustained these activities in the background and by enfolding them into people's lives. In other words, 'economic' activities were, like all others, internally and externally connected to the cultural orders of indigenous and traditional societies. Furthermore, the cultural values that guided these as well as all other human activities were qualitatively different from the new economic 'values.' The former dialectically expressed the quality of each and every relation that a person had established with everything else. As a result, such dialectical values were the opposite of the output-input ratios, which were masquerading as cultural values. The latter 'values' make everything into the measure of itself. They are self-referential in comparing one form with another of the same thing or in comparing two equivalent processes that use the same inputs to obtain the same desired outputs. The material or energy efficiency, the labour productivity, or the monetary profitability of a process is measured in terms of the corresponding inputs and outputs without any reference to anything else in its context. As such, these 'values' presuppose a mechanistic order made up of separate and distinct domains that interact with one another like the parts of a machine. As a result, these 'values' ignore the enfolded character of the natural and cultural orders, resulting in far from trivial consequences.

To the extent that technological and economic activities lost their internal and external connectedness to the cultural order of human life and society, they ceased to express something of people's lives and cultures and instead began to resemble relatively distinct functions. When joined together in a technological or an economic order, these activities superficially resembled relatively distinct structures or subsystems of a society. Evolving such functions, structures, and subsystems on their own terms linked them primarily on the basis of the power with which they could transform inputs into outputs. The metaconscious knowledge derived from this process led to intuitions that developed into 'theories' such as functionalism, structuralism, systems theory, and the metaphysics of power. I refer to these as theories because they were exact within the boundaries of particular disciplines separated from experience and culture.

In the real world, human beings and societies continued to rely on a culture. This reliance is ongoing, as babies and children still learn to make sense of and live in the world by acquiring a culture. They do not give this up when they are introduced to specialized scientific or technological knowledge or when they become engaged in carrying out a technically divided production step. The corresponding experiences continue to be internalized and symbolized in the organizations of people's brain-minds, but this necessitates a great deal of suppression of their selves and their lives. Similarly, specialized knowledge of the production step remains externally connected to the technology-based connectedness of a society and, via it, to its cultural order. Indirect as this may be, and therefore poorly understood and symbolized, no 'pure' function or structure can exist, but only highly desymbolized human activities.

Hence, *homo economicus* is a model of a highly desymbolized set of behaviour patterns that are a manifestation of profound alienation. Similarly, a relatively distinct economy separated from the remainder of society is a manifestation of a highly desymbolized cultural order, to the point that some areas of collective human life superficially resemble economic functions and structures.

In the industrializing societies of the latter half of the nineteenth century and the beginning of the twentieth century, human life became guided by three different orders: the technological order (mostly in work), the economic order (mostly in producing and procuring the necessities of life), and the cultural order (mostly in the remainder of people's lives). The extent to which people could be engaged in the corresponding activities varied from essentially being a spectator to one's life within these mechanistic orders to living one's life somewhat more fully by expressing the cultural order in an individually unique manner. The activities belonging to these three spheres were enfolded into people's lives, with the result that they interpenetrated. For example, the suppression of one's self in technically divided work made it possible to daydream about other things in life; *homo economicus* could be interrupted by ethical or religious considerations or by the ability to choose one product over another. The extensive damage to the cultural order as a result of desymbolization weakened the internal and external connectedness of family and community activities. All this influenced, and was influenced by, the evolution of a society. Its social fabric was pulled in different directions by the three orders, as implied by the well-known distinctions between mechanical and organic solidarity, between Gemeinschaft and Gesellschaft, and between the individual and society. In the latter case, the weakening influence of the cultural order caused the relationship between the individual and society to be desymbolized to resemble that between a 'part' and a 'social mechanism,' without this ever being entirely possible.

The law was also pulled in different directions by these three orders. It became more and more organizational in the technological and economic orders but retained some of its earlier qualities in the cultural order. Similarly, the political sphere extended itself into all three orders with very different effects. In other words, it seemed as though additional functions, structures, and subsystems began to appear, particularly at the interface between the economy and the remainder of society. They stretched and tore the cultural order into an ever-more distinct social structure, legal framework, political organization, individualistic morality, religion, and 'culture.' Human activities became guided on primarily their own terms, as functions, structures, and subsystems, and much less in terms of their cultures. The role of culture remained most intact in people's private family and community activities. All this will now be examined in some detail.

People's participation in the technological and economic orders could no longer be fully sustained by their lives and cultures as symbolized in the organizations of their brain-minds. Their daily-life experiences were dominated by these orders to the extent that their lives and cultures needed to be suppressed. In contrast, people's participation in their families and communities continued to be supported by their lives and cultures via the organizations of their brain-minds. However, this support was also desymbolized. Following the simple model presented in chapter 1, the clusters of experiences derived from family and community participation built up a very different kind of metaconscious knowledge than did those clusters associated with their participation in the technological and economic orders. People's divided selves were minimally present or restricted to a portion of their lives. In the technological, economic, and cultural spheres the ability of the organizations of people's brain-minds to symbolically sustain their activities was diminished, be it in different ways. This marked the beginning of the transition of the social fabric commonly found in traditional societies to that of contemporary mass societies.

Spinning off the social structure of a society from its cultural order began with work performed in the image of the machine. No longer could the cultural order assign the social roles that needed to be carried out by each new generation in order to sustain and evolve a way of life. This function was increasingly dominated by the requirements of the technology-based connectedness in terms of the categories of labour and the number of people required in each category. In addition, the rigid discipline imposed in the early factories, coupled to long working days, ensured that the rest of people's lives was little more than the regeneration of their capacity to work another day by eating and sleeping. The influence of the new order thus expanded into the fabric of the relationships of people's lives and communities. Even when working conditions improved somewhat and opportunities opened up for relaxing, socializing, and participating in the community, it simply added another sphere to their lives, which were already being pulled in different directions.

The breakdown of the extended family as a primary social group further undermined the social fabric of communities. These families could no longer provide the essential services on which traditional communities had come to rely. The resulting social problems necessitated the external provision of many of these services under the control of the Market or the state. These included day care, basic education, preparation for apprenticeship training, some health care, disability pensions, unemployment insurance, life insurance, and much more. Rudimentary alternatives had previously been built into the individual and collective life within extended families.

The social fabric of rural communities was also deeply affected. It had already come under significant pressure from agricultural reforms. Almost everywhere, this resulted in the migration of a substantial portion of the rural population to the new urban industrial centres or to the colonies. It marked the beginning of the steady growth of the urbanized portion of humanity that continues to this day. Within these centres no new traditional social fabric could be created, for the reasons outlined above. Important adaptations to city life had to be made, which, according to some observers, were cause for concern.[14]

Industrialization also weakened the internal connectedness of the law to the cultural order of a society. In a traditional society this connectedness depended on the law's reliance on cultural values.[15] It must be remembered that most members of a society never read, let alone memorize, the laws of their community. There is no need to do so because the laws that directly govern their daily lives are an expression of the metaconscious values implicit in the organizations of their brain-minds. When this is the case, these laws will be spontaneously obeyed. If, however, a law strays too far from these metaconscious values, it risks being spontaneously and widely disobeyed. When this happens, a judge in a free society will have no choice but to declare such a law to be inapplicable for the obvious reason that it is impossible to punish or incarcerate a significant portion of the members of a community.

Two factors weakened the internal connectedness of the law to its cultural order. The growing economic order created many situations that had never existed before. In the absence of legal precedence the state had no option but to become more involved in common law or to extend the scope of codified law. In either case, the law could no longer rely on culture as much as it had in the past, thereby compelling it to take on a more organizational character, as noted by many observers of that time. Second, as the role of culture in individual and collective human life weakened, many laws began to appear as somewhat arbitrary, to the point that some people began to feel justified in disobeying them provided that they could get away with it. These developments marked a transition from the law expressing the cultural values of a community to expressing the need to organize what a culture could no longer direct and guide.

We have taken note of the most significant political adjustments to an economic order expanding at the expense of the cultural order. As the cultural ordering of human activities was all but destroyed by the new emerging order, the state had no choice but to step in to make up the cultural 'deficit.' This necessitated an unprecedented growth in the role of the state. Today this development has reached the point that it is very difficult to identify aspects of contemporary human life that most people would regard as non-political. The implications of such a belief are rarely noted. If we believe that almost everything in life is political, we are in effect saying that there must be a political solution to almost all issues and that there is a role for the state to play in almost everything. Nevertheless, this is a rather inconvenient fact for most of us, and the political right continues to exploit this dilemma. Neither it nor anyone else has any solution to the problems created by the cultural deficit resulting from extreme desymbolization in contemporary mass societies. These developments are a clear manifestation that the political sphere is no longer internally connected to the cultural order of a society.

Desymbolizing the Cultural Order

As was the case for all traditional societies, the productive activities of the industrializing societies temporarily 'borrowed' matter and energy from the natural order, and human beings from the cultural order. However, this borrowing could not be symbolized for what it was. Everything that was borrowed had to be desymbolized by treating it as if it were a commodity. This commoditization was not, first and foremost, a matter of trading commodities in markets; it was fundamentally a question of transferring them from living orders to mechanistic orders. It is impossible to build up a technological order for production by symbolizing labour as activities integral to human lives. Endless repetition is what machines do and thus what human beings must be compelled to do. It involves suppressing as much as possible one's life work in the background. The more complete the voluntary or involuntary suppression of people's lives, the more their productive activities resemble the actions of a machine. The relations between these activities and the lives of the people carrying them out thus become less of a manifestation of themselves the more the corresponding experiences are desymbolized and the more the resulting developments of the organizations of their brain-minds symbolize them as resembling the order of what is real as opposed to the cultural order. The reorganization of human work on the basis of the technical division of labour, followed by mechanization, automation, and computerization, detaches human activities as much as possible from human lives so as to integrate them as much as possible into the mechanism of production.

No matter how hard we try, it is impossible for human beings to turn themselves into the cogs of a mechanism. We can think and act as if labour were a commodity, but this does not sever its bonds with human lives and communities. No matter how hard we try or how forcefully we are compelled, we cannot turn ourselves into slaves of machines. As a result, the exponential growth in labour productivity that we have come to take for granted is indissociably linked to the disordering of human lives and societies and thus their cultural order.

We have noted that symbolization has always distanced our species from its life-milieu by interposing new spheres of possibilities. Doing so led to the systematic discovery of the potential meaning and value of everything for human life and thus the realization of this potential. Everything cultural was created as a result of our being a symbolic species. However, the potential of reorganizing human work in the image of the machine as a preparation for mechanization and industrialization did not follow this pattern. It involved a considerable desymbolization of what has made us human, to build an order of what is real at the expense of an order of what is true. Contemporary workplaces reveal with growing harshness the ultimate consequences. They are not only threatening our physical and mental health but also what it is to be human.

In the same way, symbolizing the temporary borrowing of matter and energy from the natural order in terms of commodities has led to a looming collision between the technological and economic orders and the living order of which we are a part. Again, this is not, first and foremost, a question of the necessity of trading in markets what we have borrowed from this order. It is fundamentally a desymbolization of the essential participation (of whatever we borrow) in the material cycles and chains of energy transformations by which matter and energy is exchanged within the biosphere in order to sustain all life. As a consequence, this borrowing is enslaving us to these exchanges of matter and energy, and there appears to be no end in sight despite all the talk about dematerialization. Having declared our ways of life to be unsustainable by the biosphere, are we on a collective suicide mission? How many billions of people will die when our ways of life collapse for a lack of support by the biosphere? As has been the case for all previous civilizations that have collapsed, few people are willing to take the warning signs seriously, and those who do are typically dismissed in one way or another. The reasons for this will be discussed in the next subsection.

The account of industrialization and its desymbolizing effects on human lives and societies is still incomplete. We have thus far seen how a technological order within a larger economic order loosened its internal and external connections to the cultural order. Since the cultural order establishes what is true by symbolically relating everything to everything else as much as possible, it was torn into a loose federation of what superficially appeared to be structures necessary to sustain the technology-based connectedness of human lives and societies. These apparent structures were an expression of desymbolized values (including output-input ratios and market prices) at the expense of cultural values. In this way, industrializing societies moved away from strengthening themselves as culturally enfolded wholes to become something entirely different. Many observers regarded these societies as collections of interacting parts unfolded from the cultural order. However, these developments did not go as far as the theories such as functionalism, structuralism, systems thinking, and the metaphysics of power would have us believe.

Making the Collision Liveable

In order to make life liveable, the economic order and the cultural order had to be reintegrated by means of symbolization. We have thus far argued that industrialization required the imposition of a mechanistic order on a living cultural order and the de-centring of human life according to which one of these two dominated. Such an analysis would be either wrong or incomplete because this situation should have led to widespread mental illness and anomie. Our story is incomplete because it has dealt with only one face of industrialization, which may be called *people changing technology*, and we now need to look at the other face, which may be referred to as *technology changing people*.[16] Although it could not be experienced directly, the most significant component of this change resulted from people internalizing the massive and rapid changes in their lives and societies, leading to fundamental changes in the organizations of their brain-minds, which adapted them to their new situation. This process of adaptation was self-reinforcing since it worked in the background, thereby cumulatively sustaining a cultural mutation that reunified industrializing societies in a little over half a century.

The preoccupation at the time with the concept of social struggle was symptomatic of this mutation. It was symptomatic because the issue of what constitutes a social class has never been adequately resolved.[17] One of the primary reasons for this dilemma is that, at the time, the concept of culture as it is used in this work was relatively unknown, although there are embryonic beginnings of it in the work of Karl Marx and it is fully implicit in the work of Jacques Ellul.[18] Simply put, in a limited sense a social class is a society within a society during a transition from one cultural order to another. Some social classes attempt to move forward on the basis of the one, some on the other, and some on a kind of hybrid between the two. This would help to explain the substantial differences found by these analyses during different phases in a particular society and between industrializing societies. Each social class lived a temporary synthesis of sorts between the growing economic order and the weakening cultural order. This focus was gradually being eliminated by technology changing people.

In the course of several generations the changes in people's organizations of their brain-minds led to the formation of very different kinds of metaconscious knowledge, including a new cultural unity. However, it continued to sustain the usual conflicts between social groups encountered during most of human history. These conflicts were no longer over the kind of society that was to emerge but over the benefits each social group would receive and the power and influence they would wield. What was politically interpreted as a 'bourgeois' culture turned out to be the beginning of a cultural mutation and the rise of a new cultural unity that gradually began to sustain everyone's activities in the background.[19] As we shall see, the new cultural order was very different from all preceding ones because of extensive desymbolization, but it did re-establish a new kind of internal and external connectedness in human life, be it in a very different kind of society. It was not the beginning of what was later called the end of human history but a fundamental mutation in the role that culture played in human life and society in order to cope with extensive desymbolization.

Recall what we discussed earlier. Prior to industrialization the culture of any human community fully endowed the lives of all its members with meaning and value. The unknown was lived as the interpolation and extrapolation of individual and collective human life, symbolized and sustained by the organizations of people's brain-minds (which included their metaconscious myths).[20] Everything that was radically 'other' was thus banished from thought and action. Within the sphere of the possible, everything was oriented by means of a hierarchy of values anchored in the central myth or sacred. The reality as known and lived by the members of a community was transformed into reality itself, minus whatever details would be discovered and lived in the future. As a result, the greatest good known by a community was the ultimate good, beyond which nothing more important or valuable could be imagined or lived in all of reality. This greatest good was related to whatever made human life in the world possible and which could therefore also make life impossible. These powers had to be symbolized by means of a religion. It was the same metaconscious processes, and the same sustaining functions of the organizations of people's brain-minds, that reunified the industrializing societies.

Now let us examine some of the details. Imagine how the members of the industrializing societies of the nineteenth century may have internalized and symbolized the threat of the unknown. Despite the many explicit and contradictory interpretations of what was happening to human life and society and the different kinds of futures it might bring, there were many daily-life experiences that nobody doubted. Everyone knew that every year machines were becoming bigger and faster. Every year these machines produced more output. Every year there were more factories full of these machines, jointly producing an ever greater total output. The metaconscious interpolation and extrapolation of these experiences implied more of the same: every year society was coming a little closer to satisfying the needs of more people. There was nothing on the horizon of human experience to indicate that this interpolation and extrapolation of their experiences would significantly change in the future.

As a result the following kind of metaconscious knowledge would likely have developed in the organizations of the brain-minds of the people of that time. In due course everyone's material needs could be satisfied, thereby eliminating one of the most serious social problems, namely, widespread poverty. This would also eliminate the incentive that poor people had to turn to theft, alcoholism, and prostitution. The kind of human ingenuity that solved this problem could then turn to solving other social problems. Again, there was nothing on the horizon of human experience that might suggest otherwise. In other words, material progress would surely produce social progress, and all this would undoubtedly lead to spiritual betterment.

It should be noted that the above kinds of experiences, and the metaconscious knowledge built up with them, were the result of the process of symbolization being focused on the technology-based connectedness. For the people who worked in factories, this occurred through the technically divided labour they performed, which made no sense to them whatsoever. The people who organized this technically divided labour, however, focused only on the transformations of inputs into outputs and matching the output of one activity to the input of another, regardless of the meaning and value that this might have for society. Everyone's attention was essentially on the local technology-based connectedness instead of on its meaning and value for human life and society. The *sense* of this had to emerge metaconsciously through the interpolation and extrapolation of these experiences by means of the organizations of people's brain-minds, and this produced only vague intuitions of a potential but distant future meaning and value. Although this metaconscious knowledge developed in the organizations of the brain-minds of everyone involved, it grew along different pathways at different rates for the various social classes in the industrializing societies of that time. Eventually, however, the way in which people directed symbolization towards the technology-based connectedness greatly distanced work from human life, the economy from society, natural resources from the biosphere, and everyone's consciousness from their own lives.

The additional economic, social, political, and legal changes were, first and foremost, the result of the removal from the cultural order of what once had been unique expressions of it. The corresponding institutions had to be accommodated to the growing dominance of the technology-based connectedness over the cultural order. The functions and structures of this technology-based connectedness and its performance as measured by output-input ratios imposed themselves on a dialectical order and its cultural values, symbolizing the sense of it all for human life and society. As a result, the portion of society where the technology-based connectedness dominated began to resemble the domains in a machine, each having its own distinct identity and interacting with each other across relatively distinct boundaries. As noted, intuitions about these developments eventually gave rise to a sequence of theories that banished all references to meaning. With hindsight, it is perfectly clear that the influences of 'technology changing people' were not considered by those who built the new technological and economic orders. (It was, of course, expressed in literature and the arts but in an end-of-pipe fashion in keeping with the separation of culture from industry and the economy.) What *was* became separated from its *sense*, and everything became related primarily to the technology-based connectedness and, via it, to the cultural order.

The more people directed their attention to the technology-based connectedness, the more the process of symbolization was hampered.

Their metaconscious knowledge was correspondingly limited, and related only to a distant and remote future symbolized by the new cultural unities. In this way desymbolization permeated individual and collective human life to its greatest depth, and this distinguished these cultural unities from all earlier ones. They ensured the fulfilment of what was only embryonically present in people's experiences. At some future date, culture would be paramount again. The internal and external connectedness of cultural orders would be restored, and any apparent social functions and structures would be brought back to a full life for everyone's benefit, thanks to progress and hard work.

Even the best thinkers of that time experienced and lived this as the existential ground beneath their feet. For example, if the metaconscious knowledge of progress is eliminated from Karl Marx's theory of human history, it collapses like a deck of cards. Why should every society assign the highest value to its economic activities and treat them as sustaining and driving all human life? How was the enormous diversity of cultural order to be constrained in their evolution by five stages of human history? Why should each new economic base for society be an improvement on the previous one, and why would capitalism inevitably turn into socialism? The same is true for Adam Smith's interpretation. Without progress, how could the new wealth of nations possibly compensate people for becoming as stupid as possible? How could the great invisible hand of the Market compensate for its undermining of human life, society, and the biosphere? Without progress, why should the new age and the new humanity that was emerging be the fulfilment of all humanity's hopes and dreams?

A metaconscious knowledge of progress thus constituted one of the stable elements underlying the experiences of the people of that time, despite all their political, religious, and ideological differences on the surface. There was no discernible limit to progress; its material beginnings were inseparable from its social and spiritual complements. Progress was no longer limited to particular areas of human life and society, nor was it limited to a particular criterion of success. Living as if progress were everywhere and in everything transformed it into a metaconscious myth. Everyone could participate in this progress by means of hard work, thereby increasing economic output, whose benefits had no limits. In turn, progress and work would bring human happiness and as if no other future were possible.

All this could be achieved only through the accumulation and ever larger investments of capital: the life-blood of the new emerging order and the common denominator of all its values. It was as though this capital were able to create a new kind of human life in a new world. As a result, capital became the new metaconscious central myth or sacred.[21] In sum, capital led to progress, everyone could participate in this progress by means of hard work, and everyone would be rewarded with happiness. These became the new secular deities that had 'created' and were 'sustaining' the people and the world of that time. All this occurred by means of the same processes that we believe created the gods of the past. The new gods would now create a cultural order out of the emerging economic order. Faith in these gods remains very strong to this day.

The way in which technology changed people, especially through their deepest metaconscious knowledge, transformed what these people lived as myths into an explicit mythology. Even the best thinkers of that time were incapable of putting this into words. Owing to the nature of myths, explanations like the above can only be developed with hindsight. The power of myths over human life and societies is precisely due to the fact that myths reign metaconsciously, where they are out of reach of any critical scrutiny and social divisions. In other words, people lived as if the reality they had come to know and live were reality itself, thanks to these myths. Of course, there were many more things yet to be discovered and lived, but these would be cumulative and thus change nothing essential. Across the horizon of human experience in the future would appear more of the same. The unknown, thus converted into more of the known, excluded anything radically 'other' and made reality as it was known and lived utterly reliable and trustworthy.

However, the new emerging order that was ruled over by capital, progress, work, and happiness was at its core an order of what was real and not of what was true, as a consequence of the growing dominance of the technology-based connectedness over the cultural order. Attention was now directed towards most efficiently obtaining the desired intermediary outputs from the required inputs, and transforming these outputs into inputs for the next production step, and so on, and ensuring that the overall cost of a product remained as far as possible below its market price. No reference to the meaning and value of a product for human life and society was required. Even the market price was a poor indicator of this meaning and value because of the need to commoditize everything. This led to considerable market externalities, which in turn contributed to massive Market forces. The role of culture was thus banished from what really mattered to a society, as was clearly indicated by the change in the meaning of the word *culture* during this time. It now referred to the 'decoration' on the 'social cake' namely, literature and the arts. These developments desymbolized the mainstream of society in one way, and its remaining 'culture' in another.

The most important thing that a society could undertake was the development of its technology-based connectedness by investing as much capital as possible in the required technologies. Everything else that human beings in their communities had ever hoped for would follow: material progress would bring social progress, and social progress would bring true happiness. By implication, all traditional cultures had followed the wrong approach. Their cultural unities had a top-down orientation, that of first developing their cultural orders within existing material constraints. They should have followed a bottom-up approach, beginning with the development of their technology-based connectedness. For this, their cultures were of little use and therefore needed to be banished to the perimeter of human life and society. The search was now on for a replacement of what had made humanity a symbolic species until now. The technological and economic approaches quickly reached their limitations, and they would soon be replaced by a new development, which we shall examine in the next chapter. These transformations shattered all traditional ways of life along with their moralities, religions, and aesthetic expressions. One of the most significant mutations of humanity as a symbolic species had begun.

This mutation would have been existentially unbearable were it not for the fact that the new highly desymbolized cultures of the industrializing societies of Western Europe continued to create metaconscious myths. These transformed an economic order into something that had a distant future connection to the living of human life in the world. No matter how bad life was for many people, everything would turn out for the best sometime in the future. In the meantime, the 'meaning' and the 'value' of everything had to be related to what *was*, in the form of the technology-based connectedness, as opposed to the sense everything had for life. This opened the floodgates to desymbolizing experiences. Traditional moralities, religions, and aesthetic expressions had to be transformed to fit into the new order.

The reversal of the traditional relationship between the technology-based connectedness of human life and society and its cultural order began to lay the foundations for what would eventually become a new kind of consumption, one that attempted to satisfy non-material needs with material things. We only need to think of the advertisements in which lonely people in a mass society are led to associate a whitening toothpaste with new friends and possibly that special person in their lives, and to live as though alcoholic beverages, clothes, cars, computers, and other material goods have similar powers to transform lives. To attribute this change in human consumption to a sudden materialism is not to understand the spiritual forces at work here.

Along with culture, mortality and religion also had to be banned from the mainstream of society. The new secular cultural unities included sacreds that were no longer compatible with those of traditional religions. Having their metaconscious bases knocked out from under them, traditional religions found their role in the industrializing societies gradually weakening. What remained of them was largely personal: spiritual direction in this world, and personal salvation or damnation in the next. In making these sociological and historical observations, I am not in the least dealing with the question as to whether or not a God exists. A distinction must be made between a religion, which every culture creates for itself, and faith, which results from something having been introduced into a culture by the revelation of a transcendent God.

Christianity, which had been the most influential religious tradition in Western civilization, was torn in two. Its 'horizontal' dimension, dealing with the teachings about the relationships between people, had to be reinterpreted in order to play a role in society. The desymbolization of traditional metaconscious values was intuited as a unique opportunity to demonstrate the social relevance and usefulness of Christianity. Since 'technological values' (output-input measures) were guiding collective human life, such teachings had to be restricted to personal values. A dualism was thus created between human behaviour that conformed to the new economic order and the professed personal values that could be counted on to transform a person's life. For example, factory owners played a social role that was tightly constrained by the new order. If they imposed their personal values on this role in order to alleviate the suffering of their workers by supplementing their meagre wages from their own wealth, they would cease to be factory owners as soon as their capital was depleted. The option of charging higher prices to afford adequate wages was ruled out by the markets for their products. In other words, personal values became end-of-pipe solutions that were impotent against the constraints imposed by the economic order.

In contrast, the 'vertical' dimension, dealing with the teachings about the relationship between God and humanity, appeared much too abstract in the face of so much social upheaval and suffering. Feuerbach sought to 'save' Christianity by preventing what was socially useful from being dragged down by the rest.[22] With an increasingly rational and secular humanity on a mission of bringing progress and happiness, societies had little use for a God other than in their personal and spiritual lives. Even there, science was expected to diminish people's personal dependence on this God. The best way to move forward appeared to be to downplay the vertical dimension of Christianity to the benefit of the horizontal dimension. Other Christians withdrew from this worldliness by clinging to the weakening traditional cultural order as a remnant of a Christian past. The social upheavals and human suffering of that time were seen as the consequence of abandoning the past and its Christian roots. They emphasized the vertical dimension as the spiritual key to turning the situation around. This orientation made it more and more difficult to answer the question of what exactly the world had to be saved from. It turned Christianity into something that was purely personal and spiritual: one's relationship with God. In this way, Christianity was torn into its liberal and evangelical-fundamentalist branches, a split that endures to the present day.

Each branch, in its own unique way, helped to bind people to the new emerging order. The liberal branch became so deeply involved in the world that it lost all sense of the teaching that the ways of the Christian God were radically other than the ways of the world, which prevented their own intervention into the social upheavals and human suffering from being genuinely transformative. In other words, the actions of these Christians could be guided in the background by the new emerging cultural unities because the iconoclastic dimension of Christianity had been neutralized. Evangelical-fundamentalist Christians turned their backs on the world, losing all connection between the 'good news' and the kind of world to which that news was addressed. Consequently, their activities could also be guided in the background by the new emerging cultural unities, and thus the iconoclastic dimension of Christianity was neutralized once again. It can be argued that both branches of Christianity have forgotten their Jewish roots, which had always been preoccupied with the danger that any community faced with creating false gods. This can be understood as the need for any culture to create absolute roots in reality by means of a sacred and of myths. Neither branch, therefore, paid any attention to the deeper spiritual reassignment of where human beings placed their ultimate trust. Directly or indirectly, these two branches of Christianity faithfully served the new system, just as Christianity had during the Middle Ages. It functioned much like any other religion in any culture, despite its core message of liberation from anything and everything that alienates humanity. As Lewis Mumford observed, the Christian virtues of the past were quietly transformed into vices, and vice versa.[23]

The moralities of industrializing Western societies underwent the same fate. As people's metaconscious values were progressively desymbolized as they internalized the new emerging order, the organizations of their brain-minds could no longer orient human activities within the domain of the possible in the same way. In collective human life, this did not present a particular problem since output-input ratios and monetary prices all but filled the growing moral vacuum. In people's own lives, the situation was more complex. Participatory activities in their families and communities now had to be guided by whatever little remained of traditional values or by something else. If these traditional values were regarded as Christian, a reconciliation with the new emerging cultural unities was quickly reached. All along, it was believed, God had been on the side of those who worked hard and who helped themselves through entrepreneurial initiatives. His blessings were now monetary, thus explaining why the rich and the poor were what they were. It amounted to a kind of extension of the Protestant ethic as examined by Max Weber.[24]

Everyone outside of the dominant religious tradition had to adapt and evolve in relation to the changing conditions. A few generations were able to weather the turbulence and social upheavals by staying the course symbolized by the deeper metaconscious values. As desymbolization proceeded, however, an entirely new moral orientation developed that eventually would lead to a 'statistical' morality: using the organizations of people's brain-minds as a kind of radar to scan what everyone else was doing, and 'going with the flow.' This moral orientation equated the 'normal' behaviour of people with normativity. During the transition to this kind of morality, found in all contemporary mass societies, ideologies helped to fill the growing void.

The aesthetic expressions of the industrializing societies of the nineteenth century continued to represent the intuitions of what was happening deep down in the individual and collective metaconscious long before it could be brought into words.[25] It traced the emergence of a cultural order seeking to establish an internal and external connectedness between diverse functions and structures that, much like mechanisms, were made to 'work' as well as possible instead of being expressions of, and guided by, their sense within human life and society, as was the case prior to industrialization. This emerging order delineated what was possible in the domains of thought and action and arranged things according to an economic order. This was accomplished by attaching human life to functions and structures that helped to sustain this order. All direct attempts to express the meaning and value of what was happening to human life and society became increasingly incompatible with the new emerging order. As a result, the aesthetic expressions of industrializing societies were taken out of the mainstream of life and banished to museums, theatres, art galleries, and concert halls or were to be 'enjoyed' in private. In this way, people were 'free' to enjoy paintings, sculptures, music, poetry, and literature (now collectively referred to as *culture*), differentiated from everything else in human life and society. These aesthetic expressions of what was happening to human life and society might profoundly move them, but then they had to get back to 'real' life. In this way, any attempts to question the status quo were essentially neutralized.

As one of many changes in the vocabulary of the English language during that time, the evolution of the meaning of *culture* reflects the above developments.[26] The new meaning of the word *culture* symbolized what was left of human life and society after its economic core and its sustaining functions and structures had been removed. With hindsight, it would appear that all the economic, social, political, legal, moral, religious, and aesthetic adjustments that Western European societies had to make to their ways of life did not express their unique cultures. If this had been the case, the diversity of such adjustments would have been as great as the cultures involved. The limited diversity of these adjustments may therefore be interpreted as accommodations of human life and society to the new economic order. There was no unique English, French, or German economy as an expression of these cultures. This was equally true for all the other spheres of human life and society. It is for this reason that we speak of industrializing societies, because what they increasingly had in common far outweighed their remaining cultural differences.

Despite the massive changes there was also a great deal of continuity. Within the constraints imposed by desymbolization, daily life continued to be dialectical. The ability of people to talk to each other still required a dialectical tension between their being different and yet similar. The viability of human groups still depended on dialectical tensions. Societies still had to maintain a dialectical tension between individual diversity and cultural unity. Everyone's experiences continued to reflect a dialectical tension between the increasingly weakening cultural order of the past and the growing importance of mechanisms, structures, and functions within the new order. The former continued to symbolize the order of life, and the latter the order of life in the image of the machine. There was an ongoing dialectical tension between the symbolization of some entities as myths and the way these entities were experienced, particularly when reality imposed itself. Consequently, a great many, if not all, experiences were shrouded in a certain amount of disorder. Although the processes of symbolization were severely undermined by desymbolization, each moment of a person's life was a manifestation of his or her life in both its order and its disorder. Each human life was still a manifestation of his or her community in both its order and its disorder. Human communication continued to amplify or diminish this disorder. It was still through these and other activities that communities continued to live the tensions between order and disorder and between cultural unity and individual diversity. Whenever disorder grew, a desire for order usually grew along with it, and following intense collective efforts some order was frequently restored.

Despite all this continuity, the role of symbolization in establishing a cultural order in reality had been shifted by the new emerging cultural unities. Until now, such cultural unities had guided communities towards the leading of happy and responsible lives by following in the footsteps of previous generations. The new cultural unities suggested that this happiness and responsibility was to be found in a very different place. The first priority of any community should now be the strengthening of the technology-based connectedness of its cultural order, and everything else would be granted to it by means of unlimited progress. Implicit in this new cultural orientation was a devaluing of the role of symbolization, which by its very nature establishes a cultural order that dominates the technology-based connectedness. This convinced humanity that it could abandon the very creation that had made it human until this point in its history: the symbolization of individual and collective human life by means of a culture. There appeared to be less and less need for it now. The cultural and natural orders had become almost an externality of the new economic order.

These developments were the first phase of the extensive desymbolization of the cultures of all industrializing societies. As we shall see in the next chapter, highly specialized scientific and technological knowing and doing became both cause and effect of further desymbolization, completing the domination of what is real over what is true in all spheres of human life.

4 The Triumph of the Technical Order over Cultural Orders

Symbolization and Traditions

In traditional societies there was a close relationship between the organizations of the brain-minds of their members and the tradition-based way of life. The former developed as people sustained their individual lives by dialectically ordering their experiences, while the latter developed as a result of the way these lives overlapped and were enfolded into each other. Any encounter involving two or more people was symbolized by modifications to the organization of the brain-mind of each person involved. This process also contributed to the evolution of the dialectical tension between individual diversity and cultural unity, as well as the building up of a metaconscious knowledge of what was unique to each person involved and of what was shared with all other people as part of the working culture that sustained the life of the community.

The life of any member of these traditional societies overlapped with the lives of all the other members they encountered. In turn, these lives overlapped with those of others, and so on. In terms of the simple model set out in chapter 1, the clusters of experiences related to sharing one's life with close family members and friends would be highly developed, while those dealing with acquaintances or strangers would be less well developed. The metaconscious knowledge embedded in the former would be mostly personal, while that enfolded in the latter would be related to the objectivized functioning of the cultural order.

In the same vein, the dialectical ordering of the experiences of the lives of the members of the community reflected their roles and participation in the social division of labour. If someone was a farmer, the corresponding clusters of experiences would be highly developed, as would be the ones associated with encounters with fellow workers, while those associated with occasional encounters with people for business reasons would be much less developed. Still less developed would be the clusters associated with his or her encounters with strangers who occupied one of the many other social roles related to the way of life of the community. It is possible that the farmer might not have encountered a single person occupying certain social roles (such as shipbuilders on the coast), but he would still have had some knowledge of shipbuilders; the metaconscious differentiation and integration of the word *shipbuilders* would designate its meaning and value within the symbolic universe of the community. In this respect, the organization of the brain-mind of a shipbuilder might have been something of a mirror image of that of the farmer. To use the analogy of a mental map: if the inhabitants of a city each drew a street map, it would likely be very detailed for the area in which they lived, less so for the area where they worked, and still less so for areas they merely passed through on their way to somewhere else. Provided that no errors were made, all the maps drawn by the inhabitants would generally overlap but would vary considerably in the levels of detail of different areas of the city according to their experiences. The implicit common base map may be thought of as an analogue of the metaconscious tradition of a community.

Such a tradition was the blueprint for human life in the cultural order created by a community in the course of its history. It mapped the normative way in which everything in the way of life of a community was related to everything else, even though it was symbolized in the organizations of the brain-minds of the members by means of metaconscious knowledge built up from the dialectical ordering of their own experiences. Any tradition encompassed the field of individual and collective experience within a horizon that symbolized the unknown as interpolations and extrapolations of the known. The metaconscious myths by means of which this was accomplished made that life possible by protecting it from relativism, nihilism, and anomie.

In traditional societies people lived each moment of their lives as an expression of how everything was related to everything else in their lives and in the collective life of the community. Each and every day people recreated these relations by imaginatively adapting them to situations that were similar to, and yet different from, those in the past. Such adaptations ranged from a great many that were essentially routine to far fewer that required an intervention of some kind. This typically involved thinking through the situation in order to come up with a better response. The point is that some adaptation is always required. Responses to new situations may remain unique to a person's life or they may eventually become objectivized and incorporated into the working culture of his or her community. In other words, a tradition was derived from the objectivized experiences from previous generations, which were fitted together into a religiously absolutized design for making sense of and living in the world. As a result, the tradition permitted the members of a community to sustain each moment of their lives by utilizing the organization of their brain-minds as a mental map for orienting themselves in an ultimately unknowable reality. Such a dialectical ordering of individual and collective human life in a cultural order was constantly tested, confirmed, or adapted by turning any disorder into order as much as possible. Any tradition was lived by its people in a manner that was individually unique and culturally typical.

The relationships between the organizations of the brain-minds of the members of a community and the tradition they have jointly evolved can also be examined from the perspective of human knowing and doing. Babies and children acquire a great many skills that are integral to a culture-based way of making sense of and living in the world. The acquisition of these skills generally follows the five-stage model developed by Stuart and Hubert Dreyfus for adults.[1] For babies and children, the first three stages of the model (roughly characterized by the literal application of rules, the refining of these rules and the learning of exceptions, and problem solving when the rules become overwhelmed by exceptions) are largely replaced by playful interaction with the world and thus the building of a great deal of metaconscious knowledge. Intuitions based on this metaconscious knowledge open up entirely new possibilities and responses, which then modify and expand the metaconscious knowledge, thus giving rise to new intuitions, and so on. In adult learners, the marked shift in behaviour that characterizes the fourth and fifth stages is almost certainly the result of sufficient metaconscious knowledge having been built up in the organization of the brain-mind to permit a person to effortlessly recognize a situation for what it is and, later, to fluently respond to it as well. Stuart Dreyfus estimates that a grand master chess player can almost instantly recognize some fifty thousand positions and rapidly respond to them with little deterioration in the level of play, even when they are prevented from reflecting on the situation.[2]

In a traditional society some members of each new generation served apprenticeships with the aim of becoming expert knowers and doers of a particular craft, which was essential for sustaining their way of life. The many experiences of learning a skill by working under the watchful eye of a master resulted in the internalization and symbolization of these experiences, thus enormously increasing the ability of the organization of their brain-minds to sustain them in this area of their lives. In terms of the analogy of a mental map, the maps of the apprentices working towards becoming masters would have become increasingly detailed in the areas corresponding to their crafts. In other areas, these maps would have been much less developed as a result of being no more than a bystander or an occasional helper or acquaintance. In some areas, every member of the community would develop a great deal of mastery, as is the case with the use of the language or the living of the ethics implied in the metaconscious moral values.[3] In all these cases, the mastery or expertise was fully symbolized in their participation in the tradition-based way of life of their community. As a result, any tradition-based way of life enfolded within it many traditions corresponding to particular crafts or expertise that were essential for its daily functioning.

Tradition-based knowing and doing were embedded in experience and culture because they were primarily sustained by metaconscious knowledge that was implicit in the organization of the brain-mind of the person making use of them. Since both knowing and doing were sustained by the same metaconscious knowledge, they were integral to the work process and could not be separated from one another. They represented different aspects of the elaboration of the tradition of a society in a particular sphere. Technological experiences, like all others, were integrated into the tradition by means of the culture. In their own unique way these experiences elaborated and expanded the tradition in the face of both disturbances and opportunities. The technological traditions of these societies must therefore be regarded as particular manifestations of the larger tradition into which these were enfolded in a more or less culturally appropriate manner. The apprenticeship program was essentially an extension of children and teenagers being socialized, by the adult members of the community, into making sense of and living in the world. Apprentices continued to elaborate all of this in much greater detail in the areas of their crafts.

By way of an example, consider the technological tradition of carriage building. When a customer ordered a new carriage, the master craftsman did not first design it and then build it.[4] His knowing and doing that was embedded in experience and culture did not exclude anything from the context taken into account. Nor did any significant level of desymbolization make this context less effective. The master carriage builder, like any other member of the community, had acquired a great deal of metaconscious knowledge about the role carriages played in their way of life, the kinds of values people applied to assessing how well a particular carriage fit that particular role, and the personal preferences of his regular customers. In addition, his own experiences, first gained as an apprentice, then as a master carriage builder, had refined his understanding of carriages with a growing appreciation of all the technical details, choices of materials, and unique features of different kinds of carriages. When approached to build a new carriage for a particular client, the master builder began with some ideas already in his head, which would be worked out in the process of building it. For example, he might have decided that the carriage he recently built for another customer might suit the preferences of his client rather well, provided that a number of details were modified and some dimensions altered. He might also have decided to try out an idea for improving the suspension or some other detail. Once built, the carriage would be taken on the road in order to carefully observe its handling and ride with additional attention being paid to how the innovation contributed to its overall characteristics. If successful, some of these innovations might find their way into many more carriages and eventually into the carriage-building tradition.

There was a complete absence of the kind of specialized knowledge that we would use today, such as stress analysis or materials science. The master builder, watching an apprentice craft a particular piece under his supervision, might take him to a finished carriage with a similar piece to explain that in the past there had been some problems with excessive bending causing the gradual weakening of some joints, eventually leading to failure. He might have pointed out the telltale signs of trouble to the apprentice and suggested that when the carriage was taken for a test ride, the part in question should be carefully watched. If these signs appeared, the particular piece would have to be reinforced, and the master builder might explain how this had been done in other situations. Many such experiences helped to develop the metaconscious knowledge that permitted a master builder to observe a particular part and declare that it 'looked right.' In other words, based on a great deal of experience, the particular part was deemed to be strong enough, given the materials that had been used, and sufficiently aesthetically pleasing to contribute to a handsome carriage. A knowledge of the strength and behaviour of materials was thus enfolded in this assessment. In this manner, any technological tradition enfolded the tradition-based equivalent of what today would be dealt with and optimized by distinct and separate technological disciplines and specialties.

There is one more detail that we need to appreciate with respect to human knowing and doing based on tradition. Any tradition-based approach had built into it a precautionary orientation. As a strategy for making sense of and living in the world, a tradition embodied the experiences of many generations, of which the most valuable ones were objectivized and incorporated into a working culture. The community had arrived at a very workable strategy, which was transformed into the only good and responsible one by its myths. Anything new had to be approached with caution because of its potential to disturb and possibly undermine what had proved to be both liveable and sustainable.

Owing to very low levels of desymbolization, the members of traditional societies had a good sense of how everything related to everything else as a basis for making sense of something new, which explains the ambivalence that came with replacing something that had proved itself over time with something as yet unproved. Seen from the vantage point of the myths of contemporary societies, this attitude has often been interpreted as needlessly static and opposed to anything new. However, on carefully weighing the evidence, it was our civilization that found it necessary to invent the precautionary principle, on the grounds that prevention is almost always vastly more economic as well as socially and environmentally desirable than moving blindly ahead in the confidence that end-of-pipe approaches will deal with any problems. This principle has proved to be difficult to implement because it goes against our contemporary myths, which imply that anything new must be better. Generally speaking, tradition-based ways of life were capable of remarkable feats, not the least of which was ensuring that everything they did was compatible with everything else, including local ecosystems; by comparison, our civilization excels at the improvement of performance, often at the expense of such context compatibility.

Desymbolization and Tradition-Based Knowing and Doing

Tradition-based knowing and doing was embedded in experience and culture by means of the process of symbolization. Any limitations imposed on this process had repercussions for such knowing and doing.

Three kinds of limitations were imposed by industrialization. First, a range of phenomena associated with industrial technologies either were not available to the senses or became so over time, thus limiting access by means of experience and culture. Second, human groups and societies evolved the symbolization of experience and the building of cultures in order to make sense of and live in the world. Industrialization required their redirection towards building a technological as well as an economic order, to which symbolization was ill-suited. Third, the evolution of traditions and tradition-based knowing and doing was essentially accomplished by the paradigmatic growth of a culture. Industrialization overwhelmed this kind of evolution by a flood of new situations that were unlike anything that had gone before. We shall examine these three sets of limitations in detail in order to understand the mutation in knowing and doing that resulted in their separation from experience and culture.

In the pioneering industries, industrialization gradually imposed a limitation on the accessibility of technological phenomena to the senses. Initially, the efficacy of various operations of a machine could generally be correlated with observable symptoms, such as the excessive vibration of a machine tool because of insufficient support, odours or changes in the colour of a part because of inadequate lubrication, unacceptable deformation, symptoms of extreme stress prior to the failure of a material, or the frequent breaking of threads in a spinning machine because of rubbing over a pulley due to misalignment. Similarly, in the metallurgical industry the colours of flames, processed ores, or metals generally correlated rather well with what was happening. These kinds of correlations were extremely rare in the chemical and electrical industries. Brown smoke given off in a chemical reaction can mean a variety of things, and the colours of chemicals do not correlate with their basic properties. Nothing can be observed regarding the workings of an electric circuit except in cases of a short circuit or a meltdown. In other words, where technological processes occur out of sight, below the surface, far fewer if any features can be correlated with what is happening, and their symbolization would almost certainly be inadequate. Eventually these kinds of limitations also imposed themselves in textile making, machine building, and metallurgy as the increasing speed and level of technological sophistication of operations and processes required ever more subtle details to be correlated with what was happening. The building of the technological order thus revealed a limitation on symbolization and on the technological knowledge that could be built up with it.

The second set of limitations was associated with the redirection of the process of symbolization towards very different ends. As we have said, human groups and societies evolved the symbolization of experience and the building of cultures in order to make sense of and live good lives in the world. In the technological and economic orders, however, people had to think in terms of balancing flows of matter, energy, and the technological products built up with them and the efficient, productive, and profitable transformations of required inputs into desired outputs. How this affected human life and society within the cultural order was much less pressing. What mattered was the technical 'meaning' and 'value' of flows and transformations within the local technology-based connectedness of a technically divided activity so as to ensure the lowest possible cost prices, which had to be kept well below market prices to ensure profits and the renewal of capital. The interaction with what remained of the cultural order was organized by the Market, with its individual market mechanisms balancing the demand for, and supply of, commodities. These could be reintegrated into people's lives only through their behaving as *homo economicus.*

In the previous chapter we noted that the technological and economic orders did not result from a further internal differentiation and integration of a cultural order. Tradition-based ways of life were shattered into three incompatible orders, two of which were built up by the technical approach and the economic approach, which were diametrically opposite to the cultural approach. The first two approaches treated everything in a manner that was analogous to a mechanistic order built up from separate domains; these domains were characterized by one category of phenomena reproducing a particular sub-function rather than participating in endless variation and adaptation. In contrast, the cultural approach was created and evolved to deal with human life in a world in which nothing ever repeated itself in quite the same way. The technological and economic orders jointly built an order of what is real by limiting its context to the local technology-based connectedness and the markets for the associated inputs and outputs. Doing so amounted to a substantial desymbolization of what is true for human life and society. As we shall show shortly, both the technical and economic approaches were goal directed as opposed to tradition directed.

The third set of limitations was related to culture-based reasoning becoming ineffective. The symbolization of each experience had formerly been sustained by a person's entire life as symbolized in the organization of his or her brain-mind. If a new situation maintained a dialectical tension between being similar to and being different from the situations previously recognized and responded to, a person could respond to it effortlessly. If, however, a situation could not readily be recognized and responded to, a person was obliged to think about it. Such reasoning was embedded in experience and culture in the sense that it sought to understand and respond to the situation by explicitly distinguishing it from others in order to facilitate its symbolization. In other words, culture-based reasoning sought to make situations that were somewhat anomalous to a cultural order into normal ones that could be symbolized to further differentiate and integrate this order. Such situations actually helped to evolve a tradition by taking their place in the way that everything was related to everything else.

Industrialization created a great many situations that were unlike anything that had gone before and which could not be dealt with by the further internal differentiation of a cultural order. Superimposed on this development was the shattering of tradition-based ways of life into three interacting orders. Two of these were orders of what was real, while the third was an order of what was true. For the members of industrializing societies, this meant that the organizations of their brain-minds became much less coherent, as did all forms of metaconscious knowledge, including that of their social selves. The ability of their lives to sustain each new moment in the background was diminished, and culture-based reasoning could no longer be guided by tradition. Metaconsciously, everything was less well related to everything else, and this posed a growing problem for dealing with situations that were unlike anything that had gone before.

From the perspective of the five-stage skill acquisition model, people faced a growing number of situations to which they could not skilfully respond. They had to think them through, thus forcing them back to the third stage. It led to a growing tendency to treat situations on their own terms. Skilfully coping with the world on the basis of a culture gradually mutated into a rational approach. From this perspective, a goal may be regarded as what remains of a tradition when it is desymbolized to the point where only a single aspect remains. In the circumstances we are examining, this single aspect typically belongs to the local technology-based connectedness of human life and society or its organization by the relevant markets. Goal-directed behaviour desymbolizes the living of a life to the 'living' of a myriad of separate and distinct goals. The widespread emergence of goal-directed behaviour is the result of coping with a situation in which a great deal of human life and society is so highly desymbolized that experience and culture provide inadequate guidance. It is a sign of how deeply the technological and economic orders have weakened the cultural order, and this was gradually normalized by emerging metaconscious myths that eventually constituted a new cultural unity.

The role that culture played in human behaviour was completely transformed. The new goal-directed behaviour was less well sustained by a culture than was tradition-directed behaviour. It restricted the context taken into account, by excluding as much as possible whatever was not directly relevant to the accomplishment of a goal. A goal destroys something of the cultural order by separating what is directly relevant to that goal from what is marginally relevant and by eliminating everything that is irrelevant to it. In this way, a goal translates what is integral to the order of what is true into the order of what is real. Rational human behaviour is characterized by a goal that is largely substituting for a person's life working in the background. For all practical purposes, rational behaviour prevents people from fully living their lives under these circumstances. They become spectators to their lives whenever rationality takes over. Rationality thus imprisons their lives. This marked the beginning of a transformation in the role that the organization of the brain-mind had once played as a mental map used by people to orient themselves in the world.

It would be difficult to exaggerate the scope and depth of the effects of the above three sets of limitations on the process of symbolization and the role of culture in human life and society. The evolution and adaptation of these dialectical wholes had been sustained by cultures that symbolized every element of a tradition in relation to all others in order to come to grips with the element's participation in that tradition according to its meaning and value. To the extent that the cultures of the industrializing societies became desymbolized, their effectiveness in dealing with this interrelatedness was diminished.

There was no turning back. The three sets of limitations imposed on symbolization and the role of culture were greatly intensified by technological and economic competition. The advantages in the domain of performance and power that were derived from recreating human life and society were further increased by technological growth. As a consequence of the further weakening of the roles of symbolization and culture, there was a greater necessity to rely on rational responses. It was not long before technological and economic growth driven by competition reached a phase of diminishing returns, characterized by the need for ever more talented people having to spend more time to achieve ever smaller gains. In economic terms, this meant that those companies that could write off their investments in technological innovations over a large number of machines and multiple factories could continue to make a profit and renew their capital, while smaller enterprises were unable to take advantage of these economies of scale.

Diminishing returns on investment in technological innovation led to entrepreneurs rationalizing every aspect of their business to gain every possible competitive advantage. Soon these efforts also entered a phase of diminishing returns. Technological growth and the economic growth achieved with it appeared to be reaching an asymptote without the competitive pressures diminishing in any way. All this contributed to a disordering effect on capitalism, and the prediction of Karl Marx that it would soon self-destruct[5] might well have come true were it not for the growth of the phenomenon of rationality as examined by Max Weber.[6] It gradually eliminated the asymptote imposed on technological and economic growth by the three sets of limitations on human knowing and doing. The stakes were so high that a solution for overcoming these limitations to human knowing and doing was soon found.

Rationality within Culture

The desymbolization of the cultures of the industrializing societies resulted in people becoming less well connected to their social selves, others, and the world. In the first chapter, we suggested that the ordering of the experiences of people's lives implies that the meaning and value of a particular experience are dialectically related to the meanings and values of all the other experiences. Simply put, an experience is what all the others are not. As a result of the metaconscious relating of all experiences to all others by the organization of a person's brain-mind, any reliance on what would be the equivalent of a metaconscious definition is avoided. It also confirms Wittgenstein's finding that the members of a linguistic category share a network of criss-crossing and overlapping family resemblances as opposed to a fixed set of characteristics.[7]

When desymbolization began to limit how well everything could be metaconsciously related to everything else, it produced the equivalent of a reduction in both the 'resolution' of the meanings and values of all experiences and the ability of the organization of the brain-mind to sustain each moment of a person's life in the background. Symbolization now made everything less clear and more fuzzy as the density and quality of the metaconscious connections declined. All metaconscious knowledge became shallower, and this in turn affected how all daily-life skills of coping with the world were embedded in a person's life and then in the way of life of his or her community. It also affected the metaconscious knowledge of the self, the social selves of family members and friends, the social roles of strangers, and hence the fabric of social relations of individual and collective human life. Desymbolization contributed to a reduction in the intuitive sense that people had of how everything they did and thought was integral to the way in which everything was related to everything else, thus undermining the roles of tradition and culture.

As the level of desymbolization grew, it was as if the tradition of a society were decentralized into a diversity of goals, each associated with a particular activity. It created a sense that human life was full of 'problems' that were amenable to 'solutions,' provided that they were rationally approached. Such intuitive recognitions implied that a reliance on symbolization and, via it, on experience and culture, was more or less futile. This reinforced a kind of thinking that human life and society could be improved by rational problem solving. Implied in it was a sense that underneath experience and culture existed a rational order. However, the development of these kinds of intuitions had nothing in common with Greek philosophers' invention of a universal knowledge housed in a rational soul or with philosophical foundationalism.[8] Rational problem solving was not an attempt to save the cultures of the industrializing societies of the late nineteenth and early twentieth century from relativism but, out of necessity, to go beyond them in order to move forward. From this perspective, the attempts of artificial intelligence to capture human knowledge and daily-life skills in information machines by uncovering their underlying rules, algorithms, scripts, frames, or microworlds was utterly contradictory. Had it succeeded, artificial intelligence would have frozen all cultural evolution in time or, if that evolution had itself been programmed, it would have transformed human life into the image of the machine.

The phenomenon of rationality emerged in one area of human life after another. The tradition-based cultural approach was pushed back by the goal-based rational approach. Max Weber warned humanity against the long-term consequences of rationality, which he likened to shutting ourselves into an iron cage.[9] An evaluation of his conclusion requires that we pay particular attention to the fact that rationality constituted by goal-directed behaviour was conscious and deliberate and thus plain for everyone to see, while what was pushed back, namely culture, continued to operated metaconsciously in the background, be it with a diminished scope.

Any celebration or condemnation of the phenomenon of rationality would have to include what humanity was giving up out of necessity. It would also have to acknowledge that babies and children are unable to make use of rationality to learn to make sense of and live in the world, with growing implications for future generations if desymbolization were to continue to intensify. Were we abandoning the very creation that, until now, had made us human? Shutting ourselves into an iron cage meant that human beings were diminishing their abilities to sustain their own lives and to relate to others and the world, by barring access to the full interconnectedness of human life in the world as a consequence of desymbolization. It greatly weakened people's grip on their lives and their world.

The full impact that the phenomenon of rationality had on culture becomes evident when we compare dealing with a situation by the rational approach to dealing with it by the cultural approach. The rational approach requires a goal in the absence of a well-functioning tradition and culture. The cultural approach requires no such goal other than the symbolization of the situation, which reflects the contribution it makes to the adaptation of the evolution of a tradition and a culture. A goal may be mentioned, but it simply names the meaning and value of the situation for individual and collective human life.

The goal required by the rational approach salvages what it can from a desymbolized way of life and culture. A fully symbolized goal has an 'external' and an 'internal' dimension. The former symbolizes how the associated activity is integral to and helps evolve human life and society through its relations with everything else. The internal dimension symbolizes how the activity accomplishes its goal by the operations of its constituent elements. The more the goal is desymbolized, the more the external dimension is weakened to the benefit of the internal one. The less that can be salvaged from the external dimension, the more the situation must be treated on its own terms. It is now a question of doing something better. This 'better' now has a somewhat mythical sense since it appears unnecessary to clearly state by what criteria the improvement will be measured, to which areas of human life and society it will make a contribution, and on which areas it will have a negative influence. Making things better thus implies a kind of absolutization of a situation as its being good in itself and worthy of ongoing improvements. The rational approach guarantees that the positive effects will outweigh the negative ones or that only the former will occur. Initially this was sustained by the myths of progress and work.

Once a goal has thus been imposed on a situation, the context has to be recreated. Those aspects and relations immediately relevant to reaching the goal become the most important, those that marginally affect it become less so, while those that will not affect the reaching of the goal become irrelevant. Such a hierarchy of importance with respect to reaching a goal would have been entirely different had another goal been considered. The consequence of imposing a goal on an activity effectively removes it from the sphere of human life and culture, to place it in a new rational context containing the important aspects and relations. Depending on the level of desymbolization, the activity is transferred from the symbolic universe of sense to a domain of non-sense constituted by the reconstructed context. The same kind of transformation occurs with respect to the 'internal' structure of the situation itself. Implicitly or explicitly, a rational model of the activity and its context has thus been created. The aspects and relations of this model can be varied in order to determine which form of the activity will best accomplish the goal.

Contrast this with improving an activity on the basis of symbolization and culture. When the activity is fully symbolized, improving it is inseparable from its meaning and value relative to everything else in human life and society. The role of that activity is seen in terms of its participation in the adaptation of human life and society according to the appropriate cultural values. It is a tradition-directed improvement. In contrast, a goal-directed improvement compares ultimate forms of the activity in terms of doing more of what it does. Better is now desymbolized as *more*. To the extent that all these operations have desymbolized the activity and its role in human life and society, cultural values can no longer be used, and gains must be measured in terms of the desirable results produced by the activity. In other words, the 'everything being related to everything else' is now desymbolized to the inputs required by the activity, its rationalized internal functioning, and the desired output it produces. The inputs and outputs have been commoditized to the extent that they are no longer conceptualized and dealt with as an integral part of human life, society, or the biosphere. Improvements are desymbolized to become output-input ratios. The level of desymbolization correlates with the extent to which the activity is treated on its own terms.

All the above operations are carried out by thought processes that desymbolize, rationalize, and optimize the activity. To implement the improvements demonstrated to be possible in the rational domain, the actual activity must be reorganized in the form that, in the rational domain, promises the best performance. 'How much more can the activity be improved?' If the goal-directed rational approach has been competently executed, the gain in performance as measured by output-input ratios will be as great as possible for the current stage of development.

In the above comparison of the tradition-based cultural approach with the goal-based rational approach, no mention needed to be made of whether the activity being restructured contributed to the building of the technical order, the economic order, or the cultural order. Once the level of desymbolization had become significant, the activity and its goal were treated in their own right and measured on their own terms. As a result, these three orders now furnished the resources for the building of one comprehensive new order dedicated to efficiency and power. Doing so by one activity at a time in isolation from all others meant that improvements had to come from two sources: its rationalization and its endless repetition. This latter source fully opened the door to machines and, more important, paved the way for information machines in all spheres of human life. One further obstacle still had to be overcome. Whatever remained beyond people's senses could not yet be rationalized. Overcoming this obstacle completely transformed the situation once again. It involved the separation of human knowing and doing from experience and culture, and the creation of a new intellectual division of labour not unlike the one that had previously transformed manual work. It sowed the seeds that would grow into the computer and information revolution.[10]

Technique within Culture

As the rational approach began to make use of discipline-based knowledge, it mutated into the technical approach, which separated itself from experience and culture. It began to undermine the cultural approach in every area of human life and society to build a new technical order out of the technological, economic, and cultural orders. This separation of knowing and doing from experience and culture created the phenomenon of technique. It led to the birth of our contemporary civilization. We shall examine this separation first in knowing and then in doing.

We have noted that the experiences of teenagers learning physics in high school differ from their daily-life experiences in several ways. The students cannot embody themselves in the foregrounds of these experiences since such foregrounds constitute the mathematical domain to which they can only relate by their mathematical imagination. Moreover, these foregrounds are logically ordered, while the background of their daily-life world is dialectically ordered. These discontinuities between the mathematical domain of physics and the daily-life world are extended to the metaconscious knowledge that they derive from the differentiation and integration of the corresponding experiences. Just as an expert automobile driver can almost instantly recognize a great many situations on the road and respond to them, so also a physicist can recognize a great many situations in the mathematical domain and immediately know how to go about solving the corresponding physics problem. However, the metaconscious knowledge of the expert driver that makes this possible is integral to the metaconscious knowledge built up from the dialectical ordering of the daily-life experiences, while the metaconscious knowledge of the physicist is limited to his or her 'experiences' of the mathematical domain. The former kind of metaconscious knowledge is embedded in experience and culture, while the latter kind is separated from them.[11]

We have noted that industrialization shattered all traditions, with the result that much of daily life became dominated by goal-directed behaviour and problem solving. Beyond daily life, the closest thing to a tradition was now found in the many disciplines during their cumulative periods between scientific revolutions. Our inability to create a unified science would suggest that the metaconscious knowledge accumulated by each discipline is separated not only from experience and culture but also from all the other disciplines. They do not jointly develop a common scientific 'base map' of human life in the world. There is no scientific tradition separated from experience and culture that replaces the role of tradition. Each discipline temporarily elaborates its own tradition and metaconscious knowledge, which is separated from everything else. The technological traditions were transformed in much the same way, as we shall see shortly.

The situation in which high school students or physicists find themselves is much more complex than that of a detached observer objectively examining the world. Owing to the organization of the brain-mind, scientific observers are both internally and externally related to the world, from which they cannot be separated. The assumption that, under certain conditions, we can behave as detached and objective observers must be critically evaluated because of the possibility of what Devereux calls 'counter-transference reactions,' in which observers are influenced by what they examine.[12] These influences may be negligible, but this cannot be asserted without a critical analysis. This has nothing in common with Heisenberg's claim that observers affect the subatomic phenomena they study.[13]

The processes of differentiation and integration by which the teenagers symbolize their experiences of learning physics in high school will almost certainly differentiate them from their daily-life experiences. If this is the case, the organization of their brain-minds will begin to constitute a kind of bifocal mental lens. One part is used for the mathematical domain of physics, and the other for their daily-life world. There is significant evidence to support such a hypothesis. The behaviour of physicists shows that, when they ride their bikes or play squash, they rely on their 'intuitive physics,' which has been built up with metaconscious knowledge embedded in experience and culture, while their professional activities rely exclusively on the distinct metaconscious knowledge acquired from the learning and practice of their disciplines. Riding their bicycles on the basis of these methods and approaches is impossible because other categories of phenomena dominate the physical one. The boundary conditions for solving the appropriate equations would be furnished by the conditions encountered and the responses decided on while riding the bike, so the equations could be solved only after the fact. Similarly, if college students were asked about the behaviour of objects in their daily-life world, they would rely on their intuitive physics, while if they were given physics problems, they would use their school physics.[14]

The splitting of the lives of the practitioners of science into different spheres that correspond to the organization of their brain-minds functioning as a kind of mental bifocal lens has not, however, produced new forms of mental illness akin to schizophrenia or bipolar malfunctions. The reason for this is that culture continues to play a weak but critically important role in the background. The new cultural unities that emerged in the industrially advanced societies around the middle of the twentieth century ensured the integration of the practice of science into individual and collective human life.

The widespread intuition that scientific disciplines and specialties function much like intellectual silos is relatively accurate in the sense that, when a specialist is intellectually busy in his or her silo, it is impossible to see the world, and vice versa. The role of culture in science is also confirmed by the inability of science to do anything that comes close to what cultures can accomplish: the integration of the categories of phenomena examined by all the separate disciplines into an overall interpretation of human life and the world. The current intellectual division of labour in science is incapable of dealing with the world in which all these categories of phenomena work together and contribute to our lives and the world. These limitations of modern science are far-reaching whenever our civilization applies highly specialized scientific knowledge. This will become evident when we examine how the same kind of intellectual division of labour developed outside of science.

This intellectual division of labour developed everywhere else in society. It began in the German industry in the beginning of the twentieth century, rapidly making it the leading industrial power – a position that it held for decades. Recall the limitation that industrial phenomena imposed on symbolization as a result of a great deal being inaccessible to the senses. Substituting a mathematical domain as a kind of intellectual proxy for the phenomena that escaped the senses turned out to be the winning strategy by which the new technical order was built. Some of the most common ones used were a continuum and a fluid. A continuum is a mathematical model of a solid material, whose disembodied properties are assumed to be uniformly distributed in the mathematical domain. A fluid is a similar model for liquids or gases. I shall examine the effect that the use of the continuum had on a society's knowledge of the strength of materials necessary to sustain its way of life.

The technological traditions within a craft had enfolded into them a great deal of metaconscious knowledge of the strength of beams and other load-bearing elements (a kind of intuitive stress analysis). For example, master shipbuilders regularly inspected many parts of the hull of a ship under construction. On the basis of prior experience they would decide whether a particular detail looked right for the job. The situation was similar to one in which an expert automobile driver responds to a particular situation and immediately knows what to do. Such an assessment was rooted in a dialectical ordering of a great many experiences of shipbuilding and the metaconscious knowledge built up with them. However, industrialization imposed severe limits on this approach. For example, the manufacture of steam boilers with ever higher steam pressures and temperatures, or the building of bridges with growing spans and load-carrying capacities, made it much more difficult to correlate what could be symbolized with what was actually happening with the structural elements. The solution that imposed itself was the substitution of a continuum for what was out of reach of the senses. In the mathematical domain of a continuum, calculus can be used to derive the equations governing the distribution of stresses resulting from a load, by summing the forces and the moments exerted on a finite element (a tiny cube of this continuum) that add up to zero. Before these equations can be integrated to provide a model of the stress distribution within the continuum, the particulars of a real-world situation must be considered. These include the shape of the continuum that corresponds to the shape of a real structural element, the loads that the structural element will transfer to its supports (which causes the stress distribution), and the elasticity of a real material to be assumed by the continuum (which determines how much deformation of this real material would result from the stresses imposed).

The above procedure involves a triple abstraction. First, nothing of the real world beyond the structural element is retained except for the load the world imposes on it (the input) and the desired optimal configuration of the structural element to be returned (the output). The load configuration is represented in a free body diagram showing disembodied abstract forces and moments acting on the structural element. Next, the real structural element is replaced with a continuum with the same shape. The mathematical model of the continuum has the sole purpose of analysing and optimizing the stress distribution resulting from the imposed forces and moments. All other phenomena are excluded from the model. For a structural element these may include the chemical phenomena that may rust a structural element made from steel; the biological phenomena that degrade a structural element made from wood; the phenomena that produce imperfections in any material, such as those that occur during the growth of a tree or the making of steel; the economic phenomena that affect the price of a structural element; and the sociocultural phenomena that affect the uses of the larger artefact of which the structural element is a part. Stress analysis replaces the structural elements we find in the world (where everything is related to everything else) with technical domains filled with continuums in which disembodied forces and moments operate. Such a domain thus approximates the real world in only one dimension corresponding to one category of phenomena. It contributes only marginally to a better understanding of the physical basis of these phenomena.

The third abstraction derives from the fact that this mathematical model serves the goal of designing the best structural element for the job, but only in terms of its load-bearing capacity. The practitioners of stress analysis are unable to relate their work to human life, society, and the biosphere since they have no knowledge of how their discipline relates to all others and thus how everything is related. What is 'best' must therefore be desymbolized to become a goal that is accessible to their own discipline. It makes stress analysis an end in itself, namely, determining the greatest possible load-bearing capacity that can be obtained from a set of specified inputs. What is 'best' can be expressed only in terms of contextless output-input ratios.

As a result of this triple abstraction the design of a structural element has been restricted to the context of stress analysis. Any other aspect must be taken care of by another discipline specializing in the single category of phenomena associated with this aspect. The complete design of the structural element potentially involves a number of disciplines to cover all relevant aspects. Design for strength (DFS) ensures adequate strength, design for manufacturability (DFM) ensures lowest production costs, design for quality (DFQ) deals with the reliability of the element's performance, design for environment (DFE) deals with the best performance for the lowest environmental burden, and so on. In other words, the overall design involves an open-ended set of disciplines or specialties followed by a synthesis, which must be arrived at by making trade-offs.

In taking a more critical look at the family of approaches of design for X (DFX), it becomes evident that, in the strict sense of the term, no real design is involved. First, each discipline or specialty filters out all phenomena except for one category, which it analyses and optimizes for the given structural element. Moreover, the boundary conditions and parameters can be derived only from a pre-existing design concept. In other words, what is missing in the above description is the very process of design that arrives at this concept.

Second, much of what would be involved in such a design concept is also missing. It should represent a structural element capable of making the best possible contribution to the artefact of which it is a part in terms of the ability of that artefact to contribute to and help sustain a way of life. In other words, the artefact would have to strike the best possible balance between the positive contributions and the inevitable negative ones because everything is related to everything else. What is required, therefore, is a tradition-based design that relies on symbolization for an understanding of how the structural element is integral to the human and natural orders. It is then no longer a question of the best performance of the structural element in terms of one category of phenomena. With the loss of a tradition, what engineering textbooks on design deal with is only a shadow of what constitutes real design. Much of the focus has shifted to discipline-based filtering, analysis, and optimization.

Real design operates on the level of experience and cultures since it seeks to create something to meet a human need or desire and to help sustain the way of life of a community. Design thus implies a participation in how everything is related to everything else in human life and society. Cultural values must be relied on to assess how well any design will fit, and performance ratios can then determine how well it 'internally' transforms scarce resources into the desired results. It is only to this latter component that technical disciplines and specialties can make their contribution. However, this cannot happen until the design itself has been sufficiently developed that it can be subdivided into distinct but interacting technical domains, in which one category of phenomena operates at a time to produce one of the many functions required to meet the need or desire. This approach is the cause and effect of technological products, processes, and systems being organized in terms of distinct technical domains whose functions are commensurate with the technical disciplines and specialties used for their analysis and optimization. It is the only way that full advantage can be taken of highly specialized technical knowledge.[15]

The confusing usages of the word *design* may be cleared up by introducing a distinction between design exemplars and analytical exemplars.[16] A design exemplar operates on the level of experience and culture in terms of the human needs and wants it is intended to satisfy, and how it will do so as an integral part of the way of life of a society. Coexisting alternative forms of a design exemplar reflect complex trade-offs, conflicting values, and the state of technical knowledge. An analytical exemplar operates on the level of a technical discipline or specialty, using its methods and approaches to filter out, analyse, and optimize one category of phenomena in a technical domain specified by a design exemplar. There is a close interdependence between a design exemplar and its many analytical exemplars. We have noted that these analytical exemplars derive their boundary conditions and parameters from the design exemplar. In the reverse direction, analytical exemplars can be used to test particular design details. For example, the process of analysis and optimization may lead to a modification of a boundary condition.

Owing to desymbolization and the loss of technological traditions, contemporary engineering schools have great difficulties in teaching design. Much of it has been reduced to accepting what is widely recognized as a state-of-the-art design exemplar and using the latest analytical exemplars to adapt and optimize it to suit the needs of a client and the conditions on which it must operate, including the constraints imposed by legal regulations. This state of engineering design enormously impedes our ability to face the challenges that lie before us. For example, when engineering students are asked to design a production facility that a client wishes to build to market a new product, they immediately assume a design exemplar (such as lean production) and begin to concentrate on working out the specific details using the analytical exemplars they have been taught. They analyse and optimize the details of the primary subsystems such as those dealing with assembly, materials handling, information processing, and the human organization. They are simply following what they have encountered in their field trips and engineering textbooks. Their colleagues in industry do no better.

We do well to recall that the Fordist-Taylorist production system was invented in the United States to respond to a unique set of conditions, roughly corresponding to a low-skilled work force made up of immigrants speaking many languages, and an abundance of natural resources. The Japanese invented lean production when they discovered that the conditions following the Second World War were incompatible with the Fordist-Taylorist production system. They had to develop a system that could be economic with much shorter production runs, and this turned out to have so many benefits that it rapidly displaced the Fordist-Taylorist design exemplar. The Swedes invented the Uddevalla Volvo plant design to respond to conditions of nearly full employment and a progressive legislation of equal pay for equal work, which made it very difficult for Volvo to keep good workers by offering wage incentives. None of these conditions exists today.

We live in a world that over-consumes nature and under-consumes people because of our inability to balance the productivity of labour with the productivity of resources. Moreover, globalization and free trade have allowed an unprecedented concentration of production and unlimited possibilities for corporations to relocate to nations where their costs can be externalized to the greatest extent possible, because communities have lost the right to impose labour, social, health, and environmental standards as a result of a dictatorial government or the policies of the World Trade Organization. If we had faced our responsibilities to the so-called Third World, future generations, and other life forms, there would have been a flurry of design activities to discover new design exemplars for production systems that meet out present needs. Why are we unable to identify the features of the above three design exemplars and those of less well-known innovative alternatives that are appropriate to our world, in order to attempt to integrate them into an entirely new design exemplar to help us respond more effectively to our present situation? For example, we might retain the assembly process of Uddevalla, the materials handling of lean production, the organizational design of Semco,[17] the design for ease of assembly perfected by Toyota, and so on.

It really would not be very difficult to compensate for desymbolization and the loss of technological traditions by recognizing what continues to go on in the background and including it in the engineering design textbooks. Culture continues to play an unrecognized role. This has far-reaching implications. It would make it impossible for engineering faculties to continue business as usual. The current intellectual and professional division of labour based on dealing with the world one category of phenomena at a time ought to be as impossible in engineering as it is in physics. The complementary studies component (comprising the social sciences and humanities taken by engineering students) for the greater part continues to examine the world one category of phenomena at a time. The kind of diagnosis being developed in this book could be turned into a prescription for reforming the university so that its structure would no longer necessitate the levels of desymbolization that have made it into a faithful servant of the new technical order at the expense of our common future.[18] I shall return to this problem in the next chapter.

The separation of human knowing and doing from experience and culture as described above in the case of stress analysis is paradigmatic. A continuum is but one example of a mathematical model used as an intellectual proxy for what happens in human life and the world. We have noted the widespread application of this approach in science. It has also permeated technology. Although perfected in science and technology, the separation of knowing and doing from experience and culture has spread to all areas of human life and society.[19] We have shown that, as the desymbolization of experience and culture advanced in any area, a growing reliance on rationality became necessary. In the earlier comparison between the tradition-based cultural approach and the goal-directed rational approach we noted the embryonic beginnings of knowing and doing separating themselves from experience and culture. The existence of a goal imposed a rationalization of the activity and its interactions with its immediate surroundings on their own terms, independent from everything else. Widespread desymbolization then paved the way for the technical approach.

The technical approach may be summed up as follows. It begins by studying an aspect of human life, society, or the biosphere for the purpose of making it better, or to reduce the gravity of a particular problem. As a consequence of desymbolization, the study is usually undertaken by the practitioners of one or more disciplines or specialties dealing with one category of phenomena. Second, the results of the study are compiled into some kind of model. Although the next step in the technical approach can be carried out in the most comprehensive manner if the model is mathematical and quantitative in character, 'good' results can still be obtained by formally or informally establishing some correlations among the parameters that have a significant effect on performance or on the severity of the problem to be reduced. Hence, the models can range in kind from those that are entirely separated from experience and culture to those that are partially separated from them. Third, the parameters of the model are varied, and the resulting forms of the model are correlated with performance or the severity of the problem. To do this systematically and comprehensively can be a daunting undertaking unless the parameters are logically related by a formal model, because of the availability of a wide range of mathematical techniques for doing so. Nevertheless, in the absence of such a model, the search for the best form is still possible. The ideal form of the model is the one that delivers the greatest performance or significantly reduces the problem under consideration. This form will have the highest output-input ratio(s), indicating that the greatest desired result has been obtained from the required inputs. Finally, the achievement of what has been demonstrated to be possible by the use of an intellectual proxy becomes the basis for reorganizing what was originally studied into the 'one best way' for achieving the desired result.

This technical approach can be made more complex by assessing multiple desired outputs obtained from diverse inputs. For example, banks and retailers are using such approaches to identify their best-performing branches and stores and to improve or shut down their poorly performing ones.[20] Governments use input-output economic models to test and optimize their policies.[21] Business models operate along the same lines.[22] Some are proposing to adopt these approaches in 'evidence-based medicine.'[23] The methods and approaches of operations research have found their way into many areas of application, even though the problems associated with it were recognized almost from the beginning.[24]

It would appear that the limits of the scientific approach to knowing and the technical approach to doing correlate with the extent to which a particular category of phenomena can be desymbolized. Take physics as an example. In daily-life activities physical phenomena are mingled with many others, and these frequently integrate them on their own terms. This observation may be used to open up a variety of problems. For example, the well-known statics problem of calculating the disembodied force required to push a roller over a small step has little significance unless it is embodied in a daily-life activity. Doing so will turn this problem into a design exercise in which students are asked to recommend an appropriate range of wheel diameters to permit a nurse to push a patient in a hospital bed across a threshold.

In the disciplines of the social sciences that examine human life and society, there are few if any situations in which one category of phenomena dwarfs all the others. Psychological, social, economic, political, legal, moral, religious, and aesthetic phenomena mingle together, and their relative influence varies considerably in the course of the history of a society or a civilization. For example, during a severe economic crisis a lack of trust and confidence in the whole system may temporarily trump almost any other influence. In traditional societies, such a lack of trust would have resulted from people symbolizing the situation and thus relating it to everything else that was happening. In contemporary mass societies, such a lack of trust is greatly influenced by public opinion, which results from integration propaganda complementing the role of culture. A critical appraisal of the contribution that a discipline-based social science can make to our understanding of human life in the world would probably come to the conclusion that any particular discipline can at best provide some understanding of the role of one category of phenomena in isolation from all the others. As a result, a discipline-based social science is useful for manipulating human life and the world one aspect at a time with little or no regard for all the others, but it is not very useful in helping us gain a comprehensive understanding of our lives in the world. The existence of parallel and competing schools in many social science disciplines reflects the multiple ways in which the single category of phenomena under study can be separated from all others and how the remaining connections can be accounted for.

On the other side of the university campus, the professions and applied sciences have organized themselves in the same way. Their intellectual and professional division of labour appears to be working extremely well because it is the cause and effect of the way in which products, processes, and systems are built up with distinct but interacting domains within which one category of phenomena delivers a desired sub-function. However, I have shown that when it comes to a design of any kind, a very different picture emerges. Analytical exemplars tend to dominate design exemplars, making contemporary technology inappropriate for its contexts and unsustainable by the biosphere. As a civilization, we are singularly unsuccessful in designing things for a more liveable and sustainable future, which depends on many categories of phenomena adapting and evolving in relation to all the others. I shall also show that the kind of economic growth all this is producing is almost certainly *uneconomic*. The situation would have been even worse if every discipline and specialty did not benefit from the continued support in the background of a culture, despite the fact that this culture is significantly desymbolized.

In suggesting that the technical approach acts as an intellectual proxy with respect to what falls beyond the reaches of our senses, I am not implying that this is analogous to extending human experience and, via it, a technological tradition. This approach substitutes a mathematical domain (such as a continuum or a fluid) for the insights that were previously gained by experience. An order of what is real is substituted for an order of what is true. In contrast, technological traditions included a variety of successful responses for meeting particular needs, which were widely adopted and incorporated into a working technology of a society. For example, the tradition of carriage making or shipbuilding included a variety of designs to suit the needs and requirements of a way of life according to the preferences of particular customers. Such designs were very different from what we would refer to as a design today in two respects. First, they were mostly in people's heads and were made only partially explicit in the form of drawings or specifications. Second, these designs included what experience had taught regarding the strengths of the materials, the aerodynamics of the sails, the drag of ship hulls, and much more. All this knowledge was embedded in experience and culture. Both these characteristics derived from these designs being, first and foremost, the result of dialectically ordering the experiences of building, and using and repairing the technological artefacts. Much of the knowledge embedded in these activities remained metaconscious. Contemporary designs are largely the result of the application of the appropriate analytical exemplars and, only to the extent that experience and culture are permitted to work in the background, of design exemplars. Today's designs are dominated by the drive for efficiency and power, which explains both our spectacular successes as well as our tragic failures.

The developments that I have described did not occur in the same sequence in all countries. For example, the French regarded industrialization as one aspect of rationalization and 'technicization.' The vocabulary of their language does not even contain a word corresponding to the English word *technology*. The French word *technique* is somewhat analogous to our usage of that word in English when we say that a person playing the piano has a good technique or that a person's technique in downhill skiing is exceptional. The French word *technologie* is the discourse or philosophy about technique, just as we distinguish sociology from what it studies, namely society. In translations from the French to the English, the word *technology* has often been used where the French uses *technique,* and this has caused numerous misunderstandings, especially of the work of Jacques Ellul. For him, there is no 'technological society,' even though one of the translations of his works bears this title. Literally translated, the title of the French work is 'Technique, the Wager of the Century.'[25]

Rationality, and later technique, did not first develop in industry.[26] It is true that, once industrialization had begun, these parallel developments had to be rapidly accelerated. The reasons that technique was frequently first perfected in industry and the economy stem from the close coupling of the related activities by the technology-based connectedness governed by the laws of thermodynamics. Any increase in the throughput of matter and energy in any activity immediately required adjustments of the activities with which it exchanged matter and energy, which in turn required further adjustments, and so on. The laws of thermodynamics constitute a far more rigid constraint on human life than does a cultural order.

Regardless of the path that industrialization took in different societies, the levels of desymbolization steadily rose, with far-reaching consequences. A discipline-based intellectual division of labour in social science became increasingly self-evident as the many categories of phenomena previously enfolded into human life and society now appeared as relatively distinct functions or structures that could be studied as if they were independent from each other. The more the role of culture working in the background of human life and society was weakened, the more autonomous these disciplines appeared to be. Another consequence was the desymbolization of the design of anything.

The reification and commoditization that accompanied industrialization transformed human life and society into a loose aggregate of distinct elements endowed with the capacity for limitless technical improvement. The situation was somewhat analogous to our individually and collectively discovering a jigsaw puzzle without ever having encountered such a puzzle before. As a result, the constraints that the knowledge of jigsaw puzzles imposes on the 'meaning' and the 'value' of each puzzle piece are entirely absent. Each piece must now be interpreted and dealt with on its own terms. Goal-directed rationality and efficiency-directed technique essentially treat any situation and any aspect of human life and society in the same way. These are no longer integral to human life and society, which opens them up to limitless technical improvement and turns means into ends.

All this has been legitimated and justified by a new opiate of the people. By being assured that reality is socially constructed, we are led to believe that it can be brought under our individual and collective political control and that by deconstructing it, we are able to reveal how this is being interfered with by others who corrupt this social construction for their own ends. The illusion of a socially constructed reality would be unthinkable were it not for the desymbolization of our cultures. If we could really deconstruct reality, we would descend into a complete relativism, nihilism, and anomie.

The Economy within the Technical Order

It is important to recognize that the distancing of an economy from the remainder of a society in the course of industrialization was a phenomenon limited to the nineteenth century and the first half of the twentieth century. The economy had become the locus of the technology-based connectedness that dominated the cultural order. The cultural approach was ill-suited to its organization, which essentially involved transforming inputs into desired outputs and matching these outputs to the inputs of subsequent activities. A rational economic approach gradually emerged, based on minimizing cost prices and ensuring that these remained well below market prices. As is now widely recognized, these prices are a poor indicator of the value of anything for human life and society as traditionally determined within a cultural order. Besides, before anything could be dealt with by the economic approach, it had to be separated from the cultural and natural orders by the process of commoditization, which greatly contributed to market externalities and the creation of negative Market forces.

Two developments led to a reunification of the economy with society around the middle of the twentieth century. Beyond the economy, the role of the cultural approach was steadily pushed back first by the rational approach and later by the technical approach. Within the economy, the growing use of the technical approach in industry required technical planning to apply the latest discipline-based knowledge to every aspect of the running of corporations. For this technical planning to be successful, it had to make itself as independent as possible from the Market. Beyond industry and the economy, a similar kind of technical planning imposed itself wherever discipline-based knowing and doing were used to advance any aspect of the way of life of society. The resulting convergence between the economy and society gradually led to their reunification within an all-encompassing technical order.

Owing to the reigning intellectual division of labour, the above kinds of developments largely escaped being noticed in the disciplines of the social sciences, especially in economics.[27] There were some notable exceptions, among which the work of John Kenneth Galbraith is particularly relevant.[28] In the West he was the first economist to recognize the far-reaching implications of the use of highly specialized scientific and technical knowledge in industry and the economy. Although he acknowledged that these developments were an integral part of the much larger phenomenon of rationality, he decided to restrict his attention to these two spheres, with the result that his analysis was quickly upstaged by this phenomenon and its broader influence on human life and society.

Since I have already examined Galbraith's analysis from the perspective of human knowing and doing separating themselves from experience and culture,[29] a brief outline must suffice. Any bureaucracy must undergo a major organizational transformation before it can effectively apply highly specialized knowledge to all aspects of its operations. It must acquire a technostructure as a 'collective brain,' in which people participate first and foremost on the basis of the relevance of their discipline or specialty to the decisions that must be taken.

Suppose an automobile manufacturer had carefully examined demographic, social, economic, urban, and environmental trends (in addition to market trends) and concluded that these pointed to the likely success of a new kind of vehicle such as the Ford Mustang, Chrysler's minivan, or a sport utility vehicle. Such design exemplars result from a technical imagination interpreting what is happening in the world in order to determine possible opportunities for new products. It is a question not of car buyers wanting such a vehicle but of imagining the kinds of vehicles they could be persuaded to buy as perhaps being a better fit given their preoccupations, anxieties, or lifestyle changes. Such embryonic design exemplars can then be refined through extensive consumer testing, and in the end the car manufacturer may decide to design, manufacture, and market it. Highly specialized knowledge must then be applied to each and every aspect of these operations.

Highly specialized knowledge cannot be applied to any design exemplar as a whole. It must be broken down into distinct but interacting domains in which one category of phenomena performs a sub-function that is essential for the overall operation of the vehicle. A tree-like structure emerges when the overall design exemplar is broken down into its subsystems, these subsystems into their sub-assemblies, these sub-assemblies into their components, and the components into their parts. This process implicitly defines all the different domains that can be analysed and optimized by the appropriate disciplines and specialties. It also points the way to involving the appropriate practitioners of these disciplines and specialties, who must collaborate in ways that reflect how the domains producing the various sub-functions interact to create a well-functioning car.

Applying highly specialized knowledge to the design of the vehicle thus requires a fluid organization chart that 'plugs' specialists into the process wherever their expertise is required. They cannot directly evaluate each other's contributions, with the result that their collaboration resembles the workings of many committees. Their members analyse and optimize the domains in which the sub-functions are produced, as well as negotiating complex trade-offs that must be made because the optimization of each and every domain does not guarantee the optimality of the larger design exemplar. Additional trade-offs may be required as the decisions of these committees are synthesized, when the design process that began with differentiating the design exemplar into the tree-like structure of constituent design exemplars is complemented by integrating all this work to move back down the tree to the original design exemplar. This overall process ensures that the new vehicle will function as well as possible, given the current state of knowledge. The workings of a technostructure as a collective brain involve processes that are explicit and separated from experience and culture, having a rational and technical character. A constant tension is thus maintained between the design exemplar (broken down into its many constituents according to the above tree structure) and the corresponding analytical exemplars used to improve all the individual sub-functions required for a well-functioning vehicle.

The application of highly specialized knowledge is not limited to the design of a new product. It is also applied to its manufacture, sales, and financing and the web of relations with suppliers, dealers, and customers. Owing to the many complex interdependencies, the application of highly specialized knowledge cannot be done in a sequential manner beginning with design, followed by manufacture, and then moving to sales and other details. All these interdependencies must be taken into account in the overall technical plan in order to apply highly specialized knowledge to each and every detail. For example, the technical optimality of the design must include trade-offs that take into account its manufacture and its ongoing ability to persuade potential buyers. It simply is not technically feasible to first work out an optimal design and then to make whatever adjustments and trade-offs are necessary to ensure an efficient manufacturing strategy. Nor is it possible to first jointly optimize the design and the manufacture of the new product, only to discover that in the meantime some of the trends that gave rise to the design exemplar have changed in ways that were not anticipated. Everything needs to be planned at the same time, including one or two relatively insignificant model changes that can be dealt with largely on the level of the analytical exemplars. All this involves a great deal of time and the mobilization of many of a company's resources.

As the plan to bring a new product to market advances, the ability of a corporation to adjust to changing conditions diminishes. It has allocated a great many resources to the design, which cannot be easily altered without allocating a great many more resources. According to the plan, the original allocation of these resources will be recovered from a large production volume deemed feasible by the sales forecasts. As a result, the application of the latest specialized knowledge is inseparable from mass production, and mass production is inseparable from the design and construction of a dedicated production facility. It makes no sense to buy general-purpose machinery and to adapt it to a particular aspect of production. In turn, mass production requires mass consumption. Planning for this mass consumption is likely to be successful only if the original trends indicating an opportunity for a new product have been correctly interpreted and if sufficient persuasion can be exercised.

To understand the far-reaching implications of these developments, it is important to recognize that the technical order, first created within the much smaller companies of the nineteenth century, was now complete and awaiting the computer and information revolution. The technical division of labour that had opened the door to mechanization, automation, and (later) computerization had been extended by the above intellectual division of labour, now reflected in a technostructure-like reorganization of the office that was designed to take advantage of the latest highly specialized knowledge. The technical division of labour of the 'hand' and the 'brain' of the corporation was both the cause and the effect of products, processes, systems, and strategies built up from technical domains in which one category of phenomena produced one desired sub-function at a time. The committee-like collaboration of specialists had to consider many trade-offs that included, but were not limited to, the design, manufacture, and marketing of the product.

The allocation of resources, including capital, must be as technically efficient as possible, which includes, but is not limited to, the renewal and accumulation of capital stocks. Capital, like all other resources, must be efficiently allocated, recovered, and increased by the application of the latest specialized knowledge according to the overall plan. This plan was now driven by a new factor of production, namely highly specialized knowledge, which allocated and optimized all the other factors, of which capital remained the most important. This capital would not disturb the overall technical plan unless a company had poorly managed its capital stocks and was required to borrow money (to which conditions were attached) or was unable to pay its shareholders a satisfactory return on their investment. It has been very difficult to make sense of all of this because the highly specialized knowledge used by many corporations is a 'public good' in the sense that the use of this knowledge by one corporation does not diminish its value for another. Hence, this knowledge cannot be priced according to economic theory. By default, most economists and society at large continue to see this new development as little more than a slight variation on profit maximization and an economic order ruled by the Market. It is not surprising that the economic policies of the industrially advanced world, beginning with the 1970s and 1980s, have been singularly ineffective.

Much of this can be traced back to the fundamental role that highly specialized knowledge began to play in the economy and in society at large.

It is also difficult to accept that these new corporations, characterized by the application of the latest specialized knowledge by means of technical planning, could continue to operate within the economic order of that time. The considerable time it took to develop and execute a technical plan for bringing a product to market could not be a direct and simple response to market demands. Most consumers had no idea of what was happening on the frontiers of the many disciplines and specialties that contributed to these technical plans, let alone how this could be translated into the satisfaction of their needs and wants. Even if consumers had been informed to enable them to freely shape market demands, there would have been a considerable delay before the producers could respond by developing and executing a plan, and by that time these demands would have been adapted to changing conditions. At this point, the technical plan for meeting these demands would have been too advanced to be sufficiently adaptable. It should come as no surprise that the use of the latest specialized knowledge coincided with even higher levels of mass production and thus with mass consumption, all of this taking place in a mass society.

Mass societies are characterized by a relatively high standard of living, which means that a great deal of their productive capacity has been directed towards non-essential 'needs.' All of these, besides the essential ones, are closely related to individual lifestyles and a way of life that are no longer shaped by a relatively stable tradition and culture. We have noted that the desymbolization of the cultures of these societies and the destruction of their traditions necessitated a supplement to the guidance of individual and collective human life. This supplement was provided by the mass media, which enveloped everyone in a bath of images that collectively portrayed an alternative that quickly became the primary guidance for human life and society. It reinforced the phenomena of public opinion and a statistical morality. In order for people to belong and move with the time, it informed them about what they should eat and drink, how they should dress and groom themselves, what they should possess and drive, how they should have a good time, what decisions they should leave to scientists, what they should ask their doctors, how they should plan for retirement, how they should make sense of what was happening in their world, and much more.

A lifestyle was thus portrayed to substitute for what had been lost with the break-up of a tradition. Advertisements connected this lifestyle to specific products without which it could not be enjoyed. These ads were remarkably free from information about the product. Their main purpose was to build a connection in the minds of consumers between the product and a particular symbol related to that lifestyle. Advertising companies developed a portfolio of techniques for doing so. They identified, tested, and incorporated the dominant symbols of the culture and bonded them to particular goods and services, thus endowing them with powers they did not possess. It may be argued that the consumption of material things was thus turned into participation in all the good things designated by these symbols, and these were in turn anchored in the new cultural unities of that time (as we shall see shortly). Many non-material needs could thus be satisfied by material things, which were a necessity in a society where so much depended on mass production, mass consumption, and the mass media. The problem was that the ongoing desymbolization of cultures also weakened these symbols, thereby creating the necessity for the forms of these advertisements to gradually change.

The enormous importance of the mass media in the latter half of the twentieth century depended on the fact that the culture of a society continues to work in the background in people's lives. Their reliance on symbols, and the reliance of these symbols on the cultural unities of these societies, makes this clear. The destruction of traditions and the desymbolization of cultures had made it impossible for people to continue to use the organizations of their brain-minds as mental maps by which they could orient their lives in the world. The way their culture now worked in the background was more like a radar, scanning what was happening in the world and how people responded to all of this in order to fit in and to belong. The desymbolization of cultures created unprecedented levels of social plasticity and a need for equally unprecedented guidance by a technical supplement to culture. I am not suggesting that advertising agencies and their clients were motivated by this interpretation of what was happening to human life and society. What I am suggesting is that, as they searched for more efficient ways of advertising their products, they gradually developed a portfolio of techniques whose incredible success cannot be explained without reference to the *milieu* created by mass societies with highly desymbolized cultures. It is difficult to accept the explanation that advertising informs consumers in order to enable them to make more rational choices. To confirm this, all anyone needs to do is to make a checklist of all the possible information an advertisement could include regarding a product, and score a significant sample of the advertisements of that time according to this checklist, to be convinced of the almost complete absence of any real information.

It was the sociocultural conditions of human life in a mass society that created the possibility for technical planning to succeed. After the careful examination of trends of all kinds and the determination from them of the needs that were likely to arise spontaneously or could be induced, a competently executed technical plan had a good chance of managing any gap between the planned output for a new product and the consumer demand for it by adjusting the latter through advertising. However, it was very difficult to adjust a production plan. It was possible to increase an output by means of overtime, but this frequently cut into maintenance and thus productivity as well as quality. It was possible to reduce an output by eliminating a shift, but this meant that fixed overhead costs could not be recovered from sales as planned. Neither option was technically feasible or economically viable except for a short time.

As the desymbolization of the cultures of mass societies continued, their ability to generate effective symbols diminished with it. It was for this reason that advertisements began to rely on technical substitutes, of which company logos are the most common form. These are branded into the minds of children and teenagers by advertisements that associate them with what one must have and with what one must do in order to participate in the greatest good that these societies can offer, thanks to their scientific and technical advances. People who resisted this pressure would have to fall back on their highly desymbolized culture to create some meaning, purpose, and direction for their lives, which is extremely difficult.

The new corporations could not rely on the economic order of that time for procuring all their required inputs. Technical planning, with its orientation towards performance, required new materials, parts, components, sub-assemblies, and production equipment that were specifically designed for the new products and thus were not available from any market. Even when this was not the case, the enormous size of the new corporations, compared to the size of the markets for their inputs, was so large as to make these markets almost incapable of delivering the required quantities at the price and time envisioned in the technical plan. Three kinds of solutions were developed to make the technical plan as independent of markets for inputs as possible. The first was vertical integration, based on the corporation buying the supplier in order to integrate its production into the overall plan. A second solution was to work closely with a preferred supplier according to a long-term agreement. Small suppliers could be bullied because their economic life or death was essentially controlled by the decisions of large corporations. In all these cases, the role of the Market was kept to an absolute minimum.

Although these new corporations could rely on the labour market for unskilled labour, this was not feasible for their supply of specialists. Once again, technical planning had to compensate. When a shortage of a certain kind of specialist caused salaries to increase, it took a long time for universities and governments to plan the expansion of a particular program. New faculty had to be hired, classroom and laboratory space expanded, followed by another four or more years for the additional graduates to enter the labour market. Finally, except in extraordinary circumstances, most of these new corporations relied on their own internal capital, making them as independent as possible from financial markets.

All this confirms what John Kenneth Galbraith argued many decades ago.[30] These new corporations created a planning system that made itself as independent as possible of the economic order of their day. The role of the Market continued to be significant only in the economic sectors not dominated by the new corporations. Galbraith's 'revised sequence,' by which he meant the ability of the planning system to shape and influence consumer demand in order to align it with the available technical potential and thus undermine economic democracy, was unthinkable without the desymbolization of the cultures of mass societies. Galbraith also pointed out the essential role played by Keynesian economic policies in all this. To ensure a steadily growing aggregate demand, without which technical planning could not be successful, a large public sector was created to offset the fluctuations in the private sector of the economy. To make this politically acceptable, the contraction of the public sector must not directly affect the daily lives of a great many people. The result was an enormous expansion of the industrial-military complex, along with space exploration and the use of nuclear power. A great deal of the technical frontier was thus shielded from economic forces because the defence of freedom, security from attacks launched from outer space, and energy security were deemed worthy of any sacrifice, even in the face of warnings that an all-out use of our weapons systems could not defend anyone but only destroy everyone. However, it was gradually becoming evident that the energy policies of the industrially advanced nations were entirely unsustainable. Nuclear power, apart from being uneconomic, was no solution at all, given the insoluble problem of radioactive waste. Keynesian economic policies had thus been co-opted for agendas other than the public good. Given the conditions of the time, this can readily be understood, but it exacted a heavy price. These policies could have been directed towards preparing for a post-fossil-fuel era, creating more liveable and sustainable cities, rebuilding the educational system, and much more. Perhaps the most lasting legacy of those decades was the unshackling of the technical frontier from the economic order.

The expanding technical order based on the planning system plays a very different economic role in human life and society than did the economic order that it has steadily pushed back. There is growing evidence to suggest that the technical order acts as a wealth extractor, while earlier economic orders had generally been wealth creators. This economic transformation follows directly from what we have learned about the technical order: the evolution of contemporary ways of life is decisively affected by the decisions of countless specialists according to an intellectual and professional division of labour based on the manipulation of one category of phenomena at a time. These specialists belong to groups responsible for advancing and applying a body of knowledge. Jointly, these bodies of knowledge constitute a knowledge infrastructure that supports the many decisions that help evolve contemporary societies.

As noted in the introduction, most of the consequences of the decisions made by the specialists participating in our present intellectual and professional division of labour fall beyond their domains of expertise, where they cannot see them. They are suspended in the triple abstraction imposed by the technical approach to improving the performance of everything. As a result, the undesired or illegal consequences of their decisions must be dealt with by other specialists in whose domains they fall. Consequently, the 'system' institutionalizes an end-of-pipe approach to the problems it creates. Instead of getting to the root of any problem, it adds compensating techniques or services. Our present knowledge infrastructures appear to be producing undesired results at a greater rate than desired ones.[31]

There is growing evidence to suggest that the corporations using these knowledge infrastructures have become wealth extractors as opposed to wealth producers. We have been bombarded with claims to the contrary: the jobs and exports they provide, the wages and dividends they pay, the community events they sponsor, and so on. Rarely is the other side of the coin investigated, namely, what it costs our communities to have these corporations in our midst. Estes has estimated that the social costs in the United States in 1993 amounted to nearly twice the entire federal budget and nearly half the nation's gross domestic product (GDP).[32] Korten has estimated that these costs represented roughly five times the profits.[33] In other words, with accounting procedures that make sense, most or all of these corporations should have gone bankrupt a long time ago. Since the social costs imposed on the United States far outweighed the costs, these corporations have become wealth-extracting entities instead of wealth-creating ones. What this means for the global economy becomes apparent when we recognize that more than half of the one hundred largest economies in the world are transnational corporations, although there is some dispute about this because of the difficulties of calculating internal transfers.[34] Estes attributes this situation to the 'tyranny of the bottom line.' Undoubtedly, if executives and managers had more honest 'report cards' showing how their decisions affected all the corporation's stakeholders, better decisions would be made, but this is exactly what the technical approach and the current intellectual and professional division of labour make next to impossible.

As to the participation of engineers in these corporations, there is additional evidence to suggest how these problems play out in detail. A comprehensive study of a typical undergraduate engineering curriculum for North America quantitatively demonstrated the problems with the engineering practice.[35] It showed that future engineers learn very little about how technology influences human life, society, and the biosphere. They learn virtually nothing about how this understanding could be used to adjust their design and decision making to achieve the desired goals and simultaneously prevent or greatly minimize harmful effects. The same results were found when the latest methods and approaches reported in faculty research publications were investigated.[36] A study of university education revealed that these results are typical, given the current technical and professional division of labour.[37]

Daly and Cobb have shown that one of our most talked-about economic indicators, the GDP, makes no sense in relation to human life, society, and the biosphere.[38] For example, if we all did more destructive things such as having car accidents, drinking to excess, and getting divorced, the GDP would rise. In contrast, if we all started to eat better, do healthier work, walk to more places, have more face-to-face relationships, and so on, the GDP would go down. Similarly, this indicator does not distinguish between economies that produce goods and services in a sustainable relation with the biosphere and those that do so by depleting their so-called natural capital. The picture of the economy portrayed by politicians and economists by invoking statistics such as the GDP would change completely if the costs incurred in the production of wealth were subtracted from gross wealth to reveal net wealth. Doing so would provide us with a measure that makes sense.

As an alternative indicator, Daly and Cobb developed the Index of Sustainable Economic Welfare (ISEW). On a per capita basis, it showed an increase of (on average) 0.21 per cent from 1951 to 1970, 1.57 per cent during the 1960s, 0.21 per cent during the 1970s and –0.34 per cent for the 1980s.[39] Other such indexes showed even more disturbing results.[40] The General Progress Indicator (GPI) is a little closer to the way in which the economy has affected most Americans.[41] It shows an upward curve for two decades, followed by a decline beginning around 1970, to a total drop of 45 per cent.

These results converge with those suggesting significant changes in the way that technology affects human life, society, and the biosphere as a consequence of the use of the technical approach. We are now producing pollutants (products we produce but cannot sell) at a far greater rate than we are producing desired goods and services. A study from the American Academy of Engineering estimates that, of what we extract from the biosphere, 93 per cent is turned into undesired products (pollutants) and only 7 per cent into goods and services.[42] Our materials and production systems may well turn out to be among the most uneconomic and environmentally destructive ones ever created by humanity. Some time ago, Blue Cross was the largest supplier to the largest corporation in the world.[43] Apparently, physically and mentally ill workers were the company's most valuable undesired output. To deal with these and other health problems, we have expanded our disease care system. Rapidly growing health-care budgets would suggest that the rate at which contemporary ways of life produce disease outstrips their ability to deal with it. During the decades following the Second World War, population growth and rising standards of living contributed less to the environmental damage done by the U.S. economy than did the effects of technology measured in terms of burdens imposed per unit of throughput of matter and energy. Technologies were progressively becoming 'dirtier' as a consequence of the growing use of the technical approach.[44] It is easy to multiply these kinds of examples, but the deep structural economic, social, and environmental crises are obvious. The expanding new technical order leads to a decreasing ratio of desired effects to undesired effects because it trades performance for context compatibility.

Closely related to the way in which current knowledge infrastructures shift the balance between desired and undesired consequences of technical and economic growth is their blockage of the road to genuine solutions. As noted in the introduction, unless such a solution to a set of difficulties can be achieved by optimizing the transformation of the inputs into the desired outputs, the practitioners will be unable to succeed. They are trapped within the triple abstraction within which their technical approaches operate. The technical approach is relatively well suited to developing supply-side solutions to some problems, but it cannot deal with their demand-side counterparts. For example, it can increase the capacity of transportation systems, but it cannot deal with reducing the demand for mobility. Comprehensive solutions require that both be considered. This will necessitate a very different intellectual and professional division of labour. Hence, the current division of labour and the knowledge infrastructures built up with it together prevent genuine solutions from emerging when these require non-cumulative development. We have created a system that feeds on its own problems, by adding compensatory technologies and services that make it uneconomic, socially non-viable, and environmentally unsustainable.

We have argued that, prior to the nineteenth century and the onset of industrialization, economies were largely integral to the cultural orders of societies. We have now shown that the relationship between the economy and society has essentially returned to what it was prior to industrialization, except that economic activities are immersed in a technical order instead of a cultural order. This economic transformation has given rise to my formulation of an anti-economy hypothesis, arguing that we have moved from wealth creation to wealth extraction.[45] As an integral part of the technical order, the economy now significantly contributes to the consequences of a way of life advanced by a discipline-based knowledge structure.

Further evidence of this radical economic change comes from the transformation of the Market as the 'great invisible hand' creating economic order. To understand its current role, it is useful to compare daily currency turnover in international financial markets with the currency required for financing world trade and for direct investment. This latter component represents less than 3 per cent of the former.[46] Even when we deduct a significant percentage to account for the laundering of money derived from the trade of illegal goods and services (such as drugs, arms, sex, and racketeering), there remains a major portion (which could be as high as 70 per cent) that corresponds essentially to speculation. The financial sector's primary activity is the growing of a global speculative bubble. Few people appreciate its size and the inevitability of a global economic meltdown unless we make fundamental changes. The Market presides over the trading of non-productive speculation in the form of a multitude of financial instruments that make money from money without any intervening economic activities.

When people buy shares in corporations, what is the economic significance of these shares, and what do they really own? The money does not flow to the corporations in which they are investing in order to be added to their operating capital, and therefore does not result in the direct or indirect expansion of the company's productive capacity and its ability to produce goods and services. The funds invested in the purchased shares flow to the previous owners of these shares, and these shares are bought in the confidence that one day someone else will be willing to offer even more money for them. As such, these investments contribute to a speculative bubble that has only a tenuous link to economic realities. All investment schemes, including mutual funds, attract more and more money to bid up the prices of shares well beyond what they represent, instead of reflecting the abilities of corporations to deliver goods and services. It becomes simply a matter of supply and demand, until an 'adjustment' occurs because people begin to recognize the lack of reality in the stock markets. All this puts enormous pressures on chief executive officers and national governments to increase sales and exports, which results in even further wealth extraction.

Few people have the ability to work out the distinction between productive investments that increase genuine wealth and the extractive investments that decrease it. The situation has become so absurd that, in some instances, corporations could make higher profits by participating in the speculative bubble than by producing the goods and services that sustain our ways of life. Now that a complete meltdown has been narrowly avoided, governments have had to borrow enormous sums, thus compelling us all to subsidize the speculative bubble at the expense of real wealth creation. As noted, the financial services 'industry' has developed a great many techniques for making money from money without any intervening productive activities. Outside the planning system the Market presides over non-productive services, and this is far from limited to the financial sector. We see this with energy markets, for example. The role of the Market has been harnessed to wealth extraction. Junk bonds, hedge funds, sub-prime mortgages, and much else (even mutual funds) help drive an ongoing search for techniques that can create the 'virtual wealth' so characteristic of the technical order. This order has desymbolized money and capital by disconnecting them from what makes sense.

Many thoughtful people are expressing serious reservations about where our current economic trends are taking us. Many equally thoughtful people argue that we are doing the best we can and that to expect much more would be unrealistic. To be thoughtful with respect to what is real yields very different conclusions from being thoughtful with respect to what is true. In this case, the order of what is real is built up from the gathering of as many economic statistics as possible, compiling them into models, and using these models to propose and test various economic strategies in order to select the 'best' one. In contrast, the order of what is true results from symbolizing our daily-life experiences as well as those of others, especially insofar as they relate to earning a living. The relationship between what is real and what is true is not very obvious. The one is not the objective side, and the other the subjective side, of the same coin. The one does not underlie the other, nor is it a question of which approach enjoys our confidence. It is a question of critically examining the strengths and weaknesses of each one and, since these are diametrically opposite, of using them synergistically in order to arrive at a more comprehensive understanding and response. We shall show later on how this may be accomplished.

The more the economic orders first escaped the cultural orders of the industrializing societies, and the more they subsequently became absorbed into the global technical order, the less they were dominated by what made sense symbolically. This development opened up a possible course of human action that goes all the way back to the British physiocrats in the nineteenth century. They argued that economic regularities were like all other physical laws, hence nothing could be done to control the economy. Today, some speak of natural levels of unemployment and natural markets. For the sake of argument, let us suppose that these are indeed natural phenomena. The problem is that nature has never commanded us to obey it. As a result, symbolizing economic phenomena as being natural is to make a choice, for which responsibility must be accepted. However, if we have symbolized economic regularities and institutions as human creations endowed with meaning and value, we must also accept full responsibility. Current economic disagreements are not about what is real but about what is true for our lives and communities and what we must do about it. We must resymbolize what we have desymbolized.

When we resymbolize our economic situation, it becomes immediately evident that more growth is not the solution to our woes. The pursuit of growth reveals how much a society depends on its economy and how this economy depends on the biosphere. Moreover, current growth is primarily achieved by improving the productivity of labour and, to a much lesser extent, the productivity of materials and energy. As a result, a significant portion of this growth is required to undo the unemployment and underemployment resulting from increases in labour productivity. Even when this growth is sufficient to create new jobs, most of these are of the kind on which people cannot live, and they are provided with little or no social security and pensions for their old age. The present generation is still able to protect its children, but these children are likely to have much more difficulty protecting theirs, and so on. It is the consequence of economies having gone into reverse by extracting rather than producing wealth.

The present kind of growth will intensify the competition for increasingly scarce resources such as fossil fuels. It will also make it more and more difficult to deal with a variety of environmental issues, as we have seen and continue to see in the case of global warming. Moreover, making the planet safe for the technical order by virtually eliminating our ability to impose constraints in terms of labour, social, health, and environmental standards is adding to our difficulties.

As the technical order advanced in the economy and in society, it soon became bogged down in technical information. This should have been expected because the advancement of any aspect of a way of life by means of the technical approach generates a great deal of information as it is studied, modelled, optimized, and reorganized. In other words, the so-called information and computer revolution was one of the symptoms of the gradual shift from a reliance on the cultural approach to a primary reliance on the technical approach. Consequently, the role of the computer changed dramatically. Until then, it had been thought that its applications would be very limited. This led to the so-called technology paradox, because the investments in computer and information technology did not lead to dramatic improvements in labour productivity.[47] It was soon realized that computers could not simply be absorbed into the existing organizations and institutions in which each department or division used them as tools for its work. All processes had to be re-engineered in the image of this machine and incorporated in one large integrated database. In effect, these organizations became a two-dimensional 'intellectual assembly line' based on a technical division of labour, now extended to intellectual work. Each employee received the required information from the integrated database and manipulated it according to prescribed procedures, and the system sent it to all the appropriate places. The suppliers of these organizations were encouraged to undertake the same kind of enterprise integration or resource planning with a seamless interface between the two organizations. In this way, the technique-based connectedness of these organizations added flows of information to ensure the efficient transformation of all other resources. Organizations of any kind had to adapt in order to survive in a technique-dominated world. Beyond these corporations and organizations, the Internet completed the 're-engineering' of society.

People cannot participate as subjects in the technique-based connectedness of human life and society. A person must participate as a *knower* who transforms something on the basis of technique, a *doer* who executes the one best way, or a *controller* who mediates between the knower and doer through supervisory and managerial activities. A knower is thus a spectator to the execution and management of the one best way, the doer is a spectator to the determination and management of the one best way, and the controller is a spectator to the determination and execution of the one best way. Taylorism separated the hand from the brain in productive work, but technique goes much further by also Taylorizing the brain and doing so for almost every human activity. A way of life that has evolved on the basis of experience and culture thus is all but replaced by a technique-based connectedness that is elaborated by knowledge infrastructures separated from experience and culture. Although people participate as knowers, doers, or controllers through their work in the technique-based connectedness of their society, in the rest of their lives they are on the receiving end of a great many techniques to ensure their efficient participation in this technique-based connectedness. The resulting desymbolization of human life and society is almost certainly unprecedented in human history.

Possessed by New Myths

Our awareness of our own situation is enormously muted by our culture continuing to work in the background. Babies and children still make sense of and live in the world by learning to dialectically order their experiences, even in the face of the desymbolizing influences all around them. Adults continue to live their lives because the symbolization of these lives frames and enfolds each moment through the organization of their brain-minds. The quality of human relations continues to depend on a dialectical tension between individual diversity and cultural unity. Groups formed through marriage, friendship, or mutual interest continue to depend on the same dialectical tension. A high level of skill in any human activity continues to depend on metaconscious knowledge, either embedded in or separated from experience and culture. The threat of the unknown and the risk of relativism, nihilism, and anomie continue to be kept at bay by a cultural unity.

There is no question, therefore, that human life and society continue to depend on a culture, even when it is highly desymbolized. In such a culture, babies' and children's lives are inundated by the desymbolizing effects of living with television, the computer, and the Internet. These and other techniques impair the ability of their metaconscious processes of differentiation and integration to symbolically relate everything to everything else, which negatively affects the 'resolution' of the meaning and value of everything. As a result the resolution of metaconscious knowledge is also affected, especially when it becomes torn between being imbedded in experience and culture and being separated from them. These symbolic 'deficits' allow for a deeper and more effective penetration of advertisements in particular and integration propaganda in general. As noted, integration propaganda is the bath of visual images collectively portraying the lifestyle of a mass society as diffused by the media and the Internet. Gradually, a complementary relationship becomes established (through the organizations of their brain-minds) between a highly desymbolized cultural ordering of their lives and the integration propaganda that seeks to make up the deficit. All this reinforces the technique-based connectedness of human life and society. It is built up by the technical approach taking hold of all aspects of human life, society, and the biosphere to transform them into resources for efficient processes; these in turn help sustain the technique-based connectedness without any reference to sense, and thus with detrimental effects on human life, society, and the biosphere.

These developments raise the question posed by John Kenneth Galbraith of the planning system: have we ended up serving the system that we created to serve us?[48] It had been raised much earlier and much more comprehensively by Max Weber: is humanity shutting itself into an iron cage?[49] For Jacques Ellul, it concerned the autonomy of technique: is the influence that technique has on human life and society more decisive than the influence that people can exercise over technique during this particular time of our historical journey?[50] It should be noted that, almost without exception, the philosophers of technology have failed to understand the difference between technology and technique, and the fact that Ellul was writing as a sociologist and a historian and not as a metaphysician of technique. This also poses another question: what is the sense of an economy that extracts rather than produces wealth, thereby threatening everything on which it depends?[51] These developments raise still more questions: how much further can the development of an order of non-sense proceed without relativism, nihilism, and anomie endangering the lives of the next generation? How much further can it push all other life forms and the biosphere itself? We have plunged ourselves into a deep economic, social, and environmental crisis resulting from the incompatibility between the order of life and the order of non-life.

From our description thus far, it is clear that the cultural orders that humanity creates can exist without the presence of a technical order, but that a technical order cannot exist without cultural orders. As long as humanity remains a symbolic species, the relationship between the technical order and the cultural orders is parasitical; if the cultural orders collapse, the technical order will vanish with them. In the meantime, the parasitical relationship is symbiotic in some ways and destructive of the host in others. For example, a highly desymbolized culture is made liveable by the technical means of integration propaganda, and the remaining lack of sense is taken care of by technical means such as antidepressant drugs. This raises the question of how far desymbolization can proceed and still hold relativism, nihilism, and anomie at bay. In other words, how can we make sense of our lives within an order of non-sense? How can we make sense of our lives continuing despite high levels of reification?

Lewis Mumford has argued that we live in a megamachine society.[52] By this he means that contemporary ways of life and institutions have become so mechanistic in character that we have all but become cogs in the machine. This was humorously portrayed by Charlie Chaplin in the movie *Modern Times*. The impossibility of an android becoming human has been equally humorously portrayed in the behaviour of the robot Data in *Star Trek*, when the machine is endlessly perfecting its rules in the hope of dating a woman. It may be tempting to imagine that we are part machine and part emotion, in the way Sherry Turkle discovered that MIT students differentiated themselves from computers.[53]

None of these ways of making sense of contemporary human life is satisfactory. Let us go back to our analysis of the reorganization of human work in the image of the machine by means of a technical division of labour. If the workers had actually become cogs in the production mechanism, the history of human work during the last two centuries would be completely incomprehensible. All the difficulties of coping with monotony and the many attempts to mitigate the consequences by means of job rotation, job enlargement, ergonomics, human factors, industrial relations, and the quality of working life approaches would not have been necessary. What characterizes this kind of work is that people's 'hands' are integrated into the production mechanism, but the rest of their lives remains outside, from where it causes disturbances. In order to understand these kinds of human activities, we need to carefully examine the consequences of the tension that is established between those aspects that have been reified and commoditized and those that have not. This is the ultimate bluff of technique: it cannot make life more rational and efficient without disordering it. Life cannot be recreated in terms of non-life. Nor can technique recreate the dialectical tension between what sustains life and what destroys it. What it is able to accomplish may be understood in terms of our changing technique, but technique simultaneously changing us. We become so accustomed to the commoditization, reification, and alienation of human life that we accept it as normal and give up the struggle of taking responsibility for the mistakes we have made and doing something about them. Instead we look elsewhere for our delivery from what afflicts us.

Once again, we are looking for the kind of cultural unity that could possibly bring it all together. Hence, we are looking for those entities in human experience that are lived as if they were good in themselves, had no limits as to what they could accomplish, or both. Many 'clusters' of such experiences are illustrative. They are related to how we live with science, technique, and the nation state.[54]

With regard to *knowing*, we live as if science had no limits. We find it very difficult to come up with things that science may never know, and thus metaconsciously treat it as if it were omnipotent. Without such a metaconscious myth, this would be rather astonishing. After all, Western science has given up studying human life in the world because it cannot reduce this unintelligible complexity to an intelligible one by means of detached observers who are able to control the knowledge processes in an objective manner. It has parcelled out this task to relatively distinct and autonomous disciplines, each perfecting the methods and approaches for examining one category of phenomena. This form of science has abandoned any attempt to understand anything as the interplay of many different phenomena without any one of these being dominant. The efficiency of its intellectual division of labour rests on greatly simplifying things by taking whatever is to be studied out of its real-world context in order to place it in the much more manageable intellectual context of a discipline and the physical context of its laboratories. The resulting exponential growth of scientific knowledge has been undermined by the exponential growth of ignorance of how everything is related to everything else, since this knowledge has been devalued as being subjective and cultural. Of course, there are situations in which the influence of one category of phenomena far outstrips the influences of all others, but surely these are the exception rather than the rule. In most situations the gathering of scientific knowledge one category at a time represents a desymbolization that is so extreme as to separate it from experience and culture. Having thus liberated itself from any dependence on the specialized knowledge of all other phenomena as well as from human knowledge resulting from symbolization, scientific knowledge has an existence in relation to human life that can be justified only if this highly specialized knowledge is good in itself and without limits. If we examine the situation critically, it becomes evident that there may be a potential complementarity between scientific knowledge and culture-based knowledge, provided that the former is symbolized by the latter.

Discipline-based science tolerates no criticism or suggestion that it has its unique limitations, just like all other forms of human knowledge. The most common defence of its status as a metaconscious myth involves accusations of science bashing. These are founded on a confusion between critically assessing what discipline-based science can do for us and rejecting out of hand the possibility of any limitations. It is the secular equivalent of a religious response that is as old as humanity itself. Confusing a critical evaluation of science with 'bashing' it is the equivalent of believing that the thermostat 'bashes' the furnace it controls. If the goal of science is to critically understand ourselves and the world, the responsible course of action is to constantly evaluate any possible gap between this goal and what science accomplishes, in order to reorganize it to minimize the gap. In the absence of such attempts, we can only conclude that we live as if science were a myth: all-powerful and without limits for human knowing.

With regard to *doing*, our society believes that we have been delivered from the great catastrophes of the first half of the twentieth century by our technical powers. We are mesmerized by the technical virtuosity achieved through the relentless pursuit of efficiency. Our preoccupation with performance has made us and our world who and what we are. Our lives and our world are unimaginable without technique. Could we sustain the current global population without it? Can we think of any phenomenon that is more essential and decisive for human life in the present age? How would we deal with anything other than by the technical approach? Can we make a list of the issues faced by contemporary human life and society that should not be addressed by technique? These questions and our answers to them point to our metaconscious symbolizing of technique as the greatest good we know. There appears to be no limits to its powers Technique has become our only tool, because all of life is now seen in terms of technical problems that can be solved by one technique or another.

Jacques Ellul summarized the situation of human life in the second half of the twentieth century rather well in the title of the book he wrote on the subject: *Technique, the Wager of the Century*.[55] Of course, we know that many techniques are far from perfect and that some entail serious risks. However, we live as if this could be overcome by subsequent technical development. What the technical order claims is that we are able to build it by disordering the cultural and natural orders, one category of phenomena at a time. Once again, this is impossible unless each and every body of specialized knowledge is sufficient unto itself and thus without limits. No technical specialty evolves by examining how its methods and approaches affect human life, society, and the biosphere and how this understanding can be used to adjust decision making to achieve the desired results but at the same time prevent or greatly minimize undesired effects. The complete lack of any such negative feedback can be explained only if we live as if technique did not need this because it is without limits and thus all-powerful. It is our sacred, our central myth. It is the bluff of any sacred: serve it and you will live.

We live as if technique as an approach to improving human life and the world were superior to the approach that humanity has used until now, namely, the symbolization of experience by means of a culture. As we have said, doing so by one discipline (or technical specialty) and one category of phenomena at a time is extremely effective for the analysis and optimization of anything belonging to a mechanistic order or an order of what is real that can be broken down into distinct and interacting domains within which one category of phenomena makes a unique contribution to the whole. In this way, the performance of an athlete is improved by separately optimizing the physical movements, equipment, training and performance schedules, nutrition, psychological make-up, physiotherapy and medical treatments, public relations and media exposure, sponsorship opportunities, and other income opportunities. All this would make sense if it coincided with a greater quality of life for the athlete, better relations with others, and a meaningful contribution to the life of a community. Unfortunately, things are not that simple, as the lives of many athletes demonstrate. In these and other situations we live as if life could be understood and reorganized as non-life, constituted of separate domains, each populated by a single category of phenomena.

Even for the world of technical products, processes, and systems, performance comes at the expense of undermining the fabric of relations from which all life is woven. In contrast, cultures have never attempted to improve human life and the world by imposing goals on them and instead have concentrated on the living of good and responsible lives in that world, which is different altogether. The latter is guided by values, which constantly assess how well everything in human life is related to everything else and how this dialectically enfolded ecology of relations is integral to, and contributes to, a constant adaptation and evolution of everything. It is all about living a life integral to the life of a community and relating to all other life and ecosystems in the biosphere. The effectiveness of any contribution cannot be reduced to mere performance, and that effectiveness itself is but one of many values that guide the living of individual and collective human life.

Is there any place for science and technique in all of this? There is, provided that it leads to a civilization that includes them but is not dominated by them. In such a civilization, science would be guided by helping its members gain the best possible critical understanding of themselves and the world, and technique would attempt to make the best possible use of resources under the guidance of symbolization and culture. Doing so would prevent science and technique from being metaconsciously symbolized as myths, with the result that, as with any other human creation, a community would be clear about which tasks could be entrusted to them and which ones should be dealt with by other means. This would involve the symbolization of science and technique in accordance with the meaning and value they have for human life and society.

Instead, science and technique have guided the human journey in a direction that is characterized by spectacular improvements in the performance of everything without ensuring that it genuinely benefits human life, society, and the biosphere. All this is not as abstract as it sounds. Has human life not become a matter of developing what can be technically commoditized and leaving undeveloped what cannot be commoditized?[56] To put it differently, what is being developed is whatever can enhance the technique-based connectedness, and what remains undeveloped is whatever cannot be incorporated into it. What is regulated by free trade is not the well-being of human life and society; what is being regulated is our ability to interfere with technique-based connectedness in terms of cultural values and the protective standards based on them. Economists tell us that we need to give up what is most essential for our lives: a stable job that allows for the support of a family, and a community that complements our efforts of nurturing our children. Instead, when a person's job can be done much more cheaply on the other side of the world because many more costs can be externalized for lack of adequate standards, it is his or her duty to give it up for the sake of technique-based connectedness. Similarly, if a community sees its workplaces disappear for the same reasons, it must somehow convince itself that in the long run everything will work out for the best. Instead of this kind of ideology, what we really need is fair trade that makes sense for people in communities. Our trade-related problems are also rooted in technique and are part of the project of recreating human life in the image of non-life and transforming sense into non-sense. We must seriously consider whether technique does not amount to a suicide mission.

We now come to politically organizing our affairs. We live as though everything had become political. To suggest otherwise is to risk being written off as naive. Living as if this were true is to metaconsciously accept that all our problems have a political solution, that these solutions must come from the state, and that the state's power is without limits. This omnipotent state has all but taken over the guidance that human cultures used to provide before desymbolization. Of course, we know that today's government leaves a lot to be desired because it is not smartly managing the technical means at its disposal. The resulting political setbacks will hopefully come to an end with the next election. The people will have learned their lesson, and a better tomorrow lies just around the corner. Since the state exclusively relies on the power of technique, the political spectrum has been immensely narrowed. Everyone has become convinced of the intrinsic goodness of technique, which we mistakenly discuss in economic terms. As a result, everyone's political priority is to use the latest scientific and technical knowledge to stimulate economic growth in order to produce as much wealth as possible. We can then pay the bills and distribute the rest to education, health care, social security, and environmental clean-up. It has led to an astonishing political reversal, which amounts to suspending our aspirations and putting our political, moral, and religious convictions on the back-burner until the time comes to distribute the wealth created by technique. The nation state thus politically arranges everything in relation to technique as the undisputed key to its prosperity. Of course, the same holds true for the 'state nation' formed out of several nations by conquest or agreement.[57]

There is an obvious synergy between living as if technique were without limits with respect to doing and living as if the nation state were without limits with respect to politics. As noted, much of the frontier of technique must be advanced in the public sector of the economy. In this way, the nation state can mobilize all the necessary resources to sustain this frontier in the name of the ultimate values on which it is founded. During the cold war the former Soviet Union and the Western nation states could point at each other as the ultimate threat to their freedom and security and thus justify an all-out mobilization of their resources. As a result, much of the frontier of technique slipped into the industrial-military complexes. Without this relocation many technical developments would have never seen the light of day. When the Soviet Union collapsed and there was no longer any immediate threat, the technical growth within the industrial-military complexes continued unabated and at times even grew. Of course, there continued to be other threats, but were these in any way commensurate with the scale and orientation of the industrial-military complexes? Why were some of these threats prioritized over others? We must not lose track of the relationship between means and ends. To what extent do the possibilities opened up by new technical means cast various threats in a different light? It would appear that the kinds of security threats currently being prioritized over many others correspond rather well with the frontier of microelectronics, robotics, nanotechnology, and biotechnology.

It is impossible to discuss the industrial-military complex without discussing the role played by universities, both public and private. For some time, governments have been directing the research at these institutions by requiring industrial partners. The balance between pure research, applied research, and innovation has been completely distorted. Beyond their participation in the industrial-military complexes, universities are also being harnessed to help nations survive in a global economy. In the final analysis, all this is nothing but a race to build the many components of the global order of technique. With hindsight, the arms race during the cold war may be interpreted as the beginning of the all-out effort to develop this global order of technique. Those who possess the most powerful means of all kinds will gain some benefits, but humanity as a whole will reap the consequences of having its cultural orders desymbolized and our biosphere disordered. In the long run, there can be no winners but only losers, although along the way there will always be some who will benefit more than others and some who will suffer more than others. Of course, every nation believes it will be among the winners.

Although many may be uncomfortable with what they may consider an excessive level of abstraction, technique as a global order will impose the consequences of desymbolization on everyone. As a system, technique evolves in part by responding to our initiatives, but it expands even more as a result of a component that is entirely beyond our control. Each and every area of specialization acts as a transmitter and a receiver of technical information, and the permutations and combinations of all this information flowing around the globe have made the development of technique largely independent from our initiatives. All this was already noted many decades ago when technological forecasting proved to be impossible. The relationship between investments and economic benefits becomes extremely complex and uncertain because of this flow of information. Nations will run from crisis to crisis as we all learn to bow before the necessities imposed by the system that we created to serve us.

Ironically, there has been a decline in the research on where humanity is taking its most powerful creations and where these creations are taking us by their vast influence on everything. This is not difficult to understand. Even with high levels of desymbolization, the metaconscious knowledge that is developed by our dialectically relating the experiences of our lives and the world leads to intuitions regarding the 'system' shaping them. Regardless of how this is worked out through public opinion and integration propaganda, the deepest metaconscious knowledge informs our intuitions of what really matters to our lives. Within the horizon of our experience the greatest good we know is transformed into the absolute good because there can be no other. Having conceptualized certain entities in our lives and world as the greatest good that can exist, our critical attention is deflected elsewhere. Whatever goes wrong cannot come from what metaconsciously has been declared the greatest good. Why bother spending research money on these things when, metaconsciously and intuitively, we know that in the long run it will all be for the best.

I recognize that such a diagnosis of our metaconscious knowledge of the technique and nation state flies in the face of the widespread belief that we live in a secular age, because traditional religious attitudes are either in decline or in the service of personal spiritual gain that is completely dissociated from contemporary ways of life. Not finding much of interest in the traditional religious places, we erroneously concluded that we had grown up and become secular. Although this is very reassuring, it fails to recognize that we, like those who went before us, have to create points of orientation in an ultimately unknowable reality by absolutizing what is most essential to individual and collective human life. This need was particularly acute during the first half of the twentieth century, when the challenges associated with industrialization demanded very different roots in reality. One of these points of orientation was the nation state, served by three secular political religions: communism, National Socialism, and hard-line democracy.

Karl Marx, the great prophet of communism, had revealed the dialectical process of history, and the Communist Party was charged to lead the people out of the alienation of capitalism in the north and feudalism in the south to the socialist 'promised land' of freedom. No other political parties were required, because they would deceive the people, and hence true democracy could be brought only by the Communist Party. There was no room for debate or compromise. The future of humanity depended on the Communist Party, and those who stood in the way had to be sacrificed for the good of all. There was nothing new on this front. National Socialism and fascism were founded on similar absolutes. The United States conducted heresy trials during the McCarthy era, and it did not stop there, as the pronouncements of Reagan and the Bushes have made clear.

Suppose that we compared the fundamental transformations of human life, society, and our relations with the biosphere during the past fifty years with the kinds of issues we debated in the run-up to elections and that we subsequently voted on. It would be difficult to come to any conclusion other than that we failed to intervene in these transformations by imposing our values, which would have translated into various political options that could have been debated and voted on. We live as though technique were under the control of the nation state, and the nation state under the control of the people. This appears perfectly obvious until we carefully examine the reciprocal relationship that we have with our social and physical surroundings. As I have previously mentioned, as we change them, they simultaneously change us. We are perfectly aware of how we change technique, while the way in which technique changes us is at best a theoretical possibility that we cannot directly experience. We thus make existentially bearable the autonomy that the technical order has gained over the cultural order. The very existence of political processes and institutions reassures us that all this cannot possibly be a grand illusion and that we have not become commoditized resources for our own creations. Absurd as such an explanation may appear, it is simply the contemporary version of the drama of humanity.

In the Midwest and South of the United States and in western Canada, Christianity is once again helping to meet the moral and religious needs of their societies. It is ironic because this Christianity proclaims a revelation from a transcendent God, which is something entirely other than what cultures create to meet their moral and religious needs. Doing so was always portrayed as the servitude of idols in the Jewish Bible. The opening chapters of the book of Genesis include an attack on morality, religion, and magic.[58] The warnings of the danger of serving false gods continue throughout the Jewish Bible, but this message has become incomprehensible ever since Christianity was split into its 'horizontal' and 'vertical' dimensions, making it easily assimilated by the moral and religious needs of our societies. This has led to the most contradictory political alliances, with results that have been heavily criticized in the Jewish and Christian Bibles: making the poor and vulnerable even more so to the benefit of the wealthy and powerful. All this is legitimized by a Christianity that has turned itself upside down from a call to freedom and love into a morality, and from an anti-religion of liberation from false gods into a religion faithfully serving the American way of life, evolved by means of technique and the nation state.

It is important to recall the theory of our three life-milieus.[59] During the period commonly referred to as prehistory the human group was immersed in, and evolved in relation to, the life-milieu of nature. During the period of history the human group was immersed in, and evolved in relation to, the life-milieu of society. In each case, the life-milieu was absolutized, as reflected in unique forms of morality, religion, and myth. During the last half of the twentieth century, the density of technically mediated relationships in human life and society increased to the point that technique has become the life-milieu into which we are immersed and in relation to which we evolve. It should come as no surprise, therefore, that the forms of our morality, religion, and myth changed once again. Nevertheless, very few studies take account of this influence of technique on human life and society. We are no more aware of our myths than were our predecessors of theirs. To a symbolic species these myths are unavoidable. If we were truly secular, there would be no possibility of anything in our lives taking on the form of an absolute.

Finally, we come to situating our lives in time and history. We live as if 'time will tell' and 'history will judge.' It is no longer we who are making our history, but our history that is making us. Such a myth of history would have been unthinkable in earlier times because each civilization made its own history and there was no common denominator between their journeys. Beginning in the nineteenth century with the myth of progress, all societies voluntarily or involuntarily changed course to go down the path of development. For the first time, all nation states could be classified in terms of their success in doing so, which transformed them into so-called developed, developing, or underdeveloped societies. No historical journey other than the one based on science, technique, and the nation state was thinkable, let alone liveable. The myths of the industrially advanced cultures shattered those of any traditional people, with tragic consequences, particularly in the cases of indigenous peoples.

The technical approach translates every aspect of human life into a rational schema that transfers it from culture into the technical order, and thus from what is true to what is real. Here human activities are linked together not on the basis of human choices guided by values but on the basis of technical necessities. It is impossible to choose and retain the 'good' techniques and leave unexploited the 'bad' ones. When technical alternatives present themselves, no human values are required to select the 'best' one; the one with the highest performance ratios wins out. Nor do any human choices or values significantly affect the development of technique. There is a self-augmentation of technique as it feeds on the problems it creates, because these become the foci of new technical approaches as we turn to technique to resolve the problems created by it. A second component of self-augmentation results from the flow of information within the system. For example, the publication of an innovation regarding an improvement in the frequency response of a particular solid-state device will, as a result of the flow of information, trigger many other innovations, which in turn trigger others, and so on. Since this self-augmentation is accompanied by a metaconscious bestowing of the highest value on technique, any critical attention is deflected elsewhere, with the result that whatever becomes technically possible is likely to be adopted since technique has become good in itself. Once again, no human choices guided by values need to decisively intervene. The result is that technique has become a global system that functions with a great deal of autonomy with respect to human life and society. It is as if technique had its way, even though it is our own creation.

It is the sacred or 'central myth' of technique and the nation state, as well as the myths of science and history, that makes it possible for the technical approach to assimilate experience and culture. We have become possessed by different myths that tell us that science is everything we need for knowing, that technique is everything we need for doing, that the nation state is everything our community needs to guide us down the only path left open by history, and that all this is for the best.[60] There are many expressions of our new slavery. An economic fundamentalism extols the 'natural' character of the economy. For example, there are 'natural' levels of unemployment, and we must not dream of interfering with the 'natural' perfection of the Market. Is the economy not the result of human decisions and values? When we claim that something is 'natural,' like gravity, we in effect cease to assume responsibility for it. Hence, when people are underemployed or unemployed, they are either irresponsible and lazy or in need of adaptation to technique by means of education. All the problems of the economic instabilities of the first half of the twentieth century have been forgotten, including the fact that the Keynesian approach to achieving stabilization temporarily solved these problems. There are very few politicians who have escaped possession by the new economic fundamentalism, which is obvious, provided that one judges them by what they do and not by what they say. These politicians are joined by a new breed of apologists, among whom there are some great new stars.[61] The months following the beginning of the latest recession showed the depth of our possession; we behaved as if we had no fundamental underlying structural economic, social, and environmental crisis to deal with. All we face are the excesses on Wall Street, and once these are taken care of, everything will be fine again. Even people who suggest that Keynesian policies need to be revisited are silent on how the economic conditions that drove us into the arms of monetarism came about. The possibility that technique transformed the economy into an economy of non-sense remains unthinkable.

How is it possible to explain all this other than by reference to the new myths in our individual and collective human lives? Of course, we love our kids, support our friends, give to charity, and evoke our values that suggest that all is not well and that things may be sliding in the wrong direction. Unfortunately, this affirms rather than negates what I am arguing. Despite widespread feelings of this kind, what we are and are not doing about it confirms the presence of the previously discussed myths. These absolutes distort the interdependence of our lives in the world by turning complex issues into simple black-and-white ones. Suggesting that science is not omnipotent with regard to knowing can be interpreted only as an anti-science agenda and a glorification of the past. Suggesting that the price we are paying for technique is far too high is met with accusations of technology bashing. Claiming that there is very little correlation between the political traditions of our governments and the well-being of our communities and nations is dismissed as technological determinism and pessimism, yet when diagnosed with a serious disease, no one would dream of accusing his or her doctor of pessimism. Many issues that are immensely complex take on a secular religious character: either a person is a believer (which is only reasonable when one cannot experience the limits of science, technique, and the nation state) or a person is viewed as a heretic, to which we attach labels such as *liberal*, *humanist*, *communist*, or *tree-hugger*. Such people threaten everything that makes us and our world who and what we are. In other words, anything that is entirely 'other' than our ways of life that are based on science, technique, and the nation state cannot be taken seriously, which is exactly the role myths have always played in human life and society.

For example, the concept of sustainability would suggest that we are in need of ways of life that are entirely other than the ones we live. This otherness has all but evaporated in our thinking, if ever it were actually present. Every imaginable technical change and every new university, industry, or government initiative can now be declared to be 'green' and sustainable. It turns many specialists into new secular magicians, as if merely allowing something green or sustainable could change anything fundamental. Just like all the generations who preceded us, we continue to live as if the reality we know and live were reality itself, thus barring the road to anything that is entirely other. We have simply become possessed by different myths than those of earlier cultures.

As a result of technique changing people, the emergence of new myths in the cultures of industrially advanced nations has transformed them into societies of non-sense. It is not a question of having recognized how traditional cultures alienated human life by possessing them through myths. Nor is it a question of having become iconoclastic towards any myths in order to face the reality of our lives as best we can. Instead, what is uniquely human has now become irrelevant. If we wish to know anything, it is no longer a matter of symbolically relating everything to everything else in order to experience it by means of a culture. We know very well that this human knowing is limited. Living as if science had no limits implies that our human knowing is now irrelevant to our lives and our future. Living as if technique had no limits implies that any human skills depending on the metaconscious knowledge built up by symbolization are irrelevant for our lives and our future. Similarly, living as if everything were now political is to deny the dependence of any human community on its culture for ordering and directing its way of life. As a result, our historical journey now excludes any dependence on sense, thereby making it entirely dependent on an absolutized science, technique, and the nation state. All this puts us and our communities into a terrible dilemma: in spite of everything, we continue to depend on symbolization, experience, and culture.

How we learn to live within the order of technique is particularly evident in the lives of our teenagers. The metaconscious knowledge they build up as a result of living in a technically ordered universe provides them with the intuitions that they cannot live in the world they are inheriting. It appears to be colonized by something that makes life very difficult; hence the road to becoming adult members of a symbolic species appears all but barred. There is no choice but to search for other ways to orient themselves in an ultimately unknowable reality. If the development of their cultural selves appears blocked, why not use their physical selves as a 'beachhead'? Are their maturing bodies not suggesting new powers? Of these, their maturing sexual powers appear to have the greatest influence, and this is heavily reinforced by the bath of images they encounter on the mass media and the Internet. Why not create some order for their peer group and their Facebook crowd by living through their sexuality? To interpret all this in terms of a traditional morality is to entirely miss the point. It is a question of discovering a way out of the highly desymbolized culture they are busy acquiring but that does not appear to make much sense in the face of a technical order.

The acquisition of a cultural orientation and ultimate reference point proceeds more slowly as a consequence of desymbolization. As a result, teenagers have little choice but to rely on their peer group and Facebook crowd of friends. They latch on to every possible technical means to stay in touch. Lacking an orientation and roots that make sense to them, they have little choice but to go with the flow, and for that they need to be well connected if they are to ride the crest of every new ripple and wave. Much of this is lived through their sexuality, beginning with their swear words. In Western countries in their not-so-distant Christian past, these represented the transgression of what was most sacred and high – God, Jesus, or Mary – but also the threat of being separated from all this and losing your way by damnation and hell. Today, their swear words refer to what has not yet been colonized by the technical order, in order to symbolize this as a transgression of the technical order. The F-word refers to the sacred ritual of transgression around which a great deal of life revolves. It matters little whether the associated rituals are verbally celebrated or actually engaged. In one way or another, much of teenagers' lives are pulled into one sexual orbit or another.

This quest for meaning and value in an ultimately unknowable reality does not represent an escape from the technical order, despite what is claimed in music and dance. It is the exact opposite: living through their sexuality creates the sacred transgression of technique, which makes it more liveable, but at a cost. This is clear from what has happened to the love that two people may share. The sexual celebration of that love takes on a dimension of anxiety related to a person's ability to perform in bed and to fully satisfy the other. Depending on the extent of this anxiety, a quest for improving sexual performance may arise. Something in which humanity has engaged for its entire existence becomes possessed by performance. All this was particularly evident in the free-love movement and the 'happenings' in the 1960s and the 1970s in the United States. It was accompanied by an explosion of sexual symbolism in the advertisements of the time.

In a mass society in which many people are lonely and personal relations may be disappointing, the spiralling appetite for pornography is easily understood. So are the growing threats to a relatively normal development of the physical, social, and cultural selves of children and teenagers insofar as it involves the sexual maturation of their persons. Undoubtedly, the omnipresence of 'soft' pornography plays a role. It teaches a person to stare at a naked fellow human being without the expectation of any reciprocity. Staring transforms the other into an object. We know this from how uncomfortable we can be made to feel when others stare at us. It is a way in which the other denies our humanity, which opens the door to a range of behaviour that is degrading. Our embodiment, participation, commitment, and freedom in these and other relations is reduced to a level that is lower than that found among the higher animals. We know very well how all this works during a time of war, when the enemy is made subhuman in order to lower or practically eliminate any barriers to killing, wounding, and torturing. This is much more difficult if we fully symbolize one another. In other words, in contemporary mass societies a great deal of sexual behaviour extends our conformity to technique by sexually commoditizing other people.

The sacred transgression of technique includes many other kinds of rituals. There is the hacking into the places of power. The system can be made 'ill' by launching viruses and all the other ways of contaminating and paralysing computers. Shoplifting is a way of engaging and beating the security apparatus of the system. Graffiti represents a marking of the system by acts of defiance. These and other activities synergistically reinforce the sexual transgression of technique. Paradoxically, all this fights performance with performance, thereby further immersing people in the technical order.

There is an uncritical acceptance of every possible technical means. For example, when teenagers go on Facebook, they create the kind of life that is extensively separated from experience and culture. They acquire 'friends' and participate in 'social networks' or 'communities' with little or no face-to-face experience. They lay bare their private lives in an attempt to 'market' themselves and to establish their own 'brand.' It would appear that what really matters is their 'lives' on the Internet, as opposed to their embodied lives. In other words, they are learning to live in the image of technique, although they say the opposite to each other. All this is celebrated by the Web 2.0 ideology. Once again, I am not passing a moral judgment on our kids; I am simply observing how difficult it is to grow up with little cultural support because of desymbolization, and instead having to rely on integration propaganda. As adults, we cannot set ourselves up as judge and jury of a situation that we help to advance on a daily basis.

As was the case with earlier civilizations, human life would be entirely unliveable were it not for the creation of a profane that is ordered by a sacred transgression. As noted, our metaconscious knowledge of our present situation has led to intuitions to the effect that even though our cultural being may be possessed by technique, our biological being is not. With hindsight, it has become evident that sexual freedom was turned into acts of obedience to technique by seeking ever-greater erotic pleasure through the enhancement of sexual performance. Once again, this is not a question of morality but of finding our way in a civilization of non-sense. In the same way, we declare war on drugs with little consideration given to the reason so many people have difficulty coping with a lack of sense in their lives and turn to drugs as a means of compensation.

Like the sacred and the profane of earlier civilizations, all this inserts people more deeply into the sacred order. After all, none of these activities makes any sense, and some of them are directly in the service of performance. For example, highly successful hackers have been offered employment so that their abilities of outperforming the system can now be used to improve its performance. Even in intimate and committed relationships any anxiety regarding the improvement of one's sexual performance represents an assimilation by technique.

As noted, there is a profound contradiction in these developments. The metaconscious knowledge that gives rise to the sacred of technique and the nation state and to the myths of science and history is under pressure from the desymbolizing effects of technique. As a result, the clarity of our cultural meanings and values continues to decline. There is little question that our metaconscious knowledge is similarly affected. This may well make the absolutization of the reality that we know and live more fragile, but it continues to exclude anything that is radically different. The significance of all this cannot be assessed as yet. There has been no event that has been able to challenge contemporary ways of life to permit us to see whether their cultural unities are becoming more fragile, thus opening the door either to disintegration and collapse or to mutation into something radically different. In the meantime, the highly desymbolized organizations of our brain-minds seek to compensate by acting as a radar, scanning our surroundings, especially the images of the mass media. As a result, private opinions have largely been displaced by public opinion, traditional morality by a statistical morality, and traditional religions by their secular counterparts. We are thus placed in a position where we can offer little resistance to whatever violates our private opinions, morality, and values.

This vulnerability continues to be exploited to an ever-greater degree, ranging from advertising campaigns directed at young children to governments passing legislation that steadily increases the level of public immorality, as wealth is abstracted from the many to be transferred to the few.[62] We have mentioned the liquidation of democracy in favour of free trade, and the obscene compensation packages of chief executive officers. In general, those who are willing to sacrifice everything to technique (including the jobs of their neighbours, the viability of their communities, and the health of their ecosystems) are richly rewarded, while those who are concerned about the growing loss of control over their lives are punished by deepening impoverishment.

These developments have little or nothing to do with people becoming immoral because they have abandoned the morality and religion of their parents and grandparents. The relentless push of technique encounters less and less resistance from a public whose culture has been decisively weakened as a consequence of desymbolization. Many may have a feeling that their lives and communities are being destabilized, but they have nowhere else to go. Our children may not have steady jobs, our communities are losing employment opportunities, our ecosystems may be poisoned, our health and social protection may dwindle, but all this and much more cannot be avoided as long as we have faith in science, technique, and the nation state. The only response has been the advice to work much harder and smarter in the hope that you can compete with anyone anywhere in the world otherwise you will have neither a job nor a life, since the latter requires participation in consuming the goodness of technique. We are no longer shocked by the epidemic of antidepressant drugs required to keep people going mentally and the many other drugs required to keep them going physically. We are no longer shocked by being guinea pigs in the experiment conducted on us with genetically modified organisms and products containing nanoparticles. We increasingly live by fear and not by sense.

We are desperately searching for some meaning for our lives, and it is here that postmodernism performs its essential role as the 'spirit of our age.' As a philosophy, it brilliantly intuits how we metaconsciously continue to symbolize our situation by a desymbolized and assimilated culture. Recall that before this desymbolization everything in human life was centred on an absolute that was deeply hidden in people's metaconscious knowledge. When this absolute was weakened as a consequence of desymbolization, it quickly became apparent that 'anything goes' as long as it does not interfere with technique and the nation state. Cultural relativism and decentring is possible only when one's culture is so highly desymbolized that all other desymbolized cultures appear almost as liveable as our own. Modernism thus became untenable because the closest thing that people had to a foundation for their lives in the form of a cultural unity had been decisively weakened. All attempts to create a non-cultural substitute have failed. Hence, the attack on modernism was an attack on a 'straw man.'

I acknowledge that, well before postmodernism, philosophers who did not have an iconoclastic awareness of their culture's sacred and myths usually ended up working out their metaconscious intuitions of the spirit of their age by pushing to the limit what was thinkable in that cultural context. Similarly, in literature and the arts the reifying influences of technique on human life were metaconsciously intuited as the death of the subject, the disappearance of the narrative, the impossibility of any human commitment, and a complete relativization of all human meanings and values. Structuralism faithfully reflected the fact that, in human life and society, technique had reified and incorporated everything into an order of non-sense built up with networks and structures of commoditized entities whose power to obtain the greatest desired outputs from the requisite inputs had been maximized. Deconstruction as a new reductionism merely revealed the technique-based connectedness that underlies human life and society, while relegating cultural order to the domain of ideology. It thus unveiled a new foundationalism of power.

Much of what technique does to human life and society thus becomes normal and a decisive improvement on modernism. The way we live the unlimited capabilities of scientific knowing, technical doing, and political organizing makes it possible for technique-based connectedness to dominate the cultural order. People accept that they cannot make sense of their lives and their world by means of their own experience and culture and that instead they must rely on integration propaganda and, via it, on scientific and technical experts. If, however, there had been a widespread recognition that what we experience is mostly related to people changing technique and that a critical understanding must include the possibility of technique changing people, postmodernism would have been unthinkable. Instead, it would have fostered a critical awareness of what was happening to human life and society under science, technique, and the nation state.

Postmodernism is in essence another expression of our enslavement. The new cultural unity metaconsciously instructs enslaved minds as to what to think and say. Hence the enormous resonance that this philosophy has had with the public's metaconscious knowledge of their lives and world. Those who critically examined what was happening to human life and society in the twentieth century, beginning with Max Weber, Jacques Ellul, and others, had to be explained away as pessimists, traditionalists, Calvinists, or the like. Living a good life in the world by symbolizing everything in terms of meanings and values and making a history by means of them are now regarded as little more than fading expressions of everything that we are trying so hard to leave behind in our pursuit of technical power. What we have all but forgotten is that science can only take hold of repeatable phenomena and thus has no access to life beyond the extent that it is alienated and reified. Technique can imitate life only to the extent that it is able to reconstitute life in the image of non-life. The state can organize life only to the extent that people give up their freedom. In other words, science, technique, and the nation state reconstitute life as non-life incorporated into an order of non-sense. Contemporary art is the perfect expression of this situation.[63] Postmodernism, as the uncritical spirit of our age, appears to recognize one important issue: there can be no meaning, no values, and hence no culture without a reference point. When the gods of the past fail us, humanity has no choice but to create others or, failing that, to act in their place as a kind of super-being. It would appear that we have chosen to do the former and to enslave ourselves to an order of non-sense.

On the level of our daily lives, postmodernism provides a celebration of victory over modernism as well as the ability to feel good about ourselves. We get on with daily life by supplementing the desymbolized role of culture by means of integration propaganda. It reveals to us many new exciting gadgets, medical miracles, and scientific marvels. We are mesmerized by the new bells and whistles on our cellphones, by our 'smarter' computers, and by all the knowledge of the world brought to our doors via the Internet. Before we need to make any value judgments, new improvements will make them irrelevant in any case. The same holds true for any changes in our lifestyles; we are barely integrated into one fashion when the next one is already coming across our horizon. We are so busy adapting that we hardly have time to notice what is really happening, and when we can no longer suppress some vague but disquieting intuitions, we have little choice but to have a prescription take care of it. When we are 'with it,' our 'public' opinions come ready made and require no symbolic interpretation or commitment. This is equally true for our actions when we go with the flow. All this could be threatened by a little critical scrutiny, but most of us have no time to ourselves, nor does the organization of our brain-minds constitute the intellectual bridgehead for doing so.

We are the first civilization in history that is attempting to escape the order of what is true for our lives to find refuge in an order of non-life. For example, when we are out for an evening stroll with our favourite person in life, we will answer our cellphone to get back to what is real. Living on the Web is more real than daily life. Performance is more real than living a situation. What all this means is a suspension of our lives, our history, and our being integral to and dependant on the biosphere. Instead, we treat one moment of our lives, one event in our communities, and one relation of dependence on all other life as individual parts that can be manipulated and improved on their own terms, and we believe that all this will somehow create a better future.

The University as a Bridgehead?

There is possibly a silver lining in the dark clouds that hang over our civilization. It becomes visible when we compare our situation to those encountered by earlier civilizations. All of them attempted to make sense of and live in their world by means of cultural approaches that made life liveable at the cost of enslavement to a sacred and myths in order to keep relativism, nihilism, and anomie at bay. The meaning, direction, and purpose that this gave to the lives of their members were thus anchored in alienation (the secular analogue of what in the Jewish and Christian traditions was referred to as sin). Hence, there appears to be nothing new under the sun. We continue to be possessed by myths, be they of a secular kind.

From the perspective of our being a symbolic species, this is a positive sign. Desymbolization has not advanced to the point that we can no longer live our lives. Our ability to relate everything to everything else may well be embryonic, incomplete, and distorted, and largely reduced by the techniques of integration propaganda, but they continue to sustain the living of our lives. The organizations of our brain-minds continue to interpolate and extrapolate our experience into our lives, be they highly desymbolized ones. Hence, anything that is entirely 'other' to our lives remains unliveable, unthinkable, and undoable, thereby protecting our civilization from collapse. It is impossible to know how strong this protection is, but given the weakness of present-day symbols, it may well be very fragile. Our lives continue to be weakly held together, and across the horizon of our experience we still encounter what we expect. Thus far, the struggle between life and non-life endures against the powerful forces of desymbolization that we have unleashed on ourselves. This is the good news.

The bad news is that the cultural unities of contemporary mass societies sustain human life less well than did those of the industrializing societies of the nineteenth and early twentieth century. They reveal a technical order that largely excludes our participation as human subjects. We are compelled to live by what is real and not by what is true: science excludes us as symbolic knowers, technique excludes us as symbolic doers, and the nation state excludes us as symbolic people imposing our values as best we can. Our history is decided by forces that are independent of these values. We are compelled to live by what undermines us as a symbolic species.

The cultural unities of the nineteenth and early twentieth century interpolated and extrapolated the experiences of the people at that time as something that would eventually bring them happiness. The events during the first half of the twentieth century mutated their myths, but the new emerging cultural unities reinforced what had already been embryonically present. There would no longer be a promised land of happiness but only the necessities imposed by a technical order without which the global population could no longer be sustained. We had to learn to live under this technical order by suppressing our lives (just as people had to do when they began to work on assembly lines), and to consume whatever technique offered. In other words, it is no longer so much that our lives are possessed by false meanings and values but that our lives have become a collection of weakly related aspects subjected to technical approaches to achieving endless perfection. All such relations suppress our selves. A great deal of our metaconscious knowledge is now related to that technical order. This is revealed by the fact that we live as though it were omnipotent. Thus far, the organizations of our brain-minds continue to weakly work in the background to permit us to live our lives, and our children can still make use of a culture to grow up as members of another generation of the symbolic species. Everything is not lost, at least not yet. We are still able to engage the struggle against what desymbolizes individual and collective human life. What better place to start than with university reform. We can then create a different intellectual and professional division of labour and diffuse it to all other institutions to create a more liveable and sustainable future.

Before we get started, we would do well to identify the major obstacle that stands in our way. Secular religions, like their traditional precursors, bind us to a sacred and myths, and it is by means of them that we serve these myths. Today this means that everything has to be politicized to create the illusion that we remain in control. It is the result of technique changing people: technique has become our primary life-milieu and order.

With hindsight, it is becoming apparent that if we had not politicized a great many issues, we might have been better off today. Was any political solution ever found for dealing with the difficulties encountered by people who are compelled to work in the image of the machine by means of the technical division of labour? Neither the political left, nor the centre, nor the right has ever shown any concern for or devised any solution to this problem. The conflicts between management and labour have also ignored this structural issue. Today we face the same kind of issue: we are all compelled to work in the image of the computer as a consequence of the technical division of intellectual labour imposed by this information machine. Again, no political solution to this issue will be found.

Turning from work to the economy, the same kinds of conclusions impose themselves. As a result of the weak link between the prices established in markets and their value for human life and society, the Market has contributed to a number of crises. These include the underemployment and the unemployment in the leading industrial nations towards the end of the nineteenth century, the First World War, the Great Depression, the Second World War, and the failure of monetarism to deal with the present financial crisis. It was surely no accident that these two world wars were started by the most industrially advanced nation of that time, which also suffered most acutely from the two main economic scourges of industrial civilization, namely, out-of-control unemployment and inflation. The Keynesian Revolution was helpful in creating an economy that made more sense for people, but it was soon overwhelmed by the technical order. It pushed us into one of the most destructive economic ideologies of all time, namely, market fundamentalism. When we look back at how the political right, centre, and left responded to all these situations, it becomes clear that none of them discerned the threat that was posed by the technical order as a consequence of the shift from a primary reliance on the cultural approach to a primary reliance on the technical approach. In the West it was the right of the political spectrum that, throughout all the turmoil, most faithfully served the cultural unities by creating or heavily leaning on new secular political religions. The lives of millions of people were sacrificed without any solutions being forthcoming.

If we treat university reform as a political issue, we will close the door to any possibility of a genuine solution. It will have to be apolitical, in defiance of the secular myths that guide us down an increasingly unliveable and unsustainable path. Instead, we must attempt to proceed by means of round tables, operating on the premise that we share one world and one future regardless of our political commitments and economic ideologies, and this sharing cannot be based on anything but the commitment to a public good. Such a public good will have to be reinvented, which is a challenge to the current forms of science, technique, and the nation state. The round tables must be designed to help us overcome the limitations of the present intellectual and professional division of labour in the specific ways in which these play out in our lives. For example, people in professional specialties will need to enquire as to the typical undesired consequences that tend to flow from the application of their highly specialized knowledge, gain a better understanding of these consequences, and learn to use this understanding in a negative feedback mode to modify their decision making in order to achieve the desired results and at the same time prevent or greatly minimize undesired or harmful effects. This enquiry can be greatly facilitated by the careful choice of the membership in these round tables. Other round tables can be constituted to help particular disciplines or specialties to discover critically important but implicit assumptions in their domains of practice and to investigate these, as well as examining the explicit assumptions in terms of their validity as seen by other disciplines. In other words, round tables must be designed to create a synthesis of knowing and doing, which is the equivalent of creating 'windows' in intellectual professional silos to allow their occupants a view of the world. Doing so will diminish the degree to which the members of any discipline and specialty are trapped in the triple abstraction and will encourage them to explore both supply-side and demand-side solutions to the issues being faced.

None of these kinds of efforts can be mandated from above by political decision making. University administrations must recognize that there are individuals in the system who have attempted to maintain a broader view of things and thus have escaped their disciplines or specialties to some degree. These people must be provided with an administrative framework that appropriately rewards them for the above kinds of tasks and provides modest resources if required. The entire effort must be built up from grass-root levels by people who have the intellectual capability and the determination to impose a public good on their disciplines or specialties. All this was eminently feasible ten years ago, but the system is busy selecting out these people as a consequence of intense competition, which is penalizing anyone other than full professors for undertaking the above kinds of efforts. It may well be that different kinds of people have to be hired into tenure-stream positions and that the 'old guard' needs to slowly die out, as has been the case in scientific revolutions that have transformed particular disciplines and specialties. One thing is certain: it will not be easy, and politicizing this effort will bring it back into the orbit of our current intellectual and professional division of labour, anchored in our secular myths. A great deal of entrepreneurship will be required, but the next chapter will hopefully give some insights into some of the things that need to be faced in various disciplines and specialties.

5 Desymbolization and Resymbolization

Is Resymbolization Possible?

The closing decades of the twentieth century and the opening decade of the twenty-first may well stand out in human history as an unprecedented reunification of humanity. Until then, nations and communities were divided by their languages, cultures, and histories. These divisions are in sharp decline everywhere as nations struggle to help build the technical order, but they can do so only by reifying and commoditizing their cultural orders and ecosystems. As a result, the reunification of humanity and the accompanying globalization occur to the detriment of our being a symbolic species. This places humanity in a very difficult predicament because the technical order fundamentally depends on and cannot function without our lives working in the background, which requires symbolization. Our future may well be decided by how well, and for how long, technical means are able to compensate for the loss of our symbolic capacities.

As a symbolic species, we increasingly depend on two parallel modes of knowing and doing. The one required for building the new technical order is highly desymbolizing, while the other is necessary for minimally sustaining human life in a highly desymbolized cultural order. This tension also occurs in our communities and nations. We live as if mass societies were merely the latest form that human communities have created for themselves. A close examination reveals that the term *mass society* is a misnomer because it has very little in common with all its predecessors. For example, a mass society lacks anything comparable to an integrated way of life. It evolves on the basis of a knowledge infrastructure by which everything is adapted and advanced by one category of phenomena at a time. The disciplines and specialties that create and evolve the components of the knowledge base of the technical order have replaced what once were local and particular elaborations of a way of life and culture in every sphere of human endeavour. However, resulting reification and disordering of individual and collective human life continues to be limited by people's lives working in the background.

Much of this largely unnoticed role of highly desymbolized cultures must be supplemented by integration propaganda whose bath of images is as close as we can come to a substitute for a way of life. The limitations within which this integration propaganda operate result from its being an ensemble of technical means desperately attempting to substitute what is real for what is true. Once again, this important component of the technical order cannot function without grafting itself onto a highly desymbolized culture.

The reunification of humanity is thus based on a gamble with our future. We are gambling that individual and collective human life is little more than a collection of separate details and that this is also true of all other life forms. Sacrificing everything for the building of this new technical order requires the denial of life itself. It is an attempt to represent and deal with everything in terms of an order of what is real, and thus in terms of non-life. In the meantime, the suffering this causes is immense, and because of the high levels of desymbolization, we are unable to connect the dots.

Individually and collectively we have walked into a trap. By living as if science were all we needed in relation to knowing, technique in relation to doing, the nation state or 'state nation' in relation to politically moving forward, and by relying on history to tell us when we stray from the one path, we are in a collective denial of the vital but weak roles that cultures continue to play and of our need to be a symbolic species. There are people who see no problem with this at all. Life can be represented by non-life, which opens the door to eternal life on the Web. Sooner or later, machine-based knowledge, expertise, and intelligence will surpass their living counterparts.

Most people, however, do not share this faith in technique. Hence, we must address the question as to whether or not the resymbolization of human life is possible. Will we, as a highly desymbolized species, be able to make a comeback? We could fill many volumes over these debates, but the application of a 'no regrets' principle may be more effective. If we accept that the diagnosis presented thus far is a plausible description of what is happening to us, the responsible course of action would be to investigate how experience and culture (now very much working in the background) can be strengthened by concrete and practical steps that we can take today. If this is the most responsible initiative we can imagine, we should engage in it even when the chances of success appear slim. After all, humanity has undergone other transitions whose likelihood of success initially appeared to be almost non-existent.

Of all our institutions, which one would have a potential capacity to support resymbolization? The universities are the intellectual professional centres of the new technical order, but they are deeply in the grip of technique. Academic performance measured by technical means rules everything. Concerns such as research grants, publications, impact scores, and a university's rank among world-class competitors are all that administrators appear to be thinking about, at least in public. The university now constitutes an environment that selects against faculty members and administrators who seek to challenge discipline-based scholarship by concentrating on less desymbolizing approaches. Nevertheless, we must persist. The remainder of this chapter will present some glimpses of how resymbolization may be initiated.

A Starting Point

Scoping our level of desymbolization can begin by examining how a discipline or specialty restricts human life and society to one category of phenomena. It soon becomes apparent that the way in which a single category of phenomena may be abstracted is not dealt with in any introductory textbook. Any critical examination of the losses of understanding that may occur as a result of this abstraction cannot be undertaken. For example, in most disciplines of the social sciences we have strayed far from their founding thinkers who, by their methods and approaches, made it clear that human life and the world could not be understood in this way. Parallel schools of thought have emerged because of conflicting conceptions of, and approaches to, the study of their area of concentration. These areas overlap those of other disciplines.

Today the level of autonomy of the social science disciplines is very high. Have we found a way to restrict attention to those situations in human life and the world where one category of phenomena dwarfs all the others, with the result that they can legitimately be ignored? On further thought, such an interpretation is unlikely because it would have left many areas of human life and the world unexplored. A more likely explanation is that disciplines increasingly encountered human life and the world as being so highly desymbolized that they could be dealt with as if they were constituted from separate domains, each governed by a different category of phenomena that dominated its interactions with all other domains. Such an approach would in effect transfer what is being studied from a cultural order to an order of what is real. Doing so opens it up to measurement, quantification, and mathematical representation. A gap would thus open up between this approach and its earlier counterparts, which relied extensively on symbolization. Discipline-based scholarship, thus separated from experience and culture, could then proceed as if human life in the world could be grasped in terms of an order of the real. It should be noted that no such gap between quantitative and qualitative approaches existed in the past. A qualitative approach guided the research questions being asked, the data being collected, and the interpretations that appeared to make sense. A qualitative interpretation was then tested quantitatively, and if the fit was poor, one or both had to be modified.

For years we wore ourselves out in a struggle between the integrated qualitative-quantitative approaches and the technical quantitative ones because only the latter had been declared to be scientific. The result has been a discipline-based mode of scholarship that is separated from experience and culture to a considerable extent. Moreover, the transformation continues to be self-reinforcing. The more human life and society are desymbolized as they help to build a technical order, the more they can be conceptualized in the technical images of non-life, while the application of the knowledge acquired in this way in turn contributes to further desymbolization. This process appears to describe what has happened to the methods and approaches of the social sciences during the last half-century. By implication, whatever remains as the result of symbolization in life is taken to be epiphenomena. In other words, the kinds of developments we have described in the previous chapters have become the justification for the methods and approaches that today characterize the social sciences.

With hindsight, it is clear that as the process of desymbolization advanced, metaconscious knowledge of what was happening to human life and society led to intuitions that social scientists articulated in a sequence of methods and approaches that included functionalism, structuralism, social systems theories, and most recently the structures and relations of power. In all these and other cases, human life and society were implicitly deemed to be in the image of the technical order.

Those aspects of human life and society that did not conform to this image were either excluded from consideration altogether or treated as epiphenomena.

Developments such as postmodernism imply that giving up on everything that, during much of our stay on this planet, has made us a symbolic species is a wonderful advance.[1] There can be no cultural specificity, because the universal technical order continues to weaken it everywhere. Individual and collective human life can have no centre (even an alienated one) because the technical order reifies it and absorbs those elements that boost our efficiency, and also develops new techniques for dealing with the disorder that this creates. Anything goes in human life and society because they must be endlessly malleable to take on those forms required by the technical order. Whatever does not contribute to its structures and power simply has no significance; it is merely an illusion of the past, when a cultural consciousness was so alienated that it had little grip on reality. The technical order is the new reality.

Much of postmodernism appears to be an expression of technique changing people, to the point that it prepares human life and society for technique. It is portrayed as an act of freedom: we have created technique to serve us, and we have succeeded beyond our wildest dreams because we can no longer see any limits to what it can make possible. Another way of looking at our present situation, which amounts to the same thing, is the need to be 'realistic.' If our nation does not pursue this or that technical potential, others will, and we will be left behind, banished to a new secular hell. There really is no longer an alternative. These and other interpretations end up reinforcing the autonomy of technique. They constitute the ultimate technical bluff: our indiscriminate acceptance of almost everything technique offers is taken as a sign of freedom, to the point that it possesses our awareness of ourselves and the world.

The kinds of methods and approaches that continue to succeed one another in much of the social sciences may well be instances of the methodological problems documented by Devereux.[2] We have again adjusted the methods and approaches in the social sciences to hold at bay any anxieties regarding the possible loss of what has made us human until now. All this is embedded in our (desymbolized) secular myths, which make a great many things self-evident and others completely unthinkable. Bending our scientific methods and approaches to fit the building of a global technical order may be understandable, but it is hardly scientific. This situation brings us back to the need to look out of the 'windows' of our disciplines at the rest of human life and society, which amounts to resymbolizing the category of phenomena examined by them.

The natural sciences have long dispensed with the kinds of methodological issues discussed above for the social sciences. Centuries ago it became widely accepted that nature could be understood in mathematical terms. It revolutionized physics in the way we have already described, and other disciplines soon followed. Similarly, when industrialization had advanced to the point where culture-based technological knowing and doing had reached their limits, the discipline-based organization of technical, managerial, and economic specialties proved so successful that few, if any, methodological issues were raised. Industrial artefacts, processes, and systems were built up from distinct domains in which one category of phenomena contributed one subfunction at a time. As noted, all this was spectacularly successful and continues to be so with respect to improving the performance of everything as measured by output-input ratios. Ironically, the consequences have made us more aware of how interrelated everything is. Once again, there is a complete disconnect between these methods and approaches and the kind of world they reveal.

Intellectually or professionally abstracting something from human life or the world and not critically assessing what is lost in the process is clearly unscientific and unprofessional. However, since doing so would challenge the discipline-based intellectual and professional division of labour to its very core, we protect ourselves from the anxiety this would produce by simply ignoring it. An intellectually critical and professionally responsible approach demands that the kind of specialized knowing and doing resulting from any abstraction be complemented by the symbolization of whatever is abstracted in relation to everything else. In this way, any loss of understanding accompanying the process of abstraction can be supplemented, and the limitations of the methods and approaches can be assessed.

The consequences of our being scientifically uncritical and professionally irresponsible are well illustrated by the following assessments that three prominent economists have made of the likely impact of global warming on human economies. The non-sensical character of their conclusions has been well described by Herman Daly.[3] As Daly notes, the assessments could have been dismissed were it not for the fact that the three economists have an excellent reputation. The problem lies elsewhere, and I shall supplement Daly's explanation, using the present intellectual framework.

All three economists, like most of their colleagues, examine human life and the world as if economic phenomena dominated all others. Hence, the explanations of any issue are restricted to the category of economic phenomena, which is so problematic that it even contradicts economics itself, let alone many other disciplines. If society proceeded on the basis of the advice of these three economists, we would soon be in even greater trouble.

The scenario is all too typical. The three economists examined the likely impact of global warming on human economies. They concluded that only the agricultural sectors would be affected, and since these accounted for roughly 3 per cent of GNP in nations like the United States, the impact of global warming on economies had to be very small. Even if the outputs of the agricultural sectors were cut in half, the GNP would merely drop by 1.5 per cent, representing at most a 2 per cent increase in the cost of living. With these kinds of economic assessments, it is hardly surprising that governments are reluctant to act. Why make a large economic sacrifice over an extended period of time, during which per capita income would likely double?

When we interpret the effects of global warming by means of symbolization, the non-sense of the discipline of economics becomes immediately evident. Cutting the food supply in half will enormously increase malnutrition, famine, and starvation. Everyone would be willing to pay a great deal more for food. What use are all our possessions when we are starving to death? The price of food will skyrocket, to the point where it becomes unaffordable by many; for those who can afford at least some food, most or all of their income will have to be allocated to it. The irony of such a scenario is that it justifies the making of economics into a scientific discipline because economic phenomena now dominate all others in individual and collective human life.

Unfortunately for all of us, this example is typical of what happens when we examine human life and the world one category of phenomena at a time. A mathematical representation is made of this single category of phenomena. The resulting models may be used to optimize the desired results obtained from this category of phenomena or to estimate changes in these results arising from small modifications of the required inputs. The validity of such small extrapolations of current conditions related to one category of phenomena may be reasonably valid provided that all other things remain equal. Again, this may be a reasonable assumption provided that the extrapolation is made for a relatively short period of time during which there are no significant perturbations in human life or the world. Under our present circumstances these kinds of assumptions are worse than irresponsible. We have declared our ways of life to be unsustainable by the biosphere. As is the case with food, we fundamentally depend on the exchanges of matter and energy with the biosphere. As a result, what we mean by declaring our ways of life to be unsustainable by the biosphere is that they may collapse for lack of suitable inputs or of sinks for outputs. When this begins to happen, everything will change on a scale that far transcends what we have described for agriculture. If we decide to intervene proactively, we will be initiating substantial transformations in current trends. Once again, the kinds of assumptions underlying the use of mathematical models will be completely unrealistic.

Whether we decide to act or not, the discipline-based approach to evolving our ways of life is short sighted and irresponsible. Even from a scientific perspective, the kind of mathematical modelling that goes on in a great many disciplines is based on assumptions that will destroy our civilization; either desymbolization will go too far and civilization will collapse from the inside or civilization will collapse because it can no longer be sustained by the biosphere. The specialization that goes on in the universities amounts to a frenzy of optimization efforts for arranging the deckchairs on a sinking ship.

The above argument does not imply that what is currently happening within disciplines and specialties should be abandoned. On the contrary. Recall what we said regarding engineering design. It requires two interdependent and complementary approaches. Good design exemplars that fit into human life, society, and the biosphere can be derived only by means of symbolic approaches. Once found, effective use of scarce resources can best be made by means of the analytical exemplars provided by the relevant disciplines and specialties. We will be unable to create a liveable and sustainable future for the many billions of us unless we rely on *both* approaches to knowing and doing. We will then be back to creating a civilization that includes science and technique but is not enslaved to them.

Resymbolizing Economics

In the introduction I showed how a discipline-based organization could readily be adopted in some of the natural sciences and in technology.

It is more difficult to envisage how this approach to knowing spread to the social sciences. The developments described in the previous two chapters explain how, beginning with the political economy, a discipline-based approach to the social sciences not only became plausible but soon dominated all alternative approaches.

Industrialization divided societies into their economic orders (which initially encompassed their technological orders) and what remained of their cultural orders. The economic orders were regulated primarily by the Market and secondarily by culture, while for the cultural orders, the reverse was the case. This enormous transformation had to be made sense of by symbolization, and the new metaconscious knowledge led to the birth of two new visions of human history. The success of the Russian revolution transformed the writings of Karl Marx into the 'holy books' of communism. The resulting secular political religion revealed the 'truth' about human life and history: simply put, all human communities required an economic base. There were five of these bases, and, thanks to technological advances, they formed a linear progression that would lead humanity to a future without alienation. Only the Communist Party was guided by this 'truth,' and it alone could assure the future by means of central planning according to the (natural) laws of history. All this was ultimately rooted in the myths of the nineteenth century.

The second vision survived the first, but it also was initially rooted in the same myths. The Market would create the best possible world for most people, with the understanding that this economic reorientation of human life would be kept in check by the cultures of communities in general and their democratic regimes in particular. Economic phenomena were considered to be at the very core of what it was to be human, and the expression of this economic human nature was most effective through the natural Market, which would drive human history forward. This new secular religious vision of human life also enthroned economic gods, which had to be served unconditionally, that is, without the interference of cultural values.

Hence, industrialization linked together a variety of developments that jointly transformed human life and society. These transformations implied that economics should become a discipline, since in the grand scheme of things the economic phenomena were far more important than all the others, which could therefore be neglected. Few people appeared to recognize that the acceptance of economics as a discipline implied a new economic alienation of humanity, that of serving the Market for our own good. Two world wars and the Great Depression gradually created the conditions under which Keynesian economists could be heard. They affirmed that the economy was there to serve the people rather than the other way around as implied by the discipline of economics. All this changed once again when technique began to transform the economies of the industrially advanced nations into anti-economies. Governments struggled to stay abreast of the many social and environmental problems that contributed to their growing deficits. The monetarists seized their chance, blaming the situation on these governments having interfered with the 'natural' Market and the 'natural' rates of unemployment. Economics was once again turned into a discipline as our servitude of the economy trumped everything else. It enslaved the entire political spectrum, and any talk about the public good virtually disappeared. The policies that followed have done incalculable harm to humanity, especially to its poorest and most vulnerable members. Very few people recognize that the extent to which we accept economics as a discipline is a good indicator of the depth of our economic alienation. Our current global order makes this very clear. Communities have to give up on imposing their cultural values expressed in labour, social, health, and environmental standards, which might interfere with the Market. We are called to live by economic growth alone because it will provide everything else, and hence we must give up on living good and responsible lives.

What ought to have been symbolized as warning signs of an emerging economic alienation were instead taken as a green light for extending desymbolization into intellectual life by creating disciplines separated from experience and culture. Just as physics became the model discipline in the natural sciences because of the nature of the category of phenomena it examined, so economics became the model discipline in the social sciences because of the way industrialization transformed the relations between economic phenomena and everything else in human life and society. As we have examined in the previous chapters, economic phenomena took on a growing importance relative to all other categories of phenomena, so that these could be relegated to other disciplines.

Three assumptions greatly facilitated this process, but they must be re-examined in light of the current intellectual framework. It is not a question of economic phenomena being intrinsically important but of the technology-based connectedness of human life and society distancing itself from the cultural order and all the other developments associated with it.

The first assumption is that human behaviour may be approximated by the non-cultural concept of *homo economicus.* It goes much further than abstracting economic phenomena from the remainder of human life and society and placing them centre stage. It transforms human life into something that can be measured, quantified, and mathematically represented. Whatever else works in the background is thus relegated to other disciplines. Doing so does not even leave the equivalent of Robinson Crusoes, each on his own island, trading with one another. What remains are (dead) mechanisms that maximize utility or profit. It is the antithesis of human life sustained by a culture.

Living lives involves relating everything to everything else in order to gain a deeper understanding of its meaning and value for a person's life and, via it, for the community. The maximization of utility or profit reduces this context to certain economic inputs and outputs and evaluates everything in terms of output-input ratios. There no longer is any question of a human life lived in a community. What remains is a mechanism that blindly optimizes the desired outputs that can be obtained from the requisite inputs. Only the parameters of this mechanism vary from person to person. Here the silo effect of the discipline of economics reaches one of its pinnacles. Forget psychology, sociology, anthropology, political science, law, moral philosophy, and religious studies because whatever phenomena they examined would play a role that would require attention. With the messy inconvenience of human lives and societies out of the way, there is no longer any possible intervention in the optimizing mechanisms. To refer to this as self-interested behaviour or rational conduct is to miss the point. Such behaviour is no longer open to the influence of others and is not likely to change if it encounters widespread disapproval. All human communities kept self-interested behaviour in check by their values and cultures. Once these become highly desymbolized, self-interested behaviour becomes more and more psychopathic in character.

Such a transformation of economic behaviour opens the door to the replacement of living human beings (interacting via economies embedded within societies) with mathematical models of all kinds. Wassily Leontieff, a distinguished economist and Nobel laureate, put one aspect of the problem as follows in a letter written to *Science*:

> Page after page of professional economic journals are filled with mathematics formulas leading the reader from sets of more or less plausible but entirely arbitrary assumptions to precisely stated but irrelevant theoretical

> conclusions . . . Econometricians fit algebraic functions of all possible shapes to essentially the same sets of data without being able to advance in any perceptible way a systematic understanding of the structure and operations of a real economic system.[4]

In other words, the social and historical contexts of economic phenomena have no bearing on the discipline of economics. It should come as no surprise, therefore, that it is capable of developing the most counter-intuitive and bizarre explanations. For example, 'natural unemployment levels' are the result of people choosing not to work – the answer computed by the *global economicus* mechanism. Having washed away all emotions, self-respect, role in the family, status in the community, anxiety of having enough in old age, and much more, the explanation that what is left is simply 'natural' amounts to the assumption that human nature is now simply whatever suits the mathematical models. Human well-being is now reduced to consumptive throughput, and a common good is replaced with the self-interest-maximizing activities of politicians and public servants turned into bureaucrats (as typified in public choice theory). In terms of collective human life, little remains other than the mechanisms of markets. It is no coincidence that monetarism, one of the most asocial and ahistorical and thus unscientific economic theories ever invented, declares that the most important entities occupying its silo are natural, provided that we do not interfere. Apparently, our values, hopes, and aspirations that we express by means of policies are unnatural.

In defence of classical economics, there are social and historical conditions under which *homo economicus* can approximate human behaviour. The conditions of social breakdown at the beginning of the Industrial Revolution, when people had to carefully think about how to allocate their wages because their survival literally depended on it, provide an obvious example. However, as John Kenneth Galbraith has pointed out, extrapolating this to a mass society is unscientific.[5]

The second assumption that facilitated the disconnection of economic phenomena from their context is that the production of goods and services has a negligibly small effect on the biosphere, with the result that economic phenomena can be examined without reference to it. Once again, this was a reasonable approximation of the situation as it existed during the early stage of industrialization, when the size of human economies was small relative to that of local ecosystems and the biosphere. Hence, the life-sustaining capacities of the biosphere were relatively unaffected and fully available to future generations. There was no environmental scarcity, and hence no need to price the exchanges of matter and energy between a society and the biosphere. Today this assumption is untenable. We must recognize the dependence expressed in terms of the laws of thermodynamics (no longer safely out of sight in the intellectual silo of physics): an economy can neither create nor destroy the matter and energy on which it depends, and the transformations of energy are irreversible. Consequently, any economy is a subsystem of the biosphere in terms of its physical dimension. All economic activities are connected by a network of flows of matter and a network of flows of energy that jointly constitute the technology-based connectedness of human life and society. I have shown this to be the primary constraint on the process of industrialization and, by implication, on economic growth.[6] As Herman Daly puts it:

> The physical dimension of commodities and factors is at best totally abstracted from or left out altogether and at worst assumed to flow in a circle just like exchange value. It is as if one were to study physiology solely in terms of the circulatory system without ever mentioning the digestive tract. The dependence of the organism on its environment would not be evident. The absence of the concept of throughput in the economists' vision means that the economy carries on no exchange with its environment. It is by implication a self-sustaining isolated system, a 'perpetual motion machine.'[7]

Daly goes on to show that the dependence of the economy on sources and sinks may be ignored if they are not scarce, because their supply is infinitely large compared to the scale of the economy. The environmental crisis surely provides all the evidence we need to rule out such an assumption.

Many economists deny the possibility of a resource crisis because of their faith in the capacity of technology to create substitutes for resources whose scarcity has driven up the price to the point of making them uneconomic. A few economists acknowledge that the powers of technology may well be not quite as limitless, with the result that only some substitutability may be assumed. All this should come as no surprise. Economists, like all of us, still rely on a culture, and contemporary cultures have bestowed an ultimate value on technique, much like the gods of the past. The presence of these secular myths is, in my view, the primary reason that ecological economists like Herman

Daly, who have carried out a reality check on the assumptions made regarding the relationships between contemporary economies and the biosphere, have not received the attention they deserve from politicians and decision makers. Politics is also ruled by myths, thereby making realistic decisions and policies extremely difficult.[8]

Having been separated from human life and the biosphere, economic growth is the solution to everything that ails a contemporary society, according to almost all politicians. This conviction is an expression of the equally unlimited powers of the nation state, which, like the deities of the past, can deliver us from many of our problems, especially poverty.[9] Another implication of this second assumption is worth noting. If the economy is assumed to have a negligibly small effect on the biosphere, 'natural capital' is not depleted and hence does not have to be priced. The costs of resources are essentially limited to those of extraction and processing. As a result, the maximizing and regulating mechanisms will improve the productivity of labour at the expense of the productivity of unpriced natural resources and energy. Hence, economic growth over-consumes the biosphere and under-consumes the capabilities of human beings. There will be no possibility of a genuine development (even one without growth) without addressing this imbalance, which leads to (preventable) underemployment and unemployment.

The third assumption attributes the evolution of economic phenomena to mechanisms internal to the economy. These were the mechanisms of the Market, free trade, and globalization. The economic crises of the latter part of the nineteenth and the first half of the twentieth century cast severe doubt on the reliability of the first two. John Maynard Keynes reversed this thinking by insisting that the economy have no other goal than to serve the public good.[10] His policies gained a widespread acceptance and proved their ability to stabilize the economies of industrial civilization during several decades following the Second World War. At this point, difficulties arose as a consequence of a development not anticipated by Keynes. Economic growth increasingly depended on the application of highly specialized scientific and technical knowledge. As a result, the ratio of desired to undesired effects of technological growth began to decline. Dealing with these undesired consequences as well as the oil shocks soon led to ballooning deficits.

While for the Keynesians these deficits presented an anomaly, for the majority of economists it confirmed what they had believed all along. The economic mechanisms of the Market, free trade, and globalization had been tampered with by the Keynesians, and now societies were paying the price. Once again, an explanation had been found within the disciplinary silo of economics, and this led to a kind of Market fundamentalism. The monetarists accused the Keynesians of tampering with 'natural' economic phenomena. Their message appealed to the politicians, with devastating consequences. Soon the entire political spectrum bowed before the new mantra: unless governments eliminate their deficits they will be disciplined by the 'natural' Market, and unless they participate in free trade they will be left behind. Once again, the relationship between people and the economy was reversed. People now had to serve this 'natural economy' because it was in their long-term interest to do so regardless of the consequences to their lives, communities, and ecosystems.[11] It is simply astounding how long the pseudo-science of monetarism continues to reign relatively unchallenged, even after the recent economic crisis triggered by Wall Street. This also is difficult to explain without reference to our contemporary secular myths. These economists spoke in the name of science, and their barrage of facts, equations, and mathematical theories had the appearance of being as exact as physics.

As a discipline, modern economics has not come to terms with the developments of the last century, because these are invisible from the silo. The scale of the human economy has become very large relative to that of the biosphere in terms of the flows of resources and energy, and capital is no longer as important a factor of production as it was during the nineteenth and early twentieth century. Highly specialized scientific and technical knowledge has become a factor of production, and the relative influence of this factor on economic growth has steadily increased. As discussed previously, John Kenneth Galbraith clearly described how the structure of large corporations had to change to permit them to make use of this highly specialized knowledge, which led to a new form of organization that he called a *technostructure*.[12] He also recognized that, both on the input and output sides, these new corporations could not rely on the Market because they had to plan the entire technological cycle with each product, using the latest specialized knowledge. These and other developments created a planning system that is now increasingly integrated by enterprise systems linked by satellites and the Internet. Despite Galbraith's and other voices, modern economics continues to assert that the mechanisms of the 'natural market' have an unlimited capacity to produce growth as long as we enslave ourselves to it. Richta and his colleagues in the former Czechoslovakia brought the analysis of Karl Marx into the twentieth century by concluding that the influence of scientific and technical knowledge had changed everything.[13]

Economic heretics, including Keynes, Galbraith, and Daly, have had to leave the silo of the discipline of economics because in one way or another they are all convinced that the economy is there to serve people as opposed to people serving the economy. Convictions of this kind are incompatible with achieving intellectual satisfaction within a disciplinary silo. Windows have to be created that have to look out on human life, society, and the biosphere in order to understand how we contribute to a particular category of phenomena and how these phenomena in turn contribute to human life and society. At minimum, this requires disciplinary silos with a great many windows. A body of knowledge cannot remain separated from experience and culture. Without such a conviction, life in a disciplinary silo may be intellectually satisfying. The mathematical models of modern economics are elegant and persuasive. Also, the higher the level of desymbolization of human life and society, the more the autonomy of economic phenomena and hence of the discipline of economics will appear plausible. However, this will be acceptable only if we are willing to give up on ourselves as a symbolic species.

As economics gradually turned itself into a discipline separated from experience and culture, it became incapable of noticing how human life and society were passing it by. The conditions that made *homo economicus* a plausible model have not existed for a century. The environmental crisis as a sign that natural capital is being depleted has been intensely discussed since the 1960s. It soon led to a global consensus that contemporary ways of life (based on technological and economic growth) are unsustainable. The markets described by Adam Smith, in which the size of any producer was small compared to the markets for its products, have been the exception for nearly a century. The mutual advantages of free trade disappeared when capital became internationally mobile. In response to the recent crisis on Wall Street, economists offered little but a deafening silence. When we put all the evidence together, it appears reasonable to come to the conclusion that the examination of economic phenomena in isolation from all others is not only unscientific but also tied to half a century of bad advice, with devastating consequences for the lives of billions of people. Worst of all, there is mounting evidence that, for the first time in human history, economies are being transformed (or have already been transformed) into anti-economies because they now extract wealth as opposed to creating it.[14] The most important contributing factors will be summarized as follows.

In the previous chapter we noted that the global economy is largely uneconomic as a consequence of speculation. Less than 3 per cent of daily currency turnover in international financial markets corresponds to the currency required to finance world trade and foreign direct investment.[15] After the deduction of 30 per cent to account for the laundering of money obtained from the exchanges of illegal goods and services,[16] the conclusion imposed itself that these financial markets are almost entirely in the service of non-productive speculation. We have all known for some time that the financial services sector creates financial bubbles,[17] but few seem to appreciate their enormous size and the associated risk of a global financial meltdown.[18]

It may be tempting to interpret this development as an extension of a well-known pattern of events that accompanied industrialization. As a result of the growing productivity of land and labour, the people who became redundant in agriculture were absorbed into the growing manufacturing sector of the economy. When automation and computerization vastly increased its productivity of labour, the manufacturing sector shrank to the benefit of the service sector. Since it was difficult to imagine that society's need for services could be satisfied any time soon, the future appeared very reassuring.[19] Financial services became an important component of the service sector. It is tempting, therefore, to interpret one of the most significant economic changes of recent times as the creation of a fourth sector in the economy. The employment of capital in this fourth sector of the economy may well be more than twenty times greater than that in the other three sectors combined, according to the above figure of daily global currency flows. Such an interpretation is problematic because this fourth sector is entirely unlike the other three. It is based on making money from money, without any intervening production of goods and services. Since such speculation is a zero sum game, and since this game has been most effectively played by hedge funds, it amounts to a massive extraction of wealth from the poor and powerless to the benefit of the wealthy and powerful. The very least that can be said about this new fourth sector is that it is entirely non-productive and non-economic.

Until recently, almost all moral, religious, and political traditions would have condemned such activities. This has largely disappeared in the United States. A powerful evangelical elite appears to have forgotten that in their Bibles the Jewish kings and religious leaders were judged on how they dealt with the poor, the vulnerable, and the weak.[20] Among the elite is a group referred to as the Family, which has completely reinterpreted the Jewish and Christian traditions in terms of efficiency and power.[21] This group appears to be unaware that such a reinterpretation represents the servitude of the gods of our time as much as the service of the golden calf did in an agricultural society.[22] As was the case in the industrializing societies of the nineteenth century, Christianity has been reinterpreted according to the necessities of the time, legitimating everything that this tradition condemns but cannot live without. Individual salvation is emphasized in order to cover over the role being played by the evangelical community in the United States under the guidance of utterly corrupt political and religious leaders. This also is a situation about which the Bible has a great deal to say.

Continuing the above explanation a little further, the growth of the fourth sector in the economy has made the role of capital completely abstract, all but severing its ties from real productive service and activities. The production of money from money represents the triumph of technique over capital.[23] Techniques such as operations research backed by powerful computers are used to monitor and intervene in financial markets. An enormous technical ingenuity has gone into designing a host of financial instruments to, as efficiently as possible, make money from money. It represents a thermodynamic impossibility and an enterprise of non-sense. It is a first step in the creation of anti-economies that extract rather than create wealth.

The creation of this fourth sector in the economy has many historical antecedents. It is surprising how quickly the Market was co-opted by speculators of all kinds, to the detriment of producers and consumers.[24] The transition from the traditional role of the Market to its being harnessed as an instrument of speculation occurred during the period from 1960 to 1990. Today, almost all of us have, voluntarily or involuntarily, become speculators. It begins with our buying contracts for the supply of gas or electricity and ends with the necessity of putting some money away for our old age.

A second factor in the creation of anti-economies becomes evident when we examine the tiny portion of the global economy that is still involved in the production of goods and services. This is becoming increasingly uneconomic under the influence of technique. In chapter 4, we noted that as far back as 1993 the social costs that corporations imposed on the U.S. economy had already amounted to twice the entire federal budget and nearly half the nation's gross domestic product.[25]

Another investigator reported that these costs represented roughly five times the profits earned by these corporations.[26] In other words, if we modified our accounting procedures to reflect genuine wealth production, these corporations would have gone bankrupt a long time ago.[27] We also noted that these trends are confirmed by an examination of national economies. If the costs incurred in the production of the gross domestic product were subtracted, net or real wealth would be stagnant or declining, depending on the assumptions made (and this has been going on since the 1970s).[28]

A third factor in the creation of anti-economies is exemplified by the federal, state, and local governments' largely abandoning their roles to protect and nurture the common good. The vast scope and magnitude of this decline have been documented by David Johnston.[29] It represents a massive redistribution of wealth by taking from the many to give to the few in the name of helping communities survive in the global competitive Market. For example, Wal-Mart Corporation owes much of its success not to entrepreneurship and creative corporate strategies but to concessions extracted from different levels of government: deals involving free land, long-term leases below market rates, tax reductions or exemptions, and workers trained at government expense. Johnston reports a spokesperson admitting that Wal-Mart commonly seeks subsidies for opening new stores, in about one out of three cases by having local and state governments compete to obtain the supposed economic benefits for their communities.[30] This scheme is hardly limited to Wal-Mart and is widely practised by automobile manufacturers and others.

In order to justify the subsidies, these governments uncritically use the claims made by corporations that they provide jobs and economic development. No performance standards are negotiated during these deals, and follow-up studies are rarely done to ensure that these benefits did indeed materialize. The problem is equally acute for the building of sports facilities and the sponsoring of official games.[31] A spectacular Canadian example was the building of the SkyDome (now the Rogers Centre) in Toronto. When citizen groups that are defending the common good seek to obtain some evidence of these government and corporate claims, the results usually show another instance of taking from the many to give to the few.[32] Many corporations know full well that the building of government relations by lobbying represents one of the most lucrative business strategies.[33]

The problems are even more acute in the U.S. private health-care sector and in government drug procurement.[34] It is not surprising that the corporations involved did almost anything to prevent the creation of a public health-care system. This included the spread of a great deal of misleading information regarding the Canadian public health care system. The role that the Republican Party played in this drama should have been embarrassing to the majority of the American people. How can one of the richest nations justify not being 'your brother's keeper' when it comes to basic health care that most cannot afford? One might have thought that evangelical support for the Republican Party would have vanished forever. The privatization of other public sector services such as education, security, incarceration, and municipal services would open up even greater opportunities for transferring a society's wealth to the very top. Despite the rhetoric of free trade, these activities are so deeply woven into the U.S. economy that they substantially contribute to its uneconomic character.[35] Canada is rapidly catching up with the United States under the leadership of a government of Market fundamentalists, with the same support from evangelicals.

Another contributing factor in the creation of anti-economies is the transfer of wealth from the many to the few. It goes under the name of deregulation, which redistributes costs and risks. The official explanations are well known and are once again rooted in Market fundamentalism: for example, regulation interferes with the 'natural' workings of the Market, thus opening it up to corrupt bureaucrats and politicians; corporations know their own business best, including what their customers demand and expect; and government should get out of their way since this kind of self-regulation is far more efficient than what governments can achieve. There is no point repeating all the other usual mantras, but the result is once again that the poor, vulnerable, and weak are exposed to growing risks, with a court-sanctioned impotence to defend themselves.[36] The results of all this deregulation are, for the greater part, uneconomic and represent huge losses of productivity and substantial increases in the costs of running the economy.[37]

To sum up, technique appears to have transformed economies into anti-economies that extract rather than create wealth. The traditional sectors of the economy have growing uneconomic components, the fourth sector is completely uneconomic, and the fifth sector of illegal activities is also entirely extractive. The economic emperor has new clothes, all persuasively described by the monetarist economists and their political disciples. Many religious leaders have also joined the pack. There really is nothing new under the sun other than that technique has replaced capital. All we need to do is to reread Max Weber's study of the relationship between capitalism and the Protestant ethic.[38] We now have a new Protestant ethic that is an expression of and a submission to technique.[39]

If the worship of the economic emperor were limited to the above groups, there would have been a political and religious rebellion decades ago. There is a complete absence of a political and moral imagination across the entire political spectrum. There remain a few dissenting voices, but almost no one pays any attention to them. The warnings of those who are invited by the mass media rarely have much effect, because what they say will be decontextualized from their message in order for it to be recontextualized so that the show will work for the intended audience, with the result that their voices are encapsulated in a very different message. In the face of this situation and the almost complete disregard of what is really happening, we must face the possibility that we also, like all those who went before us, are swept along by the currents created by our new myths. What I am suggesting is that just as capital was the lifeblood of the industrializing societies of the nineteenth and the first half of the twentieth century, so technique is the lifeblood of our age.

It is not a question of simply condemning irresponsible corporations, greedy chief executive officers and their lobbyists, corrupt politicians and their underlings, or religious leaders in sheep's clothing. They play the social roles necessary to evolve the technical order. There are no good or bad ways of doing so but only differences in the technical competence with which efficiency and power are pursued. After all, an efficiency of 85 per cent is always better than one of 84 per cent, and anyone who systematically chooses the less efficient options on the grounds that they are economically, socially, or environmentally responsible will be fired or will fail and disappear from the technical order. Experts and specialists can be distinguished only by their levels of technical competence, and certainly not by their moral, religious, or political convictions. The pursuit of efficiency and power has become the only way of life. Even if everyone were to wake up as saints tomorrow, the order of technique would constrain people's knowing and doing in such a way that much of the destruction of human lives, societies, and the biosphere would continue. Over and above this 'structural' component exists the component resulting from actions that are deliberately in excess of all of this. The structural component is the systematic disregard for the way in which everything is related to everything else in human communities and the biosphere. The intellectual and professional division of labour limits our attention to improving the desired outputs that can be obtained from the requisite inputs, as if they were not integral to everything else. Consequently, the individual and collective pursuit of technique is not only uneconomic, because of a rapidly deteriorating ratio of desired to undesired effects, but also psychopathic. The problem is integral to the technical approach to life and the technical order it creates. Good intentions and the invocation of values lead to inefficiencies that will not be tolerated very long and that in many instances are merely naive smoke screens.

In order to understand this, recall our reciprocal relations with our surroundings: as people change and evolve technique, they are simultaneously changed and evolved by it. Without the latter's influence, we might have had the moral, religious, and political resources to offer a great deal of resistance. Instead, the influence of technique has been such that, like all those who went before us, we are possessed by the spirit of our age and the gods that rule over it. All this is both good and bad. For example, our current economic recession would instantly have turned into a complete economic meltdown if most people had had a realistic view of what the economy has become; if this were to happen, the human suffering would be unimaginable.

Almost without exception, analyses of our contemporary situation ignore the 'technique changing people' component of our relations with our surroundings, which causes cultures to be realigned by the system and transforms a great deal of human behaviour in its image. This has given rise to an economic vision of a Market liberated by free trade and globalization to maximize growth through unbridled competition. It is peculiar at best. It amounts to regarding the evolution of the biosphere in terms of the survival of the fittest members of any species without accounting for the whole.

In the course of industrialization, human life became increasingly regarded as essentially economic, in the sense that economic phenomena dominated all the others in human life and society. However, their further evolution by mean of technique gradually made these economic phenomena uneconomic. Making anything efficient by maximizing its power to produce a desired output from requisite inputs amounts to detaching it from everything else and reorganizing it without any consideration to everything else. Within the biosphere this helps to create a process of de-evolution, and in society the process of desymbolization. It is the antithesis of all life, both biological and cultural. By being separated from experience and culture, the scientific knowing and technical doing, as well as the economic growth based on them, have been allowed to largely disregard their contexts (human life, society, and the biosphere), with the result that they undermine them. In other words, the growing evidence that the costs incurred in economic growth lead to a reduction in net wealth production represents the economic dimension of this undermining, and this economic extraction will inevitably lead to collapses of one kind or another unless we decisively intervene. Thus far, our attempts at intervention amount to little more than concentrating on one symptom or another and discarding all the rest. There has been a proliferation of good causes, many of which are taken on by non-profit and non-governmental groups who all imagine that the world would be better off dealing with our situation one issue at a time. Unintentionally they contribute to the situation, which is a disaster in the making.

Bringing these considerations back to the discipline of economics, it follows that its shortcomings are but a single instance of a widespread pattern. When we know and evolve one category of phenomena with little or no regard for the context of human life, society, and the biosphere, we place ourselves among the most foolish of all human societies and cultures. Dealing with economic phenomena outside of their context is to make it impossible to understand the influence of this context on people's confidence in the economy. Not considering how consumer and investor confidence is affected by the things we do is not to understand something on which everything economic ultimately depends. After all, this confidence is embedded in experience and culture and thus beyond the reach of the discipline of economics.

Thus far we have encountered five economic sectors: agriculture, manufacturing, services, financial services based on derivatives, and illegal activities. To this must be added three more: a sixth sector made up of the 'underground' or 'shadow' economy; a seventh sector comprising everything in human life and society that sustains economic activities without remuneration; and an eighth sector made up of all the services rendered to economic activities by the biosphere (such as those of pollinating insects) that are technically compensated for as a consequence of their unsustainable use.

The first three sectors are almost entirely in the grip of technique. The planning system has become highly successful at reducing its dependence on the Market to an absolute minimum. The desymbolization of cultures and the necessary technical supplement of integration propaganda have largely undermined the sovereignty of the consumer, and, to the same extent, economic democracy has been disabled. The fourth sector employs a host of sophisticated techniques that harness the Market to speculation. The fifth, sixth, and seventh sectors are, like the remainder of human life and society, under enormous pressure from the technical order. The eighth sector constitutes the economic dimension of the environmental crisis.

If this very simplistic categorization of economic activities and their contexts is of any value, it follows that various proposals for addressing our current economic woes have a very long way to go. For example, imposing a reduction in the scale of the global economy as a subsystem of the biosphere is surely a necessary condition, but it does not address how the technical order must be mutated to make this possible. The environmental crisis is directly produced by the technical order, which cannot grow and evolve other than by disordering human life, society, and the biosphere. Hence, there appears to be no alternative other than resymbolizing the technical order in terms of what it is doing to human life and the world. This will give us a better understanding of how economic phenomena evolve in relation to all other phenomena within the technical order. We must accept full responsibility for this since there is absolutely nothing natural about it. If we are ready to do so, we can put this understanding to work. Subjecting technique to human values will be a long and difficult struggle, but it will gradually establish a balance between human values and efficiency as measured by output-input ratios. Eventually it could bring technique back into a greatly strengthened cultural order that would compel its mutation. All this assumes that the desymbolization of our cultures is reversible. No one can know this, but it is our responsibility to undertake this task.

Industrialization created the conditions that permitted economics to transform itself into a discipline separated from experience and culture. Its examination of human life, society, and the biosphere in terms of one category of phenomena has been an intellectual disaster. From the disciplinary silo it was impossible to see how the relations between economic phenomena and everything else continued to undergo rapid changes, with the result that the assumptions that were perfectly valid for much of the nineteenth century are no longer valid today. Bringing these assumptions in line with today's conditions is the equivalent of removing the foundations of the entire intellectual structure of economics. Rebuilding it will not be an easy task. Moreover, resymbolization will not be the only option for doing so. Perhaps the best we can hope for is a growing consensus that the study of human life and society one category of phenomena at a time has let us down. Such a consensus would bring a growing recognition that much more dialogue with others from beyond our own disciplines is required. At first this dialogue may be informal with kindred spirits from within and beyond our departments and disciplines, but it will not thrive until university reforms create a sustaining institutional support.[40]

Any resymbolization of economics could begin with a critical and detailed examination of the hypothesis that we have created anti-economies.[41] The approach should be similar to that taken by the founders of the discipline: economic phenomena must be examined in the context of all other phenomena that make up human life, society, and the biosphere. Doing so requires intellectual round tables where members from different disciplines, having come to similar conclusions about the limitations of their own approaches, begin to examine what has happened during the last two centuries. This will involve intellectually 'mapping' the interactions between all these different phenomena, including their joint evolution. There are two constraints on this mapping. First, the economy operates within thermodynamic constraints. The second is that human beings live lives and that jointly these lives contribute to the historical journey of a community, no matter how desymbolized and alienated it may be. As a result, every category of phenomena is a unique manifestation of this individual and collective life. At the same time, it is subject to a reciprocal interaction. Careful attention must be paid to the balance between the extent to which people can change their world according to their needs and aspirations and the extent to which this world affects their consciousness and way of life. Some may object that this introduces a value judgment about human life, and they are correct. Fortunately humanity is now united in regarding enslavement as an unacceptable form of life. This, however, brings us into the analysis as observers open to counter-transference reactions – a possibility that must be carefully monitored.

There will almost certainly have to be a number of these intellectual round tables proceeding in parallel. They would be distinguished by their intellectual foci, each concentrating on how everything relating to everything else unfolds a particular category of phenomena. Their interpretations will undoubtedly lead to different 'intellectual base maps,' and from these differences parallel schools of thought may emerge. In this way the social sciences can avoid the problems of intellectual monocultures, provided that these schools of thought operate as an intellectual ecology. After all, they are engaged in the same task: understanding an unmanageably complex reality by simplifying it in some way. Doing so inevitably involves value judgments as to what is important enough that it must be retained, what is less important but must still be considered, and what is so peripheral that it may be ignored. Anyone who believes that making theories or models can be done without these kinds of value judgments has not taken into account the ideological baggage that has come with discipline-based scholarship, which avoids this problem by creating many others. If each school of thought recognizes that making human life and the world intelligible is an exceedingly complex and open-ended task, it may also recognize the essential role for dialogue with all the other schools of thought. Given the above constraints on the intellectual process, this certainly does not mean that anything goes, especially in an intellectual ecology where every participating school of thought evolves in relation to all the others and jointly in relation to the reality they seek to understand. Every school of thought must be open to the possibility that it has overemphasized, underemphasized, or neglected certain aspects. It may make academia a much more collegial and stimulating place to work.

There will be no question of returning to the past, when each discipline was constituted by multiple schools of thought and the notion of discipline-based scholarship precluded meaningful dialogue. Once disciplines no longer examine human life and society one category of phenomena at a time, the situation will change completely. Any category of phenomena now becomes an intellectual focus for understanding different aspects of how everything is related to everything else in human life and society. It will no longer be a question of knowing more and more about less and less until eventually a sub-specialty knows everything about nothing. Nor can it be a question of abolishing any intellectual division of labour in order to create a 'holistic' approach that knows less and less about more and more until nothing is known about everything. The suggested approach takes a middle course: it seeks to establish a dialectical tension between breadth and depth by symbolizing a category of phenomena in relation to all others and at the same time understanding its unique 'internal' features. It may be regarded as the explicit and conscious equivalent of the metaconscious processes of differentiation and integration.

Some details of the task that awaits us may be illustrative. Karl Polanyi has argued that the emergence of markets and a money economy constitutes the Great Transformation.[42] In any case, the corrosive effects of money and capital on traditional cultures is well known but less well explained. Before anything can be traded in markets, it must be commoditized, and before anything can be commoditized, it must be desymbolized. As a result, markets substitute monetary values for cultural values, with far-reaching consequences. A Market economy requires that almost everything take on a monetary value. When money becomes the value of values, it cannot only take on the status of a metaconscious myth and possibly that of a central myth or sacred. When this happens, either traditional religions must be reinterpreted or secular religions must be born, with the well-known consequences for human life and societies.

If we were truly convinced that monetary values are a poor substitute for cultural values, we would use markets much more prudently. They are a human creation like all others and the best means we have found thus far for regulating the production and distribution of goods and services. We get into trouble when the next step is taken. Seeing no alternative to markets, we declare them to be natural. Instead of using our invention carefully and critically by realizing that, like all other human creations, it is good for certain things, harmful for others, and irrelevant to still others, we use them indiscriminately.

A more realistic approach is based on a critical evaluation of what markets can and cannot do for us and to use them accordingly. An awareness of the inherent limitations of markets and monetary values would help us make clear and effective decisions as to what should be kept in the public sector of the economy and what can best be dealt with by the private sector so as to yield the best possible ratio of desired to undesired effects. It is but a first step towards reharnessing our economies to our needs and values. The planning system must also be modified. This will require a different kind of professional division of labour to complement the restructured intellectual division of labour under discussion.

Markets would perform their roles much better if they were proactively protected from speculators. The equivalent of a Tobin tax (named after the Nobel laureate who proposed it) would be a simple means for vastly reducing speculative bubbles by levying a modest tax, with the income being used to benefit the public good. It is astonishing that the last crisis produced by Wall Street has not led to the implementation of such a tax. It shows the enormous clout of the fourth sector in the economy, which is engaged almost entirely in uneconomic activities. Such a tax would return markets to their role as the best means for organizing the production and distribution of goods and services.

The ratio of desired to undesired effects of these markets can be enhanced further by appropriate regulatory frameworks. For example, the labour market was much less destructive of society when children were put out of its reach. Technique must be confronted. Since technique evolves everything by means of contextless output-input ratios coupled to highly desymbolized cultures, ever more regulation was required to technically compensate for what cultures used to perform free of charge. Such an end-of-pipe approach has turned out to be so ineffective and costly that we have thrown up our hands and abandoned our future to the global Market, which is 'liberated' by free trade and globalization. It is not doing anyone much good. Corporations compete in a global race to externalize as many costs as possible because all so-called trade barriers such as labour, social, health, and environmental standards are constantly being weakened. This places chief executive officers in the role of having to take every possible action to justify growing share prices, by imposing ever more relentless ways of externalizing costs. It is a classic case of positive feedback that is driving our lives and the world in the wrong direction. It is the primary mechanism by which our economies have been transformed into anti-economies under the influence of technique.

These trends must be reversed by imposing adequate labour, social, health, and environmental standards on the global Market. The arguments against imposing such standards amount to an extreme case of 'short-term gain for long-term pain.' It is true that, until now, higher standards have been almost exclusively achieved by adding end-of-pipe technologies and services. The higher the standards, the more crippling are the costs on the system.

There is an alternate way, however, in which standards can be raised. As I will explain shortly, a change in the intellectual and professional division of labour opens the door to preventive approaches, which can achieve a much better ratio of desired to undesired effects. Prevention is almost always cheaper than end-of-pipe compensation, as we have recognized by creating the precautionary and no-regrets legal principles. Many have doubted the potential of preventive approaches, arguing that it amounts to a free lunch. This reaction is understandable because it is based on past experience; higher standards were achieved by end-of-pipe approaches. With a different intellectual and professional division of labour, not only does this objection evaporate, but feasible alternatives open up. It will permit economies to deliver goods and services with a fraction of the burdens currently imposed on human life, society, and the biosphere. As preventive approaches spread and grow in sophistication, standards can gradually be raised, and this in turn will significantly improve the ratio of the desired to the undesired effects of markets.

What has erroneously been called free trade must also be rethought. Genuine free trade is supposed to benefit all parties as a result of the comparative advantages they bring. Such advantages occur only when there is no international mobility of capital. Investments withdrawn from particular activities would flow to other sectors of a nation's economy to create alternative employment, thereby protecting a society from unemployment and underemployment. The invention of the theory of mutually advantageous free trade came during a time when the international mobility of capital was minimal. Free trade is genuine only when both parties benefit, otherwise it will be *forced* trade for one of the parties. The assumptions underlying the possibility of mutually advantageous free trade were invalidated by the developments of the twentieth century, particularly following the Second World War.

This raises the question of what free trade is under the current circumstances. Freedom is granted to flows of goods, technical services, and people with an advanced education, and it is withdrawn from communities that can no longer object to these flows when they produce unacceptable harm according to their democratically established standards and regulations. Freedom makes the world safe for the indefinite expansion of the technique-based connectedness at the expense of cultural orders, thus permitting democracies to operate as long as they do not interfere. It also permits an almost limitless expansion of the speculative bubbles associated with a larger component of global currency flows. Hence, free trade is forced trade for people and their communities.

We must also resymbolize many daily-life activities. What happens when we deposit money in a bank, buy shares, contribute to our pension funds, and use paper money? Herman Daly, referring to Frederick Soddy's theory, points out that such deposits in effect result in banks' printing money.[43] Except for a small portion, banks are permitted to lend such deposits to third parties. Such lending has no commonality with people lending to others, because they in effect give up the use of these funds in return for a payment of interest or for a personal favour. Banks, on the contrary, give up nothing, because they do not own the money they lend. If bank loans are not repaid, the depositors have not lost their money, because they have given no permission for these loans.

The money supply has in effect been increased by permitting banks to borrow against a very small portion of deposits kept in reserve.

As a consequence of the global speculative bubble we must ask, when people buy shares in a corporation, what they really own. Their money does not flow to the corporations in which they are investing in order to be added to their operating capital, resulting in the direct or indirect expansion of a company's productive capacities. The funds invested in the purchased shares flow to the previous owners of these shares, and these are bought in the confidence that someday other people will be willing to offer more money for them. These investments contribute to a speculative bubble that has only the most tenuous link to economic reality. All investment schemes, including mutual funds, attract more and more money to bid up the prices of shares beyond what they actually represent. Instead of reflecting the ability of corporations to deliver goods and services, it becomes simply a matter of supply and demand until an 'adjustment' occurs, when people begin to recognize the lack of value in the stock market. In the meantime, this puts all the more pressure on chief executive officers and governments to increase sales and exports, which results in still greater wealth extraction.

When people make contributions to their pension funds, they directly and indirectly stimulate wealth extraction and accelerate the coming of the next adjustment, and possibly an eventual financial meltdown. People help to ensure that the world in which they will be retiring will be less financially sound, even if adjustments do not take back a significant portion of the value of their pensions. Few people have the necessary background to work out the implications of the distinction between productive investments that increase genuine wealth and extractive investments that actually decrease it. The situation has become so absurd that corporations can frequently make higher profits by participating in speculative bubbles than by producing the goods and services that sustain a way of life. In addition, when there is a meltdown of public confidence, governments usually have to borrow enormous sums, which is how we all subsidize the speculative bubbles at the expense of true wealth creation.

Herman Daly has argued that the significance of paper money may be explored by imagining that the real economy could be separated from the extractive economy and observing what happens at the interface.[44] The real economy involves exchanges of people's productive and creative capacities for earned incomes as well as exchanges of portions of these incomes for expanding and improving this production, which depends on exchanges with the biosphere. The extractive economy includes costs that corporations externalize into human life, society, and the biosphere, and government subsidies to these corporations as well as loans required to restore confidence in their currencies and the international speculative bubbles. Owing to information that flows across the imagined interface, corporations are obliged to make enormous amounts of money in order to increase the value of their shares, and this can best be accomplished by externalizing as many costs as possible. Similarly, governments are compelled to implement monetary, fiscal, and trade policies that maintain the confidence of international speculators, and thus they reinforce and amplify the extractive component of their national economies. The making of money from money in the global speculative bubbles increases the claim of this virtual wealth against the real wealth of a society. Hence, any investment in these bubbles is extractive. Individual recipients of their newly made money are able to exchange it for real wealth as long as most of the others reinvest it in the bubbles. If all the money created within the bubbles were used by their owners to claim real wealth, it could not be done because all the assets are already owned by others. As Herman Daly explains, the money created within the speculative bubbles thus becomes a national debt owned by the individual but owed by the community.[45] In his discussion of Frederick Soddy's theories, he explains that money is transformed from an exchange medium into an interest-bearing debt as long as the bulk of that debt is not liquidated.

Since the Second World War there has been a complete transformation of the role of money and capital in the economy in particular and in human life and society in general. The extractive component has grown to the point that our current systems have nothing in common with what was referred to as capitalism in the nineteenth and the first half of the twentieth century. It may be argued that these transformations have created anti-money as well as an anti-economy because of their destructive effects on true money and true economies. All this is perfectly invisible to the discipline of economics; hence, resymbolizing the economy in parallel with other efforts (outlined in the subsequent sections) is becoming more urgent every day if we are to avoid an economic meltdown.

Resymbolizing the Social Sciences

Once economics had been turned into a discipline, the other social sciences soon followed suit. Their transformation was greatly facilitated by the kinds of changes described in the previous two chapters. Intellectually I lived through these changes, not because I am that old but because of how I came to the social sciences. I did not intellectually grow up in these sciences although I have taught them all my professional life. While completing a doctoral thesis during the early 1970s, I decided that engineers needed to learn a great deal more about the social and environmental consequences of their professional activities if we were to effectively confront issues such as the environmental crisis. It led to my first encounter with the discipline-based approach to understanding human life and the world. After I had examined a number of introductory textbooks to the social sciences, it quickly became evident that they were able to describe much of human life and society with very few references to the influences of science and technology. Typically, there were the occasional acknowledgments of these influences, but the principle arguments and theories remained almost entirely unaffected. Many years later I was asked to contribute a chapter on technology and social change to an introductory sociology text, and it ended up near the end of the book. I may well have had an exaggerated sense of the importance of science and technology, but it has always struck me that it ought to be impossible to give an adequate explanation of human life in an industrially advanced society from the perspective of any social science discipline without intellectually bumping into science and technology. I was also convinced that on my side of the campus we faced the mirror image problem. We learned very little about the way technology influences human life, society, and the biosphere and even less about the use of this understanding to ensure that we achieved the desired results and, at the same time, prevented or minimized harmful effects (I will say more about this in the next subsection).

While writing a research proposal for a postdoctoral fellowship, I came across Jacques Ellul's book *The Technological Society.*[46] It intellectually changed everything. Two things struck me. First, implicit in this book was the most accurate description that I had ever encountered of how my technical mindset worked. Second, this description helped me to understand that this was a structural and not a moral issue. As engineers we all use the same methods and approaches, the results of which are integrated with those of all other specialists within an institutional framework. This 'wiring' together of our knowledge and skills causes many unintended, undesired, and unforeseen results that no individual practitioner could understand, let alone control. Whatever the moral, political, or religious intentions of the practitioners, there is a knowledge system with characteristics that depend more on its structure and organization than on the intentions of the participants. Good intentions and noble convictions are at best a necessary but insufficient prerequisite to bringing about essential changes. It became clear to me that the problems of our civilization are both within and all around me.

My postdoctoral mentor gave me, as an outsider, almost complete freedom to learn and make sense of the social sciences. Deeply influenced by their great thinkers, I sought to follow in their footsteps by imagining what Karl Marx, Max Weber, Emile Durkheim, and many others would have found if they had observed the human life and society of my days. There was nothing particularly radical about this, because the European scene was not yet as much in the grip of the discipline-based approach as was the North American. These influential thinkers might temporarily place a particular category of phenomena in the foreground of their analyses, but they never abstracted them entirely from its context. I wrote a first draft of what was to become *The Growth of Minds and Cultures*,[47] which sought to explain how the many different phenomena that were examined by the social sciences mingled in individual and collective human life, beginning with the life of babies and toddlers.

Along with a better understanding of the way in which many different categories of phenomena mingled in human life and society came the realization that some kinds of knowing and doing did not participate very much in all of this. It led to the theory of scientific knowing and technical doing being separated from experience and culture, thus requiring a distinct sociology of science and a sociology of technology. Furthermore, societies in which this knowing and doing dominated the daily-life knowing and doing embedded in experience and culture had little in common with traditional societies, thus requiring a sociology of industrially advanced mass societies. The chair of my sociology department agreed with this assessment, and my students from the social sciences as well as engineering were enthusiastic in their support and, fortunately for me, were very vocal about it. Thirty years of research have amply confirmed these early findings.

We find ourselves largely in agreement with Horowitz,[48] who has examined the decline of the 'big five' social sciences (economics, political science, sociology, psychology, and anthropology). Problems with disciplinary boundaries were traditionally remedied by 'hyphenated' subdisciplines such as political sociology. Although the creation of such sub-disciplines was helpful, it did not stem a deep-seated concern that individually and collectively these disciplines were losing their grip on understanding human life in contemporary society. There was a minority of generalists who continued to practise the approaches of the founding thinkers, examining a category of phenomena against the background of all the others. A growing majority of 'specialists' asserted that these approaches lacked scientific rigour and that only quantitative ones had a future. Data were collected by means of surveys in order to make the actions of various groups, organizations, and institutions more effective. When a victory of this approach appeared certain, entirely new developments emerged.

According to Horowitz, during the intellectual turmoil the 'little five' sub-disciplines arose: communication, environmental studies, urban affairs, criminology, and policy studies.[49] These developments may be interpreted in a somewhat different way, namely, as responses to the many new developments in human life and society. The communication sub-discipline attempts to understand the ways in which all relations between human beings in a society are being transformed by computer and information technologies. Environmental studies seek to understand the ways in which, by using science and technology for economic growth, our civilization is changing the biosphere, and how this in turn changes human life and society within it.

Urban affairs includes cities but goes well beyond them to the patterns that are making the old dichotomies between urban and rural, city and farm life, traditional and modern disappear under the influence of universal technical and economic growth. These are pressing everything into the service of efficiency and performance, thereby creating the megalopolis on the national level and globalization on the international level. All this is strongly linked to the effects that it has on the biosphere, and how this in turn will affect urban affairs.

Criminology is closely linked to the transformation of traditional societies into mass societies. The latter evolve much less on the basis of traditions, customs, and cultures than on a scientific and technical knowledge infrastructure. As a result, the guidance traditionally provided to individuals has been greatly weakened, to the point of necessitating new ways of organizing societies, including other ways of dealing with behaviour that threatens the social order. Weakened and desymbolized cultures offer a much lower resistance against antisocial behaviour of all kinds, which now appears to be as common at the top as at the bottom of the social hierarchy.

Policy studies attempt to evaluate what went right and what went wrong in our frantic efforts to improve everything by means of scientific knowing and technical doing. It also seeks to evaluate the myriad of new technical possibilities and the risks they represent. Custom and tradition are being replaced by the compulsion to do almost everything that becomes technically possible. Ironically, these developments in the social sciences bring them into the orbit of making everything better by means of the technical approach. This is driven by the conviction that we urgently need better strategies in almost every sphere of human life. Big or little, the social sciences are taking on a technical aspect at the expense of the pursuit of knowledge for the common good.

Interesting as these interpretations may be, they do not really get to the root of the problem. The fact of the matter is that governments and university administrators feel that the public university in general, and the social sciences in particular, are not delivering enough value for the investment of public funds. However, a solution has been found. University reform would bypass the institutional structure of faculties, departments, and disciplines. Research chairs have been established to head networks of researchers to pursue issues deemed by others to be of great importance. In professional faculties, research funding to individual researchers increasingly requires ties to industry or government ministries. Whatever reasons are provided for these developments and whatever grounds are invoked to support or oppose them, the result has been to draw the public university into the orbit of the technical order and the many challenges it creates. Examining human life and the world one category of phenomena at a time has been displaced by doing so one issue at a time.

From a social science perspective, does the rise of the 'little five' disciplines and the research networks represent ways of transcending the limitations of discipline-based knowing? The answer to this question is painfully obvious. Academics are cross-appointed to the 'little five' or informally affiliated with research networks, but they are granted tenure and promotions largely based on their contributions to a particular discipline represented by their 'home' department. I know of no university that has successfully resolved this dilemma. Unless we are willing to tackle how the contemporary university is 'wired,' in terms of the knowledge components (disciplines, departments, and faculties), the 'little five' and research networks will continue to experience all the frustrations of discipline-based knowing. The knowledge infrastructures of our societies will remain largely unchanged in terms of their inability to support decision making that creates high ratios of desired to undesired effects.

It would appear, therefore, that the emergence and the evolution of the 'little five' disciplines and research networks do not represent a triumph over the discipline-based approach to scholarship and instead represent a triumph of technique in this domain. We therefore need to tackle head on the problems of this kind of intellectual and professional division of labour.

It may be argued that the above discussion has missed the point. Discipline-based approaches emerged as a result of adopting more objective and rigorous approaches. Hard data are collected about a particular situation by means of survey-based or other research instruments. This data can be statistically evaluated and placed on a mathematical foundation to varying degrees.[50] Quantitative approaches appear to be very objective and scientific until we attempt to establish their sensitivity to the insights implicit and explicit in the process of formulating the questions for the research instrument. For example, if we asked members from different schools of thought within one discipline to prepare what they would regard as an appropriate research instrument, the results would be very different. Furthermore, the research instrument puts respondents into a hypothetical situation that may or may not correspond to their own metaconscious knowledge of what is happening. If there is little correspondence, there will be a certain level of confusion and possibly some suspicions or even distrust as to what they are contributing to. Finally, what they would really do if they were confronted with a situation in which they would have to live with the consequences might well be an entirely different affair.[51]

The problem is not so different from the one encountered in physics, where the theory largely determines what can be measured and which results will be accepted because of their dependence on the theory. The presence of parallel schools of thought means parallel vantage points, theories, and methodologies, which would obviously lead to the creation of different research instruments, data, correlations, and conclusions. There simply exists no scientific approach for deciding between them, any more than we have been able to determine which school of thought is the most scientific in its methodological approaches, theoretical framework, and interpretations. It would appear, therefore, that the quantitative approach obscures the underlying methodological issues, which have never been resolved.

It is impossible to avoid the kind of methodological issues so well laid out by Devereux,[52] to which I would like to add the possible influence of myths. Since we are all intellectually divided along parallel modes of knowing, dealing with anything beyond our domain of expertise presumably takes place in terms of the mode imbedded in experience and culture. No school of thought, or any discipline, is able to entirely sever its connections and dependence on the knowing and doing embedded in experience and culture, particularly in its vantage point, its implicit and explicit assumptions regarding the nature of what it is studying, and how this is embedded in the larger world. If, on the one hand, we accept that all human cultures are anchored in myths and that as a result we are alienated, the methodological issues raised by Devereux make all of us vulnerable; it demands humility and collegiality. If, on the other hand, we regard contemporary cultures as being secular and free from myth, we face the troubling problem of why science and culture do not increasingly converge and overlap. I will develop the former interpretation by accepting the existence of secular myths.

The presence of myths enormously decreases our ability to achieve a high level of detachment and objectivity. The reasons are both intellectual and existential. The more we develop a critical awareness of our myths and how they influence everything we do, the more we expose our existential and cultural roots in an ultimately unknowable reality. Uprooting ourselves in any way opens the door to relativism, nihilism, and anomie. Doing so would lead to even deeper anxieties than those described by Devereux, and we may be unable to keep them in check by unconsciously adjusting the interpretations of what we observe. The reason that these possibilities have been largely ignored is most certainly related to the intellectual difficulty of searching for the myths of our own culture. It requires an understanding of how everything is dialectically related to everything else and how the absolutized exceptions are anchored in myths. This cannot be done by means of discipline-based scholarship. The findings of the many disciplines cannot simply be synthesized; they must be transcended. As a result of this combined intellectual and existential challenge, it is almost certain that, to a very high degree, we uncritically accept the myths of our culture.

Let us compare Karl Marx and Max Weber on this point. Karl Marx's work was one of the best critiques of the assumptions of his day. However, with a great deal of hindsight, we can see that his sociological and historical understandings collapse like a deck of cards when the nineteenth-century myths of progress, work, and happiness are removed.

We have shown how the technical division of labour, followed by mechanization and industrialization, separated the technology-based connectedness from the cultural order of his society. One possible way of interpreting this development is by the emergence of a distinct economic base and a superstructure composed of the remainder of society. However, this development was unique to the nineteenth century and cannot be extrapolated back to the beginning of humanity and forward to encompass its entire future. Only the myths of progress and work make this possible. A development unique to the industrializing societies of the nineteenth century was absolutized as an instance of the dialectical process of history. This absolutization opened the door to the creation of the first secular political religion: communism.[53]

From a methodological perspective, Marx never solved the problem of how a scientific approach and a scientific socialism would be possible if everyone were alienated by the capitalist system. Again, this issue has far-reaching consequences. How would the revolutionary class shed its false awareness of itself and the world in order to guide humanity towards the socialist period, free from alienation? Why was the proletariat in a better position to do so than were all the other social classes? Only by means of the myths of progress and work could certain features of human life and society that were unique to a particular social epoch be transformed into more or less permanent features of human history. In these and other respects the work of Karl Marx represents what I would regard as the most brilliant intellectual elaboration of the cultural order of the industrializing societies of his day. However, it is entirely uncritical of the cultural unity of these cultural orders that represented the gods of his time.

Max Weber sought to understand the beginning of another far-reaching transformation of human life and society. With hindsight, I believe the phenomenon of rationality to be the beginning of the widespread separation of knowing and doing from experience and culture. From a sociological and historical perspective, this phenomenon became widespread first in Germany after the death of Karl Marx. Max Weber did not absolutize this development and did not transform it into a key for the understanding of all of human history. He simply warned that the consequences of permitting the phenomenon of rationality to develop would be disastrous. His interpretation did not elaborate the cultural unity that was beginning to emerge and which would eventually dominate the second half of the twentieth century. Max Weber exposed something of the roots of human existence at that time and place and the way in which humanity, with all the best intentions in the world, was really creating another form of enslavement. It made him very depressed for a time, to the point that he could not teach. Since Weber did not intellectually elaborate the new cultural unity, no absolute theory resulted and no secular political religion could be based on it.

What I am proposing is the possibility that the different interpretations of the most influential thinkers during the last two hundred years were based on their critical or uncritical elaborations of the cultural unity of their day. A strong case can be made that a great many scientific concepts and theories have been brilliant, but uncritical, elaborations of various aspects of myth and that the social sciences (and also the physical sciences, to a lesser extent), like all other human activities, have great difficulty in escaping our dependence on myth.

The simplistic views of our relations with reality have long been shattered, as I have illustrated by means of several quotations from some eminent physicists. In the same vein, the idea of emerging secular societies made up of individuals liberated from alienation implies a similar highly simplistic view of our relations with reality by means of symbolization. It is now becoming evident that this idea became thinkable only when desymbolization had advanced to the point that our myths had become very weak and much less able to provide human life with meaning, direction, and purpose in an ultimately unknowable reality. Were it not for the technical amplification of these myths by means of integration propaganda, the current epidemic of nihilism, relativism, and anomie would have been even more devastating. It would appear, therefore, that myths continue to work in the background of individual and collective human life.

If myths, the deepest forms of metaconscious knowledge, continue to play their roles in human life and society, the task of coming to grips with what is happening is a great deal more difficult than can be assumed under discipline-based scholarship. Resymbolizing a particular category of phenomena in terms of the contribution it makes to the way in which everything is related to everything else must be expanded. Anything that is exempted from this relational character of human life and society, and thus lived as if it were absolute, creates a new form of enslavement. Since enslavement is an unacceptable form of human life, any threat to freedom must be clearly understood. In the past, that threat came from the alienating character of cultural orders. Today it comes from the technical order and its permeation of the highly desymbolized cultural orders, including their cultural unities. By examining human life and the world, one category of phenomena at a time, discipline-based scholarship has made all of this intellectually invisible. It continues to have disastrous consequences for our understanding of where we are taking science and technique and where these creations are taking us through their vast influence.

If human life and society evolve primarily on the basis of the technical order and secondarily on what remains of the highly desymbolized cultural orders, the threats to human freedom will be unlike those of the past. Living in mass societies has very little in common with living in culturally ordered societies. We continue to make the same mistakes as the one that some of us lived through in the 1960s. The supposed freedom to dress in any way a person wanted, the explosion of sexual liberties, and the ability to choose the morality of the day would have indeed constituted an extraordinary liberation had these taken place in a traditional society. Instead, this happened in a mass society with a highly desymbolized culture under pressure from an emerging technical order. Seen from the perspective of this order, the so-called liberties were simply irrelevant for its maintenance and evolution. What mattered was a new technical conformity: who cared how you dressed or with whom you slept as long as you were technically competent as a specialist. Of course, almost anything non-technical 'goes' in a technical order with a highly desymbolized culture. We are certainly more decentred than people were in traditional societies. When human knowing and doing have been extensively separated from experience and culture, we are at best knowers, doers, or controllers in our work, which represents only one portion of our way of life. In all of the remainder we are on the receiving end of endless technical interventions to make everything as efficient as possible. Such reification goes hand in hand with the decentring of our lives. All kinds of cultural elements can be gathered into an eclectic mix because their ties with the cultures and histories of other peoples have been severely weakened by desymbolization.

The following correlations in the evolution of different kinds of activities within the most industrialized societies of the twentieth century are difficult to explain without the desymbolized cultures' continuing to work in the background of people's lives. Take the example of the correlation between developments in physics and those in art.[54] An explanation to the effect that Picasso was inspired by reading physics papers and that physicists were inspired by Picasso's work is difficult to sustain. However, if we accept that secular myths worked in the background of these people's lives, correlations between events in physics and those in art would be exactly as expected. Similarly, there is an extraordinary correlation between the development of art and the process of industrialization[55] and between the emergence of the phenomenon of technique and the development of various new art forms.[56] Another correlation exists between the emergence of debates in physics dealing with the nature of reality (including the extent to which it can never be understood)[57] and the weakening of our roots in reality as a consequence of the desymbolization of cultures. Similar correlations can be found in literature and poetry, which go well beyond the utopian and dystopian novels. For example, in hermeneutics there emerged the structuralist approach to interpreting a text.[58]

The above correlations between the developments of art, of science, and of technology (and later of technique) illustrate what we have referred to as technology changing people and technique changing people. The many daily-life experiences gradually externalized the consequences of this influence. Science and technology (and later technique) sought to know and change the external world each in their own way, while art interpreted the experiences of that world. All three unique kinds of human activities contributed to, as well as expressed, the evolving cultural unities. In this way, all three human activities were joined by the ongoing efforts to symbolize human life and the world in an ultimately unknowable reality. These activities also registered the profound consequences when this symbolization was increasingly undermined by the separation of knowing and doing from experience and culture.

What is true for the evolution of science, technology (or technique), and art is almost certainly a microcosm of the way in which all human activities contribute to, and are a reflection of, evolution of a technical order. It represents an important consequence of our being a symbolic species. Our lives work in the background of each and every experience, and our community's technical order and desymbolized culture work in the background of our lives. When levels of desymbolization were low, this was evident to every great thinker, none of whom examined human life and society one category of phenomena at a time. When the levels of desymbolization rose, the gradual imposition of the technical order resulted in human life and society appearing more mechanistic. The intellectual division of labour based on disciplines rapidly gained influence, and the generalists all but disappeared. Much of this amounted to an alienated working out of the cultural unities of the time.

The philosophy of technology also provides clear manifestations of the secular myths working in the background during the 1960s, 1970s, and 1980s. Any scholar who was not uncritically elaborating the new cultural unity that was enfolded within technology was dismissed as a pessimist, a determinist, or worse. The problem was not limited to the excesses of the few people who reacted as if a nerve had been touched. The issue was systemic in character. Why did the reviewers not insist on the removal of such unscientific terms and their replacement by a proper scholarly argument? Why did the editors and publishers not insist on the same thing? Why did the readers not protest against this deterioration of scholarship into a secular religious polemic between the believers in the new cultural unity that was enfolded in technology and those who were critical of it? It was an incredible experience to sit through one conference after another, hearing speakers summarily dismiss colleagues as common heretics, with few if any members of the audience challenging this travesty of scholarship. These problems were not limited to the philosophy of technology. The few people who challenged the scientifically unfounded claims of the artificial intelligence community were also treated as heretics. Even today a publisher associated with a prestigious institute of technology systematically refuses to publish works that are critical of technology, even when its referees recommend otherwise.

An extensive debate took place during that time over the kind of society or civilization that was beginning to emerge. Was it a post-industrial society, an advanced industrial society, a post-capitalist society, a global village integrated by the mass media, a consumer society, an information society, a 'technotronic' age, a spectator society, a megamachine society, or a new industrial state?[59] Generally speaking, the proponents of these theories recognized that a new and very important phenomenon had emerged whose influence on human life and society was far-reaching. Such an observation ought to have posed a serious challenge to discipline-based research. It did not. Apparently, the researchers relied on their metaconscious knowledge to go beyond their disciplines, in which case (judging by the results) the cultural unities of their societies were working in the background. Almost without exception, this led to the conclusion that technology would make tomorrow even better than today. It was not long before 'technique changing people' caught up with these findings, quickly making them less and less relevant. None of the experts foresaw the major developments that soon followed. Very few of these gifted researchers recognized that the core of their theories represented an important part of a much larger pattern. This should come as no surprise because that pattern contradicted the cultural unities of their societies. Another methodological approach amounted to interpreting everything in terms of intuitions based on a metaconscious knowledge of technique, thus making much of human life resemble functions, structures, systems, or relations of power. A complete critical analysis should have included a component of 'people changing technology (or technique),' a component of 'technology (or technique) changing people,' and an evaluation of the balance of their influence on individual and collective human life. Such an analysis requires a transcendence of the discipline-based mode of scholarship, and the introduction of the insights of cultural anthropology, the sociology and history of religion, and depth psychology, to achieve a greater critical awareness of how myths work in the background.

We have not adequately probed the methodological implications of our being a symbolic species. Despite high levels of desymbolization, individual and collective human life continues to rely on symbolization to a non-trivial extent. As a result, the different kinds of human activities that contribute to a technical order continue to be a manifestation of individual and collective human life working in the background. Only in those activities where people are knowers is this background separated from experience and culture and, secondarily, embedded in experience and culture. Probing the extent to which this is the case and how this is affected by desymbolization ought to be an integral part of the study of any sphere of human activities. It will deepen our understanding of the way in which any sphere of activities is dialectically enfolded into the larger wholes to which it contributes. A corresponding dialectical tension should be reflected between the analyses of these wholes. It is impossible to separate them according to the conveniences of a particular intellectual division of labour and its institutionalization. The macro-level studies question the micro-level analyses and vice versa. In addition, comparative analyses of the evolutions of different spheres of activities can reveal the extent to which they are manifestations of the larger wholes and, in the final analysis, of the cultural unity of a society. If the levels of desymbolization are high, such correlations may be weak, but they are not likely to be altogether absent.

To the extent that we remain a symbolic species, humanity and its life-milieu are indivisible. This cannot be understood by a discipline-based intellectual division of labour. It makes intellectually invisible to us technique's desymbolizing pressures on human life as well as the ways in which technique excludes us from our lives by its mediation and reification. In order to create a more liveable and sustainable future, methods and approaches will have to be found that give us a clear understanding of the role of technique in human life and society. Such approaches will also help us to understand many specific issues such as how work has become one of the primary sources of physical and mental illness,[60] how the urban habitat as a technical artefact imposes many stressors,[61] and how our food is being depleted of vital nutrients.[62]

Many among us are convinced that there is nothing to fear; humanity is simply undergoing another mutation in its historical journey. This mutation is stressful, but the epidemic use of antidepressant drugs will be a temporary aberration. A few have even predicted a glorious future for humanity.[63] Here I can do nothing else but take a stand. I believe that these kinds of visions of a new humanity are so alienated and reified as to be almost unrecognizable. If everything can be reduced to bytes as the new 'atoms' of the 'new creation,' all is lost. The problem may well be that everything we do fundamentally depends on our lives working in the background.[64] Since these lives are lived in communities, our cultures are an integral part of all of this.

The optimistic futures suffer from the same flaws as do the predictions that have accompanied almost all new technologies for over a century.[65] When a new technology is invented, imagining how it may change the future is, almost without exception, limited to 'people changing technology.' The reasons stem from the fact that 'technology changing people' depends extensively on metaconscious processes and thus lags behind 'people changing technology.' In addition, any causal relations between 'people changing technology' and 'technology changing people' are at best indirect. In addition, we usually examine 'people changing technology' one invention or innovation at a time, but 'technology changing people' cannot be understood in this manner, any more than the erosion of a beach can be understood in terms of any particular wave. Until these methodological issues are addressed, predictions as to what lies in store for us as a result of a new technology will remain almost comical when read with hindsight.

Modifying the methods and approaches in the social sciences may bring them closer to our perceptions of the natures of human life and the world. When the levels of desymbolization were low, these natures were perceived as being so highly interrelated as to preclude the possibility of discipline-based scholarship. When the levels of desymbolization rose, this changed. What I am suggesting is that we take the next step in this development of the social sciences. We have uncovered too many aspects of human life and the world that are somewhat anomalous with respect to the present discipline-based approaches. If this resonates with my readers, a growing sense may develop that discipline-based approaches are ill-suited for gaining a better understanding of human life within a technical order.

There almost certainly exists a dialectical tension between the prejudgments that help constitute the vantage point of any scientific observer; the limits of what can be observed from that vantage point, and their effects on the process of interpretation; and the methods and approaches that appear most appropriate for the advancement of human knowledge, along with the findings that result from their use. Such dialectical tensions may be latent during certain periods, but they will never go away. Sooner or later, the difficulties of dealing with an ultimately unknowable reality tend to impose themselves. Every human community (and this includes the 'invisible colleges' of the sciences) will develop a unique interpretation of human life and the world. Doing so will be accompanied by intuitions regarding their natures. Whether or not these are worked out as formal world views or theories matters little. When such intuitions are widely shared, the supposed natures of human life and the world suggest the most appropriate approaches for advancing human knowledge. When these approaches become widely recognized as the best available, thus leading to their widespread adoption, they will sooner or later unearth certain aspects that will appear somewhat odd or in other ways different from what could reasonably be expected given people's intuitive understandings. When such experiences are no longer isolated exceptions to what was to be expected, people's responses may gradually lead to a reinterpretation of the natures of human life and the world.

In the case of cultures, this may take centuries as a transition from one historical epoch to another. In discipline-based science that is separated from experience and culture, such reinterpretations take much less time because they have been put on a logical and mathematical foundation. Also, the 'world' of a discipline is restricted to one kind of phenomena. As noted, we may well be suspended in language, with the result that we do not know what is up or down, but any community has no choice but to decide what is up and what is down, even though sometime later this is almost certainly going to change. The certainty of what is up or down can only come from myths. The dialectical tensions of the vantage points of the scientific observers in the social sciences, what can be observed from these vantage points, the methods and approaches being used, and the resulting findings have all become too great. What is the point in continuing with endless intellectual fashions succeeding one another in almost every discipline?

I am not arguing that the methodological hypotheses that we have briefly sketched above should be adopted. Fortunately, there is not a chance that this will happen, and if it did, it would be a disaster. The last thing we need is another totalitarian theory dominating the social sciences. Naive as I may be, and having come to the social sciences as an outsider, I am hoping to make a small contribution that may lead to an 'ecology of knowledge' within which different approaches, each with their strengths and weaknesses, would be involved in collegial dialogue in order to discover where humanity is taking our most powerful creations and where these creations are taking us. Relying on symbolization in this endeavour is unavoidable. The great thinkers of the past symbolized everything in relation to everything else, which worked in the background when they focused on particular phenomena. Since their time we have gained a great deal of understanding of the process of symbolization and the role of culture in human life, and we should put this to good advantage. Implied in this is the acknowledgment that even our very best scientific efforts can never be free from alienation. The only colleagues who will not be able to join this new ecology of knowledge will be those who claim to be free from all alienation as truly detached and objective observers.

In their present forms the social sciences are ideally suited to the building of a technical order that requires no other knowledge than that derived from human life and the world one category of phenomena at a time. We have come to the same kind of methodological crossroads as did some physicists who saw no other way out than by accepting an indivisible reality. The subject matter is so complex that it will require an ecology of knowledge guided by the shared acceptance that none of us is free from alienation. At the risk of being dismissed as a pessimist, I interpret the situation as one in which symbolization makes human life possible, but it is ever so fragile. If the myths that ultimately sustain the life and historical journey of a community in an ultimately unknowable reality fail, the community will immediately be plunged into chaos. As best as I am able to understand, symbolization weaves a fragile fabric of meaningful and valuable relations over an abyss where human life as we know it is impossible. As a result, there can be no human freedom apart from alienation, and this includes all our social science endeavours.

One way in which we could all contribute to the resymbolization of the social sciences is by reflecting on the effects that science and technique have had, and continue to have, on the category of phenomena examined by our discipline. We could start with the texts commonly used in first-year university and check their indexes for entries to these phenomena. Doing so will immediately reveal that science and technique receive little or no attention in the social sciences. Surely this ought to be surprising, given the fact that there is hardly any daily-life activity that does not directly or indirectly depend on one or more technologies. Besides, all of them take place in surroundings that are essentially technical artefacts. If technologies of all kinds permeate almost all human activities, and if the experiences of these activities are symbolized by means of neural and synaptic changes, will technique not also permeate the organizations of our brain-minds and hence our cultures? If this is a reasonable working hypothesis, the category of phenomena examined by our discipline is likely approached with the following prejudgments regarding technology and technique.

First, by our growing up in societies whose ways of life are primarily evolved by means of the scientific approach to knowing and the technical approach to doing, the organizations of our brain-minds will almost certainly contain metaconscious images that reflect this situation. Second, it is highly likely that metaconsciously we have assigned a value to science, technology and technique. What these values might be may be imagined by means of the following thought experiment. Suppose that we woke up tomorrow morning to a world from which science and technique had been removed. Who would we be, how would we live, and what would our world be like? Surely our lives and our world are unthinkable without them, in which case science and technique have made us and our world who and what we are. These human creations are unlike anything else in our experience. They are so fundamental and essential that we (metaconsciously) value them very highly, and it may well be that nothing more valuable can be imagined. Does this mean that science and technique are the equivalent of the cultural gods in the past in the sense that they created and now sustain us and our world? These human creations may well play the same roles as myths did in the past, in which case we cannot begin to fulfil our roles as relatively detached and critical observers of the phenomena we study without, at the very least, having some understanding of the effects that such myths have on our work.

A third prejudgment that could interfere in our work as social scientists is a corresponding lack of awareness of the limitations of science and technique. It reflects the fact that they are used almost everywhere. Moreover, the mass media report almost daily on the future benefit of the discoveries in our laboratories. Just think of all the cures we are being promised. Over the years I have asked audiences of engineers to point out issues that cannot be confronted by means of some technology. They, like all of us, find it difficult to come up with such issues, which puts all of us in a situation described by Maslow to the effect that if your only tool is a hammer, all your problems look like nails.[66] Would any of us hire a contractor to make repairs to our house if he or she believed that only a hammer was required? Banging away at a leaky toilet, a cracked window, a binding door, or anything else would simply create destruction. Nevertheless, this is exactly the situation in which our civilization finds itself. When was the last time that we decided that the use of technology in a particular situation was inappropriate, and that we reached for a different approach in order to deal with it? We are back to the possibility that science and technique constitute secular myths in our cultures.

Finally, there is a widespread tendency to confuse the regulation of technology with technology bashing. The dilemma is the following. If our civilization lives with its most powerful creations by treating them as possibly the greatest good and if it is very difficult to conceive of some limits to their goodness, how do we deal with experiences that appear to contradict this as a consequence of harm done to human life, society, or the biosphere? One of the most common ways of dealing with these situations is to dismiss them as science or technology bashing engaged in by subversive people, who are in turn dismissed as communists, liberals, tree-huggers, reactionaries, pessimists, and more. It is the secular equivalent of the way in which heretics were treated in the past. We would not dream of accusing the thermostat on the wall of 'bashing' the furnace even though it constantly assesses the effect that the furnace is having on the room temperature, in order to 'criticize' it, based on the set point as the symbolic representation of our comfort zone.

The other common way of dealing with the prejudgment that is creating an incompatibility between the many problems we encounter and our metaconscious images of our most powerful creations is to 'be realistic.' For example, every technology exhibits a variety of problems during the early phases of its development, but these are later resolved. If this were not the case, we would not be where we are. Hence, we feel that there is no need to worry. Besides, if we do not develop and adopt this or that technology, others will, and our nation will start falling behind and could even begin a slide towards backwardness. Such a 'realistic' approach leads to the same conclusion: every technical possibility must be pursued, and whatever problems are associated with it must be accepted and dealt with. There can be no question of asserting ourselves. In other words, there no longer is something equivalent to a set point on the thermostat in contemporary cultures, something that reflects our values and aspirations and by means of which we independently guide our most powerful creations. These cultures, through their myths, have been permeated by such creations, thus making their regulation impossible.

These four prejudgments, which we bring to whatever we study as social scientists, confront us once again with our alienation and the impossibility of coming even close to being critical and detached observers. If we would all seek a greater critical awareness of the influences that these four prejudgments have on our work, I believe that the social sciences would be much more aware of the technical order we are building.

Resymbolizing Engineering

How would most of us react to an instructor who taught young people to drive a car by focusing on its performance as indicated by the gauges on the dashboard and only occasionally glancing out of the windows, especially when strange noises are heard? I suspect that the most common response would be that this question is so silly as not to merit a serious response. Nevertheless, as I discovered early in my career, this is a perfectly valid analogy for how we as engineers individually and collectively 'drive' technological innovation when solving a problem or accomplishing a goal.

As noted in the previous chapter, undergraduate engineering education as a process of secondary socialization into the profession was examined in the light of two questions. How well do we teach future engineers to understand the influence that technology has on human life, society, and the biosphere (the equivalent of looking out of the windows of our profession)? To what extent do we teach them to use this understanding in a negative feedback mode to adjust design and decision making in order to achieve the desired results, but at the same time prevent or minimize harmful effects (the equivalent of making steering and pedal corrections based on what is seen out of the windows)? These two questions were converted into extensively tested research instruments to permit the quantitative scoring of every element of a typical undergraduate engineering curriculum. Since the results were potentially controversial, great care was taken to adhere to the most rigorous standards possible. This included testing every single course for which the required information had been made available to a curriculum review of the Canadian Engineering Accreditation Board. The results showed that the answers to both questions were virtually nothing. The research instruments are shown in table 1, and an overview of the scores is shown in table 2.

It should be emphasized that the scores reflect the upper limit of what was achieved, since any course material relevant to the above two questions was evaluated in a non-critical fashion; that is, it was not assessed in terms of what was known in the relevant discipline or disciplines. Doing so in a few instances showed that a critical evaluation would have yielded even lower scores. For a full discussion of this study, the reader is referred to my earlier publications.[67]

To some of my colleagues, these results will undoubtedly appear counter-intuitive given the endless curriculum reforms, which have included courses on human factors, life-cycle analysis, environmental engineering, energy, and public policy. However, from the perspective of the above study, such courses do not alter the structure of the methods and approaches of any discipline; they simply add disciplines. What is required is a complete reorientation of the curriculum to make it preventive with respect to the human, social, and environmental consequences of technology. Engineering design and decision making must be modified on the basis of both the intended and the unintended consequences.

The ability of faculty members to adjust their decision making according to what they observe by 'looking out of the windows' of their specialties was evaluated by applying the same research instruments to their publications, which were listed in the same database that was prepared for a review by the Canadian Engineering Accreditation Board. The results are included in table 2. It would appear that many years of research and consulting produced even lower scores than those of the undergraduate curriculum. It may be argued that this is the result of the high levels of specialization of the journals in which they publish.

Table 1
Research Instruments

Scoring System	Technical Courses
0	no reference to context issues.
1	minor reference(s) to context issues, which remain peripheral to the thrust of the paper or course. Usually, this amounts to little more than outlining the context in which the problem arose, but once the problem is cast in engineering terms, little or no reference to context is made.
2	some reference to context issues with some consequences for the thrust of the paper or course.
3	major reference to context issues with substantial consequences for the thrust of the paper or course.
4	substantial consideration of context (as in 3) plus evaluation of consequences to adjust or reassess methods or theories (i.e., 4 includes some negative feedback; 3 does not).
Context Issues Include	
a)	implications of technology for human life, society, or nature.
b)	ethical considerations and relationships to values.
c)	non-technical aspects of engineering education and the professional paradigm (following T. S. Kuhn).
d)	implications of engineering theories and practices, including the consequences of quantification and mathematization, particularly of a qualitative sociocultural human reality.
e)	implications of engineering decision making, including the implicit and explicit values, beliefs, assumptions, and models which guide it.
Scoring System	**Complementary Studies Courses**
0	no reference to technological issues.
1	minor reference(s) to technological issues, which remain peripheral to the thrust of the paper or course. Usually, this amounts to little more than outlining the problem, but once the problem is cast in social scientific terms, little or no reference to technology is made.
2	some reference to technological issues, which have some influence on the thrust of the paper or course.
3	major reference to technological issues, with substantial consequences for the thrust of the paper or course.
4	substantial consideration of technology (as in 3), plus evaluation of consequences to adjust or reassess methods and theories used in engineering and technological practice or in the social sciences and humanities.

Table 2
Faculty Scores

	Year	CHE	CIV	ELE	ESC	IND	MEC	MMS	Mean
Core courses	1	0.5	0.5	0.4	0.1	0.8	0.3	0.5	0.4
	2	0.3	0.7	0.3	0.6	0.5	0.2	0.3	0.4
	3	0.7	1.2	0.7		1.6	0.7	0.9	1.0
	4	1.8	1.4				1.3		1.5
	N	30	38	29	18	25	32	26	
Core and technical electives	1	0.5	0.4	0.4	0.3	0.8	0.3	0.6	0.5
	2	0.4	0.6	0.4	0.6	0.5	0.3	0.4	0.5
	3	0.7	1.1	0.6	0.7	1.5	0.8	0.9	0.9
	4	1.5	1.5	0.8	1.1	1.3	0.9	0.5	1.1
	N	57	51	67	94	43	67	47	
Publications	Score	0.2	0.6	0.1	0.2	0.6	0.2	0.3	0.3
	N	367	237	306	204	121	290	169	

Key to abbreviations
CHE – Chemical Engineering; CIV – Civil Engineering; ELE – Electrical Engineering; ESC – Engineering Science; IND – Industrial Engineering; MEC – Mechanical Engineering; MMS – Metallurgy and Materials Sciences
N – Number of courses scored.

This is undoubtedly the case, but it is equally true that if their methods and approaches were more context-sensitive, professorial career paths and technical journals would have co-evolved very differently. Since this did not occur, there is a strong structural bias against any faculty member, department head, or dean who takes seriously his or her responsibility to protect the public interest above all else. In the meantime, many societies continue to permit their engineering professions to regulate themselves, on the assumption that they protect the public interest.

These findings are exactly what should have been predicted, given the almost complete reliance of the curriculum on disciplines in both the technical and complementary studies (social science and humanities electives) components. The present intellectual division of labour in engineering is both the cause and the effect of artefacts and processes being built up from distinct domains in which one category of phenomena at a time contributes a particular function to the ensemble. As a result, the discipline-based intellectual division of labour is ideally suited to engineering analysis and optimization. This distinguishes modern technologies from their traditional counterparts. The latter were created by means of a technological knowing and doing embedded in experience and culture, with the result that these technologies were appropriate to their sociocultural contexts and were generally sustainable. The traditional division of labour did not interfere with these essential qualities of technology. It was not until the discipline-based approach, separated from experience and culture, became dominant that concepts such as appropriate technology and sustainable technology had to be invented in order to address issues that could no longer be taken for granted.

There is an additional and equally serious problem with current undergraduate engineering education. For decades the engineering professions in the United States and Canada have recognized that they were doing a poor job of teaching engineering design. In the United States substantial efforts were made to remedy this situation. The results were extremely disappointing, but this could have been predicted given the discipline-based structure of the curriculum. Engineering analysis and optimization do little else than create highly efficient artefacts and processes, but unlike their traditional counterparts these are not effective in any comprehensive sense. To be genuinely effective, artefacts and processes must make efficient use of resources, and this can adequately be assured by output-input measures. However, they must also fit into human life, society, and the biosphere; that is, they must at the very least be compatible with their surroundings. Good design goes much further. The better the design, the greater the positive contributions to the surroundings, and the less the surroundings will be undermined. Overall effectiveness must therefore trade off what can be achieved through analysis and optimization, with synergistic interactions with the surroundings. Such efforts cannot be sustained by knowing and doing separated from experience and culture, one category of phenomena and one discipline at a time. They require parallel modes of knowing and doing.

Engineering analysis and optimization is a necessary but highly insufficient prerequisite to good design. It requires the additional components of craft and art. The current curriculum does not crowd out these components, but its 'hidden curriculum' devalues them as being subjective and arbitrary. The hidden curriculum comprises everything that is communicated to the students implicitly through the formation of metaconscious patterns.[68] There is little question that craft and art can be taught. All traditional societies did so very effectively by means of a master-apprentice relationship based on learning by doing.

In order to overcome the powerful desymbolizing influence of the discipline-based knowing and doing separated from experience and culture, a curriculum must be devised that develops both modes of knowing and doing (those separated from and those embedded in experience and culture). Doing so involves a resymbolization of engineering, where engineering analysis and optimization serves the larger process without dominating it. This may not be as difficult as it appears.

As a consequence of its discipline-based intellectual division of labour, engineering analysis and optimization constitutes the core of all engineering teaching and practice.[69] Design and decision alternatives are adjudicated in terms of efficiency or similar output-input criteria. All this translates into a strategy for increasing the power of anything it touches to extract desired outputs from requisite inputs, without paying much attention to the consequences for human life, society, and the biosphere. Engineering is a microcosm of technique and, in some of its branches, possibly its most developed form.

In other words, if the engineering profession were imagined as a collective person, he or she would have a psychopathic personality. Worse, this personality trait would be considered essential by most employers, themselves engaged in extracting wealth by participating in the technical order. Here my work joins that of Joel Bakan.[70] He reminds us that, under U.S. law, the corporation is treated as a legal person with the same rights as other Americans but with few of their responsibilities. His research shows that, in a comparison of the personality traits of this legal 'person' with those of a human being, the corporation behaves like a psychopath. The corporation shows no remorse for illegal behaviour, stops at very little to avoid getting caught, pays any fine as just another business expense, and shows no feelings for any victims born or unborn. All this is carefully hidden behind a mask of public relations 'speak' that forms an integral part of the integration propaganda found in mass societies. It is simply a question of doing business, with the new brand of chief executive officers being the greatest beneficiaries (as we shall see in the next subsection).

The people working for these corporations are compelled to live a kind of schizophrenic lifestyle, which prevents them from becoming psychopaths themselves. By means of the techniques developed within their specialties and disciplines, they contribute to the many decisions their corporations make, without feeling much or any personal responsibility for them. When they go home, they turn into different people. They become loving spouses, caring parents, and friends on which you can count. When they fail in their personal lives, they often feel remorse and regret. There is little new here. Dictatorships, totalitarian movements, and criminal organizations have always depended on this kind of behaviour. What is new is that highly desymbolized, contemporary cultures make all this much easier, leaving little moral or religious resistance in defence of the public good. It usually requires a whistle-blower with extraordinary courage and superhuman persistence to hold any corporation responsible for its actions, because the law offers very little protection, especially after recent deregulation.[71] Both the corporation and the engineering profession serve the cult of efficiency with little regard for human life, society, and the biosphere, which makes them psychopathic in character.

Engineering, along with economics and the other social sciences, helps build a technical rather than a cultural order. When discipline-based specialists participate in corporate decision making, they contribute to the corporation becoming uneconomic because the costs that it imposes by increasing its power to extract desired outputs almost always outstrip the wealth it creates. In turn, the kind of economy that these corporations help to build is an anti-economy, for the same reason. Social scientists hardly notice these developments because the prejudgments they bring to their studies make it next to impossible to understand how the scientific approach to knowing and the technical approach to doing have changed everything in individual and collective human life as a consequence of desymbolization. As a result, the psychopathic spirit of technique possesses all of us specialists.

For example, the failure of an engineer to use the most efficient methods and approaches is taken as a sign of incompetence, and doing so systematically is taken as grounds for dismissal. Few, if any, organizations can permit their employees to assert their humanity by selecting methods and approaches that are less harmful so that performance is balanced with context compatibility to create better ratios of desired to undesired effects. Any organization that faces intense competition will have no choice but to externalize as many costs as possible. As noted previously, our laws and accounting procedures place few restrictions on doing so, with the result that corporations and the economy as a whole have become wealth extractors. To be an engineer in this 'system' is to play a particular social role, which requires the suspension of most of our values and convictions. The profession has little choice but to fit in. The same is true for all the other related professions. It is a deep structural problem inherent in our knowledge infrastructures, which are based on disciplines separated from experience and culture. These knowledge infrastructures are the antithesis of civilization. As a result of 'technique changing people,' all of this has become perfectly normal and acceptable. There are very few remaining points of resistance.

An acknowledgment of these deep structural issues with our knowledge infrastructures and our desymbolized cultures amounts to a recognition that we have trapped ourselves in the labyrinth of technology. In order to find our way out, we need to resymbolize engineering education and much else. As noted, in the province of Ontario we began to organize such an initiative under the former Premier's Council in 1995. In 2002 preventive approaches for reorganizing our current knowledge infrastructures were recognized as one of twenty-five recent leading innovations by the Canada Foundation for Innovation. In 2003 the Canadian Natural Sciences and Engineering Research Council and the Social Sciences and Humanities Research Council drafted the Society, Technology, and Science 21 project to redirect research and teaching at Canadian universities, to tap into this potential. However, all these initiatives fell victim to economic fundamentalism.

Engineering education can be resymbolized, beginning with design. It depends on two parallel modes of knowing and doing (one embedded in, and the other separated from, experience and culture). An affirmation of this dependence raises concerns that objective scientific and applied scientific methods and approaches will be abandoned in favour of subjective ones. Tinkering with the teaching of engineering design failed to critically examine the dominance of analysis and synthesis. Good design is at the heart of engineering practice. As a creative process it is deeply metaconscious and hence personal.[72] However, the end result is not personal. It is very much open to engineering analysis and optimization.

On the receiving end of engineering design, most of us have experienced that, although the many things we have purchased may efficiently perform one or more functions, they frequently do not fit very well into our daily-life activities. This lack of fit cannot be addressed by adding disciplines (such as human factors), because the issue must be addressed on the level of experience and culture. First-year engineering students can be made aware of these kinds of issues by having them analyse how well certain designs fit into our surroundings, beginning with how people encounter them in their daily lives.[73] Doing so examines these designs 'from the outside in,' while engineering analysis and optimization operate 'from the inside out' by examining a succession of domains, each producing one sub-function by means of one category of phenomena that is mathematically modelled by one discipline.

For example, I asked my students to examine the design of the current Toronto streetcars from the outside in. This involved listing all the connections between the streetcars and their physical and social surroundings, followed by an evaluation of how well these worked for people and the biosphere. They were then to develop alternative concepts with a better ratio of desired to undesired connections. They quickly realized that the design was ill-suited to moving people around and that it was only marginally compatible with its social and physical surroundings. The floor height of these streetcars was designed for a European-style rail station platform or a subway platform. Getting on from street level or getting off back to street level is needlessly time-consuming and difficult, if not impossible, for very young children, parents with strollers, people in wheelchairs, elderly people with walkers, people with luggage or shopping carts, and so on. Similarly, trains are designed to run at constant speeds as much as possible, with the result that their weight is not a critical issue. For streetcars the opposite is the case. The excessive weight of these railroad-style streetcars makes the job of the operator much more difficult because they are unable to respond effectively to busy traffic. In many areas the weight of the streetcars caused so much vibration that the rail supports had to be redesigned and replaced; it also caused excessively high levels of track maintenance and repair. Excessive quantities of energy are required to accelerate these streetcars, of which only a little is recovered during deceleration. The increased noise and vibration interferes with a variety of activities such as people talking on the sidewalks and nearby residents working, sleeping, or relaxing at home. The students quickly concluded that from almost any perspective this streetcar design represented a very poor fit with its surroundings. At the same time, they learned a great deal about how it could be improved.

When the same first-year students were asked to examine the common household refrigerator from the outside in, they also had no difficulties in arriving at superior design concepts. A much better insulated space could be integrated into the usual kitchen cabinetry. If this space were located on an outside wall, cold air from outdoors could be used whenever available. The compressor and exterior heat exchanger for releasing heat could be advantageously located to make use of the best seasonal temperatures available, and this would also prevent the compressor from reheating the cooled space. In addition, the ease with which items could be put into or removed from the refrigerator could be greatly enhanced by giving the insulated space different shapes.

The same kinds of analyses were undertaken for buildings. The students had to pay particular attention to how a building appeared as they approached it; how well it fit into the street and neighbourhood; how easy it was to enter and find one's way within it; what it would be like to work, live, or shop in it; how well it related to other facilities in its surroundings; and how well it connected to various modes of transportation. They were then shown how these approaches could be used to arrive at a design concept for a new building to be constructed on a vacant property. The students were also asked to evaluate a number of streets in terms of how well these worked for people and the contribution that these streets made to the liveability of city blocks and local neighbourhoods. It was followed by examining the connections with the biosphere, with particular emphasis on making buildings as energy self-sufficient as possible. This shed light on the importance of orienting buildings and how this in turn affects street layout and infrastructure design. Exercises of this kind can be multiplied almost indefinitely.

By means of these analyses, students learn to distinguish between design exemplars and analytical exemplars.[74] The former correspond to a person's level of experience and culture, while the latter correspond to discipline-based analysis and optimization separated from experience and culture. To return to the example of the streetcar, once the current design exemplar had been examined from the outside in, it could be further examined from the inside out by reducing it to a number of domains within which a single kind of phenomena produces a subfunction. For example, the load-bearing structure of the streetcar can be examined by means of a continuum having the identical geometry and material properties. Similarly, the electric motors contain several domains in which electromagnetic or mechanical phenomena contribute to the propulsion and deceleration of the streetcar. In other words, once a design exemplar has been agreed on, the details of its inner workings divided among distinct domains can be examined by analytical exemplars associated with the appropriate disciplines. It should be noted that this will almost certainly provide opportunities for improving the design exemplar; however, it is impossible to design anything by moving from analytical exemplars to design exemplars because the boundary conditions required for the equations involved in mathematical modelling and optimization derive from design exemplars.

Unfortunately, whatever these first-year students learned in the above kinds of exercises was quickly drowned out by a curriculum almost entirely focused on the analytical exemplars of the disciplines relevant to a particular branch of engineering. Almost without exception, the design textbooks that the students read make a complete muddle of the process of design by blurring the differences between knowing and doing embedded in experience and culture and those separated from experience and culture, as well as those between design exemplars and analytical exemplars.

When, several years later, I asked some of the same students (now in fourth-year industrial engineering) how they would go about designing a new production facility for a client wishing to bring a product to market, they did so in the way implied in their curriculum. They immediately jumped to working out the details of an assembly line, the materials handling system, the information system, the organization, and so on. When I stopped them to ask what kind of production system they were designing (Fordist-Taylorist, lean, Scandinavian, or other), for most of them the choice was so obvious that they had difficulty understanding what I was asking. I had to remind them what I had taught them in first year: all these systems were invented as a response to very different economic, social, and environmental conditions. When I asked them which of these conditions existed today, they quickly began to realize that the current ones were very different. I then returned to my question as to what kind of production system would be appropriate for the client and our world. It certainly was not the system they were beginning to design as if this could be done by means of analysis and optimization.

By carefully examining previous design exemplars in terms of how well they fit into and contribute to their surroundings, it is possible to choose elements that will fit rather well into our present world. For example, conditions of unemployment and underemployment would greatly benefit from the kind of assembly process pioneered by Volvo in its Uddevalla plant.[75] It was based on the principle of autonomous working teams who assembled cars with the support of sophisticated information systems. Work cycle times were eighty or a hundred minutes, depending on the workshop. In the one, the car was moved once, and in the other not at all. To further improve labour productivity, this assembly process could be modified by incorporating features from other production systems, including the design for ease of assembly and the just-in-time materials handling of lean production. Possibly the greatest improvements could be realized by incorporating these autonomous working groups into a Semco style of organization.[76] This organization (which has been visited by many people from all over the world) ensures that the values of capitalism benefit everyone in the company, with spectacular economic and competitive benefits. In this way, more appropriate design exemplars can be developed that are best matched to the prevailing conditions. We should also implement the findings of social epidemiology that show that healthy and creative work always balances the demands placed on workers with the control they have over their work.[77]

The objection will be that Volvo was compelled to shut down its experiment because it was not sufficiently competitive. This shutdown was partly due to dramatic changes in Swedish society, the European market, and the global economy.[78] Some of these changes could have been addressed by a comprehensive adaptation of the design exemplar to the new conditions by means of preventive approaches. Volvo had only responded to conditions of full employment and strict legislation requiring equal pay for equal work, which did not permit the company to hang on to Swedish workers by offering pay incentives. It must also be pointed out that, in the rest of Europe, traditional assembly-line work was carried out mostly by 'guest workers' rather than their own citizens. My point is that a superior design exemplar could have been worked out that might well have made the company competitive through the full use of preventive approaches.

Such a design exemplar would undoubtedly permit further development. For example, the environmental impact of automobile-based transportation could be significantly reduced by selling transportation services as opposed to the automobiles that provide these services. A customer would then be able to buy a certain number of kilometres of transportation with a chosen level of luxury, safety, and fuel efficiency. Turning automobile manufacturers into transportation-service providers has enormous potential benefits for the biosphere and hence for our future. Materials and energy could then deliver far more services, and this would translate into greater competitiveness. Developing such preventive approaches could offer an automobile manufacturer a highly competitive niche in the current global market. This is undoubtedly what the future of the automotive industry will look like, because comprehensive preventive approaches will gradually demonstrate their superiority as we are forced to make changes in the face of issues such as global warming and the sustainability of cities.

Once a design exemplar has been established by these kinds of approaches, the details of all the sub-systems can be analysed and optimized on the basis of discipline-based approaches, as in the problem sets that students encounter in their undergraduate education. These problem sets examine one category of phenomena in a mathematical domain, of which a continuum and a fluid are previously noted examples. It is important to emphasize once again that these phenomena are not analysed in the real world where they mingle with many others, and consequently the results obtained with discipline-based methods and approaches are distinct from those embedded in experience and culture. Our lives and our world are unthinkable without an endless mingling of many kinds of phenomena. Despite this obvious fact, almost the entire undergraduate curriculum is devoted to the solution of problem sets that can be tackled by analytical exemplars. As a result, students come to believe that the products, processes, and technologies they encounter in these problem assignments are the current design exemplars, which they only need to adapt to specified conditions using the appropriate analytical exemplars. How these design exemplars were invented in response to a unique set of conditions and how they evolved with these conditions are never discussed. The design exemplars in the problem assignments thus appear as givens, and this is exactly the message students receive.

What is lacking in undergraduate engineering education is the recognition that the design process relies on symbolization to grasp a dialectically enfolded world. After all, what matters about a particular product, machine, process, or system is how well it contributes to human life and society and how well it minimizes any harmful effects to its surroundings. Most of these kinds of investigations must occur on the level of experience and culture, and it is this that can most effectively guide the processes of analysis and optimization through the lens of a particular mathematical domain of a discipline or specialty. The current undergraduate engineering curriculum and the engineering practices based on it have this backwards, with far-reaching consequences.

A first consequence is that of analytical exemplars largely steering the design exemplars. As a result, engineering practice is dominated by endless optimizations of the sub-functions provided by a particular category of phenomena that correspond to a particular discipline. Technological development is dominated by whatever modifications to design exemplars are derived from analysing and optimizing the performance of one aspect at a time. The evolution of design exemplars is thus a response to technical advances measured in terms of performance as opposed to an improvement in the overall effectiveness of the way they contribute to human life and society. Engineering design and decision making thus reflects the influence of technique as life-milieu and system on the minds of engineering practitioners.

A second consequence of analytical exemplars dominating design exemplars is our inability to make the major non-cumulative adjustments that will transform the building of a technical order into the creation of real economies and into ways of life that can sustain communities and that in turn can be sustained by the biosphere. Attempting to do this with noble intentions, but with the same intellectual and professional division of labour, will only work for the very few instances in which increasing output-input ratios constitutes a genuine solution. Dealing with the issues we face is almost never a question of optimization. Optimization ignores the way everything evolves in relation to everything else in living orders, and thus builds a non-living order. We must go in the opposite direction, and the introduction of preventive approaches can begin this process of reversal.

The third consequence of the dominance of analytical exemplars over design exemplars is that we engineer products, processes, and systems that are dysfunctional from the perspective of experience and culture. Take modern buildings as an example. Being composed of optimized elements assembled in optimal ways, they externalize any consideration of context compatibility. For example, a concrete that is optimized for load-carrying capacity and cost may have acoustical properties that negatively affect the privacy of the spaces that it helps to create and may also increase the ecological footprint of the building. Analysis and optimization based on disciplines produce a great many externalities, as we have seen in economics. All the things that specialists (trapped in the triple abstraction) cannot consider contribute to the fact that engineering design and decision making produce very poor ratios of desired to undesired effects.

The resymbolization of engineering education and practice could begin with the introduction of preventive approaches, in parallel with the transformation of the professional division of labour.[79] These make use of the knowledge we have regarding the influence of technology on human life, society, and the biosphere in order to adjust design and decision making to ensure the desired results, while at the same time preventing or greatly minimizing harmful effects. It involves a collaboration between two parallel modes of knowing and doing, one operating on the level of design exemplars and the other on the level of analytical exemplars. The former corresponds to looking out of the windows of our specialties to see what else is out there, to ensure that a genuine service is being rendered and that 'collisions' are avoided as much as possible, while the latter ensures the optimal use of resources.

These approaches can contribute to restoring a better balance between technique and culture. Preventing collisions is much more economic than first producing them and then dealing with some of the negative consequences by means of end-of-pipe approaches, which is essentially what current practices accomplish. Preventive approaches can steadily improve the ratio of desired to undesired effects of design and decision making, thereby addressing the roots of our current economic, social, and environmental crises. It will then be possible to cost-effectively raise social, health, and environmental standards. In the current regime the bulk of competitive advantages are derived from the externalization of costs.[80]

The launching of these kinds of initiatives is easy. I have never had any difficulty in convincing undergraduate engineering students that such an approach to engineering is vastly more interesting and rewarding than is the current one. Our profession could be in the forefront of helping society to address some of the most fundamental issues, and this also could bring its rewards. Probably the most effective way to unlock the potential of preventive approaches is to withdraw the right of engineering professions to regulate themselves. In North America they have steadfastly ignored the evidence that there is an alternative to the current approaches, one that could protect the public interest in a much more comprehensive manner. It is time that the engineering profession undergo public scrutiny. The kinds of questions that have been asked of undergraduate education could be used as a measure to track the extent to which engineering education and practice is being reoriented in a preventive direction. I know that this will be seen as a threat by many of my colleagues, who are so highly specialized that the demands for a new trade-off between breadth and depth will cause a great deal of anxiety. Nevertheless, they, like their students, must appreciate the tremendous benefits that can be realized. Furthermore, the first industrially advanced nation that systematically exploits the potential of preventive approaches will gain such huge competitive advantages that everyone else will be forced to change.

In an age in which we face deep economic, social, and environmental crises, it is more than a little depressing to watch some of our brightest minds spend their talents and time on optimizing design exemplars that are part of the problem instead of the solution. The whole exercise is not much better than rearranging the deckchairs on the Titanic, knowing it might go down. We are acting as if a little more analysis and optimization will bring the kinds of qualitative changes that are required. We do everything except question the structure of the curriculum itself; we sprinkle the term *sustainability* everywhere, add optional courses or minors, and roll it all out in a new academic plan. Unfortunately our current academic planning is all about performance, with the result that it has adjusted the entire university to technique. All this prevents bright young minds from creating a liveable and sustainable future.

The first step in resymbolizing undergraduate engineering education is to put the regulatory boards, which accredit engineering schools, under public scrutiny. They must be able to demonstrate that the kinds of scores shown earlier in this subsection are steadily improving. In response, engineering faculties can redesign their hiring, tenure, and promotion criteria to encourage a rebalancing of breadth with depth and then begin to overhaul the curriculum as bright young minds step up to the plate. Once this initiative begins, student pressure will quickly mount to do more, because to be engineers who are as socially and environmentally literate as they are technically competent involves far more interesting, meaningful, and rewarding work. The making of designs and decisions that have a far more positive ratio of desired to undesired effects will then spill over into their workplaces upon graduation, and it will not be long before their clients will simply demand the higher standards that can be achieved with a preventively oriented engineering process.

Resymbolizing Management

Management science and business administration were the result of business schools adopting the discipline-based model of teaching and research that was borrowed from operations research and economics.[81] The resulting problems are in many ways similar to the ones we have discussed for economics, the social sciences, and engineering. The complex and uncertain issues faced by managers are transferred from the order of what is true to the order of what is real, thus opening them up to measurement, quantification, and mathematical modelling. Knowing and doing separate themselves from experience and culture, and discipline-based rigour trumps relevance. Mathematical models of all kinds are devised to represent data sets. Examining the world of business one category of phenomena at a time (including production, marketing, sales, finance, accounting, research, and human resources) does not gradually accumulate into more comprehensive models that can eventually approximate the behaviour of large complex organizations.

In addition, given the five-stage human-skill-acquisition model discussed earlier, mathematical modelling within a discipline-based context is rule based and hence cannot simulate the fourth and fifth stages. Even then, it remains restricted to one mathematical domain into which has been admitted a single category of phenomena. Real management has more to do with messy, contradictory, and incomplete understandings of a variety of issues, of which only the simplest aspects can be measured and modelled.[82]

What physics is to the sciences, operations research is to some professions, including management. Data are collected to be represented in the mathematical models. Simulations of various scenarios can thus be run by means of these models in order to determine the best way forward. It is a clear instance of the technical approach. In the world of business it led to strategic planning separating the knower (the strategic planner) from the doer (the manager). The rise and fall of strategic planning has been well analysed by Mintzberg,[83] and much of it resonates with our findings in previous chapters. In a later analysis, Mintzberg extends these criticisms to the masters of business administration (MBA) programs and why these fail to prepare their students for real management.[84]

As a rule of thumb, operations research works well for machine-based systems of all kinds, including those that 'produce' information. The reason is obvious: these machines and the larger systems built up with them are designed, built, and operated according to the order of what is real and are constituted of distinct and separate domains within which one category of phenomena dominates all others in the production of one sub-function at a time. Operations research is generally a complete failure in simulating a dialectically enfolded order of what is true. Since even the most modern corporation is far from limited to its technical infrastructure, it is clear that operations research can contribute very little to the overall understanding of the running of these organizations and their interactions with the human world and the biosphere. In other words, operations research is like any other human invention: good for certain things, irrelevant for others, and harmful to still others. Skilled practitioners know the limitations of their tools so as to choose the most appropriate ones for the tasks to which they are well-suited. None of this is taught in operations research, and the broader understanding that is required for such judgments is not even acknowledged in most cases.[85]

A variety of developments contributed to the pressures on business schools to adopt the discipline-based professional division of labour. The symbolization of management experiences became increasingly difficult as the corporation grew in size. We have noted that when the limitations of technological knowing and doing embedded in experience and culture were reached, businesses were compelled to grow in size in order to derive further benefits from technological innovation through economies of scale. Once knowing and doing had been separated from experience and culture in all areas of the corporation, every aspect of its operations had to be planned, and this necessitated further increases in scale. All this posed enormous challenges for the abilities of senior management to clearly symbolize their experiences of the organization and to derive the metaconscious knowledge necessary to sustain these skills, as an executive vice-president of General Motors put it in the title of his book: *On a Clear Day You Can See General Motors*.[86] Eventually the limitations on managing by experience became so extensive that complementary forms of knowing and doing had to be found.

In addition to the problem of scale, there were other difficulties that impaired the ability of managers to symbolize their experiences. These were related to the re-engineering of the corporations in the image of the order of what is real at the expense of the order of what is true. As noted, the technical infrastructure of these organizations had always been organized in the image of what is real. Initially the people using this infrastructure continued to do so according to their cultural order, but this quickly proved to be unsatisfactory. Goal-directed behaviour led to the emergence of a new kind of organization, which Max Weber called *a bureaucracy*.[87] After the Second World War the tops of these bureaucracies were temporarily reorganized into what Galbraith referred to as *technostructures*.[88] These technostructures began to use information machines, which led to the 'technology paradox' that resulted from investments in these machines not being commensurate with productivity increases.[89] At this point, business process re-engineering began to recreate the organization in the image of the information machine, and enterprise integration completed this task.

Today, these organizations are run as if they were intellectual assembly lines. The integrated database presents each worker with the required information inputs, on which he or she performs a production step, of which the outputs disappear into the system. As noted previously, the symbolization of the worker's experiences of the local technique-based connectedness was ineffective and thus required a desymbolized technical approach. We are now dealing with business organizations that have almost entirely been recreated in the image of (information) machines.

These developments further reinforced goal-directed behaviour and its translation into technical approaches. They have permeated each and every business unit: in the morning its members assemble to hammer out a mission statement, and after lunch a strategy to achieve it, and later the planning of how to do so. The report is then put on a shelf because the participants know full well that the day-to-day reality is far too unpredictable and messy for the plan to be of much value. Nevertheless, MBA graduates have been taught that everything can be reduced to business models, value chains, and other simplifications that reduce everything to the local technical order.

What cannot be translated into this order of the real goes into the background or disappears. A decision to make any change whatsoever in a business unit will affect all the others, the corporation, and the world beyond and in turn will be affected by these changes. What is 'hard' will trump what is 'soft' because it can be quantified and modelled. It will get most or all of the attention, while what is soft (because it cannot be treated in this way without extreme reductionisms) will carry less weight. If the consequences of any business decision under consideration are reasonably small and cumulative changes, they may be predictable by some of the models, but if this is not the case, the predictions may be of little value, no matter how accurately the models depict the current situation. Underneath lies a more fundamental issue: the dependence of the technical order on desymbolized cultural orders cannot be taken into account, even though much of what will happen revolves around this dependence.

The above transformations put enormous pressures on managerial skills that are embedded in experience and culture. Also, the university is not equipped to effectively transmit any professional knowing and doing that is mostly embedded in experience and culture. This kind of knowing and doing is largely metaconscious and can be transmitted only by apprenticeship-based learning by doing. The enormous success of operations research during the Second World War appeared to hold out an answer to the issues faced by business schools. They could adopt a discipline-based model of teaching and research and thus fall in line with their host institutions. Anyone who protested was simply swept aside and ignored.

With hindsight, it is now clear that the few critics should have been taken more seriously. Unfortunately their case was difficult to articulate at that time. The intuitions of these critics based on their experiences of what worked and what did not work could not easily be turned into 'hard' data. The mainstream would not listen to anything but 'hard' or 'scientific' evidence. Such evidence requires the kind of data that can never be obtained from a dialectically enfolded human world without intellectually destroying it by transferring various aspects to a mathematical domain. Once we subscribed to a scientism that uncritically assumed that the order of what is real was objective and trustworthy, and the order of what is true for people's lives and communities was subjective and unreliable, a great deal was lost. Our civilization has paid a high price for this scientism.

It is tempting to regard times of war as unique situations where performance is all that matters – kill or be killed – and this depends on the efficiency of the means and strategies employed. Hence, tools such as operations research ought to be ideally suited to the task. We soon learned that this was not the case. The classic example of this is the Vietnam War in which Robert McNamara played such a decisive role.[90] He extended what he had learned in the Harvard Business School to fighting the war in a new way. It was a monumental disaster because the 'soft' factors won the day. Military strategists, with all their discipline-based techniques (including operations research), appear not to have learned this lesson in Iraq and Afghanistan.

The limitations of operations research as the pinnacle of the technical approach are even more serious in peacetime, when ignoring the context is inexcusable. The following account is illustrative. As a leading scholar in operations research at that time, Stuart Dreyfus was asked at a party to explain what he did. In response to his description, the person asked if this was how he decided when to replace his old car with a new one. Dreyfus replied that he would not dream of it. Reflecting on his response, he eventually came to the conclusion that if he would not trust operations research to make good decisions in his own life, how could he expect others to do so?[91] This led to a significant professional reorientation.

Operations research as a paradigmatic discipline in management education has done a great deal of harm to our lives, communities, and workplaces. It has contributed to the emergence of a new kind of management style that Mintzberg refers to as 'heroic.'[92] Before it could emerge, other developments had to take place. A discipline-based approach to managing transfers all situations into the domain of what is real. Anything that cannot be calculated is largely ignored. According to these calculations, plants are relocated, mergers are attempted, and budgets are assigned to each and every initiative. Discipline-based MBA programs produce calculating managers. The problem is that new ideas and initiatives cannot be produced by means of calculations. A creative act is required. Nor can these ideas or initiatives be calculated into existence. It comes as no surprise, therefore, that these managers have tended to adopt one management fashion after another. So-called high technology, restructuring and downsizing the corporation, re-engineering the corporation, or maximizing shareholder value all amount to executing the myths of corporate cultures. They certainly do not constitute genuine strategies. When all else fails, managers can always pressure their staff to 'massage the numbers' as a kind of secular magic.

Calculating managers face other problems as well. It is impossible to calculate to what extent, when confronted with daily-life situations, they mingle phenomena belonging to production, marketing, sales, finance, accounting, and human resources from within the corporation and all the many phenomena from without. Discipline-based management education and MBA programs avoid these problems by inventing simplistic cases backed by data in which one category of phenomena dominates all the others, with the result that it can be assigned to a discipline, course, and classroom. This educational system will never compel the real world of management to follow suit. It is true that corporations have been restructured to increasingly resemble a technical order rather than a part of a cultural order. Nevertheless, the former ultimately depends on the latter even in business, as we are gradually discovering.

Calculating managers also have difficulty dealing with new trends. Nowhere in human life and society do new trends or developments come ready made. They all have embryonic beginnings that are inevitably ambiguous, incomplete, and even contradictory. It is next to impossible to translate these beginnings into facts, let alone trends. It is here that people who have developed a great deal of metaconscious knowledge are able to symbolize their experiences of these embryonic beginnings into something meaningful. They may have a sense that something is going on, which other people with less experience may dismiss as soft and subjective. Gifted leaders stand out by their ability to see something significant in the works where others see nothing at all. Such leaders may pay very careful attention as things evolve in order to confirm or reject what, until then, are little more than intuitions. They may simply get it wrong, but this is not at all unique to soft data. Anyone who has ever attempted to design an indicator will know the kinds of distortions that may result from getting a hard handle on some issue. As a result of these weaknesses, it is essential to pay attention to everything, whether it is amenable to becoming hard or whether it will remain soft.

Calculating managers are frequently very smart with respect to the hard aspects of issues and also frequently very ignorant when it comes to the soft aspects. In other words, they may be lacking in 'people smarts' and have difficulty understanding that everything is not reducible to a technical order. They rely primarily on their cognition and intellect and do not trust their feelings and intuitions. We find these people in all professions. In engineering they are the practitioners who believe that any issue is amenable to rational problem solving.[93] In computer science we have the hackers.[94] Most MBAs want to become leaders of teams, but because of their common lack of people smarts, few want to play on these teams.

The calculating management style is the perfect incarnation of technique in the business world, with the same consequences for human life, society, and the biosphere. Performance now has little to do with producing better products or services for people, with improving customer service, or with doing creative research that responds to new trends and opportunities. All this is rich in context. Instead, performance is reduced to output-input ratios taking the form of raising shareholder value by increasing stock prices and externalizing costs, rushing products to markets, exploiting customers, increasing employee workloads to permit more layoffs, abandoning communities, and manipulating finances. Enron was simply the tip of the iceberg. In order to make technique existentially bearable, we transform everything into a political issue in order to reassure ourselves that we remain in control. It is by means of our secular political religions that we faithfully serve technique.

When more and more of these calculating managers became chief executive officers, the above trends reached new heights. How could calculating-style CEOs make themselves stand out? The pressure was on to come up with bold new ideas to prove their leadership abilities. Simply improving output-input ratios by means of techniques of all kinds hardly inspires people to follow such leaders. It was not long before a solution was found and, having responded to a need, it spread rapidly. Chief executive officers had to become heroic leaders who *were* their corporations, much like kings and queens were their nations before constitutional monarchies. These 'heroes' have become single-handedly responsible for every success of their corporations, and their inflated self-worth is reflected in the obscene salaries and bonuses they demand. Power has become concentrated in the CEOs to an extent not seen since the nineteenth century's captains of industry. To make this palatable to employees, lip service is paid to empowerment, teamwork, and decentralization. When troubles arise, boards of directors can now point their finger at the heroic leader, who can always exit with a golden parachute. Everyone gains except the employees.

The pattern has reoccurred over and over again. The new heroic CEOs come on stage, pronounce their bold ideas, temporarily increase share prices, and then exit to repeat the cycle elsewhere. The boards of directors sit through cycle after cycle but appear to be impotent to do something about it. The press makes polite protests, public opinion follows, and politicians murmur that something ought to be done. There seems to be no one defending this circus other than the CEOs.

There are deep structural reasons that little is being done, even after the near meltdown initiated by Wall Street. The intellectual and professional division of labour based on disciplines selects for calculating managers. From among these, those who are willing to play a heroic role are usually the ones to rise to the very top. It is little different from what nations had to endure from their rulers before their heroic actions were constrained by constitutions.

Mintzberg puts it very well when he asks how a CEO can single-handedly add billions of dollars to share value, bring out a revolutionary new product, merge two very large organizations, and (I will add) do all the work that used to be done by the people he or she laid off. What is really happening is that they perform an essential role in the system of technique. They help create the appearance that we are still in control and the illusion that everything that happens is for our own good. They try to give a heroic face to an anonymous system that influences them much more than they can influence it. Are they able to oversee, let alone control, the flow of information within the system and the fallout of innovations that it produces? Are they able to oversee, let alone control, the speculative bubbles of the anti-economies? Surely the power they exercise is the power of technique, and if they fail to exercise this power by in any way curtailing it in the name of anything human, social, or environmental, they will soon cease to have this power. It is the tyranny of the technical bottom line.

Imagine what all this does to the morale of the workforce when the CEOs sit on pedestals and claim all the credit and most of the rewards. Anyone who has carefully followed all the soft information coming out of workplaces cannot help but notice the war that has been declared on people. All this is confirmed by the demand-control and reward-imbalance models based on the socio-epidemiological research related to healthy work.[95] For employees, the re-engineering of the corporation has led to unprecedented monitoring, an even greater loss of control coupled to increasing demand, and a growing imbalance between responsibilities and rewards.[96] Ever fewer people are able to enjoy or even derive some satisfaction from their work. The re-engineering of the corporation necessitated by information technology continues to reduce the skill levels required to carry out many jobs, while at the same time every effort is made by society to raise educational standards and provide continuous education and retraining. Meanwhile, many of the industrially advanced nations continue to haemorrhage jobs to other parts of the world, and this trend is making inroads into middle-class jobs. No economist or politician appears to have any idea of what to do about these trends, which in a relatively short time can spell social disaster. There is indeed a war on people, but I do not believe it has much to do with left- or right-wing politics. It is deeply rooted in technique and slavishly executed by those who wield technical power.

What from the perspective of MBAs would undoubtedly appear as one of these soft issues bringing on left-wing ideologies can be a serious issue to anyone with children entering the work force or for anyone attempting to understand what is happening to work. When economists calmly tell us that no one should expect to have a traditional career or depend solely on a single source of employment, I can think of no stronger evidence of enslaved minds. What are economies for if more and more people are excluded from fully participating in them? What does democracy mean when employment issues are taken off the table by secretly negotiated free trade agreements?

The days of heroic CEOs are undoubtedly numbered. They will likely go down in history as temporary aberrations, but the damage they have done to the people with whose careers, lives, and communities they have gambled will be difficult to repair. Besides, below this summit of greed the system of technique continues to advance, partly in response to the damage that has been done. The few wake-up calls that have been sounded have been largely ignored, and the Goldman Sachs of this world, as bastions of the anti-economy, have survived rather well. Mintzberg reports that during the 1990s the salaries of the CEOs of American companies rose by 570 per cent, corporate profits by 114 per cent, and average employee salaries by 37 per cent. In 1999 median shareholder returns dropped by 3.9 per cent, but the direct compensation of CEOs rose by 10.8 per cent.[97]

Equally disturbing is the behaviour of many deans and faculty members of the business schools. During interviews on the media we are left with the impression that the whole world should follow their business models. It is as though they can no longer conceive of any limits to these models, and because they talk and act as if these models were omnipotent, their arrogance is equally limitless. The only reason their message resonates with public opinion is that the public has been deeply affected by the transformation of human life and society, first in the image of the machine and then in the image of the information machine, as well as by the belief that ultimately human life is economic

Fortunately not all executives are of the same mind. A report produced by the Centre des Jeunes Dirigeants d'Entreprise critically examines the role of the corporation in the twenty-first century.[98] It warns that the present system may well be on the way to collapse because of the social problems it causes. For at least twenty years corporations have been profitable at the expense of societies. Moreover, unemployment and underemployment cause many dislocations and tensions, which in the long run can threaten everything. Our problems have become social and spiritual, but we continue to focus on the material. Ironically, the report also insists that our working lives, careers, and communities must become more adaptable: lifelong careers are unrealistic; people should pursue multiple activities and thus reduce their reliance on one employer. Corporations should be held responsible for all the damage they do to society and the biosphere. Possibly the most surprising, the report insists that the present system is incapable of sustaining human lives and communities and that therefore these must change in accordance with the system. Despite everything, it appears that we human beings must remain 'just in time' resources for the technical order.

In order to get to the root of the problem, it is essential to examine the processes of symbolization and desymbolization in the life of a manager. Regardless of the level of desymbolization, the manager's experiences of what is happening in the corporation will create metaconscious images of it: how things work or do not work within the organization and beyond and how all this forms the equivalent of a corporate culture. This metaconscious knowledge provides a basis for hunches and intuitions in terms of how a particular situation or problem is a routine expression of the whole or something that is out of line. In the case of the former, the usual kind of response immediately suggests itself without a moment's hesitation; in the case of the latter, the situation has to be given a great deal of thought, and the result will most likely be non-cumulative in character. As far as the reactions of colleagues are concerned, the manager's metaconscious knowledge may also lead to hunches or intuitions that provide further feedback, including assessments of success or failure.

In following up on these hunches and intuitions, varying levels of creativity may be involved. The lowest level of creativity will result in responses that barely go beyond a manager's experience. It may be very skilfully crafted, but it does not go any further. On the highest level of creativity the manager's experience is a jump-off point for devising something highly original and yet symbolic in the deepest sense of being artful. Another altogether different range of responses is based on a critical and sceptical attitude towards one's own experiences as potentially being too subjective and unreliable. In such cases a manager's metaconscious knowledge will at best guide a careful analysis of the situation in an attempt to simulate some kind of mathematical model. The manager trusts hunches or intuitions based on metaconscious knowledge only to the point that they can be translated into hard features, with the result that they are no more than a jump-off point for a calculating approach.

Artful, craft-like, and calculating responses all have their limitations. An artful response may generate an idea that is wonderful on its own terms as opposed to the terms of the situation. A craft-like response may be highly skilful but too routine to cope with rapid or complex changes that demand more non-cumulative responses. A calculating response risks being so analytical that it may seem cold and dehumanizing.

Each approach has weaknesses associated with its strengths. An artful approach may lose touch with the situation, a craft-like approach may become too rigid and fail in all but the most routine situations, while the calculating approach may be so reductionistic that it misses the essentials. Each one can thus benefit from the other two. An artful approach can be kept in check by a calculating approach. A craft-like approach may benefit from being questioned by an artful one. A calculating approach may be less vulnerable if, at the very least, it is questioned by a craft-like approach and, for unusual situations, by an artful one. As a result, artful and craft-like approaches may be dysfunctional. Craft-like and calculating approaches may be functional but lacking the virtuosity necessary for success. Artful and calculating approaches risk becoming impersonal and disconnected from the soft aspects of a situation.

An alternative way of interpreting all this can be based on the recognition that the symbolization of the experiences of managing an organization is affected by what we derive from our senses, our feelings, our intuitions, and our thoughts. Sensing and feeling primarily depend on the particular experience. Intuitions primarily depend on the metaconscious knowledge of all the experiences that have contributed to the organization of the brain-mind. Thought depends on reason or rationality, where the former is embedded in experience and culture while the latter is separated from them.

Still another way of interpreting all this is to regard visionary styles of management as artful but kept in check by science and craft. Scientific (discipline-based) styles may be effective when kept in check by art and craft. People-oriented styles based on experience are affected when kept in check by art and science. The results are more or less equivalent to the research on personality types and the way they depend on sensing, feeling, intuiting, and thinking. Given this topology, the reason the calculating styles of management are so destructive is clear. By dogmatically assuming that everything that is relevant must be hard and that everything irrelevant must be soft, these styles fall short of being visionary, comprehensively analytical, and people-oriented.

The complementarity of the three approaches becomes even more apparent when we examine the daily business of management. It has no common denominator with what happens in the classrooms of business schools. During the running of a corporation there are no 'pop-up' problem statements that include only the relevant data, nor are there any single right answers. First, a manager notices that something is not quite right or that a unique opportunity has presented itself. Initially all this will be very embryonic and probably driven by hunches and intuitions. The manager may decide to keep a close eye on whatever has attracted his or her attention in order to seek some kind of confirmation of the hunches or intuitions, or an assurance that things are not what they initially appeared to be and no further attention is required. If subsequent events confirm the hunches and intuitions of the manager that something is in the works, the situation may warrant a careful analysis, leading to a confirmation that indeed a problem has presented itself. The nature of the problem is itself part of the challenge. The kinds of phenomena that are relevant are not given, the way they are in a business school problem. It will not be easy to assess which of them play a primary or secondary role or such a marginal role that they can perhaps be neglected. As the formulation of the problem advances, it is essential to constantly verify that the balance of influence of the different categories of phenomena is what it first appeared to be. Some considerable time may elapse before anything close to a classroom definition of a problem can be approached. Even then, such a problem definition must be verified to ascertain that all the relevant information has been collected. The final problem definitions are likely to have little in common with those encountered in the disciplines taught in business schools. The corporate world simply cannot be dealt with effectively, one category of phenomena at a time.

Once the problem has been defined, what to do about it is not likely to resemble anything encountered in school. Real situations are much more multidimensional. They simply do not respect the discipline-based separation of the categories of phenomena related to production, marketing, sales, finance, accounting, human resources, public relations, and more. Even when the 'world' of an organization is highly desymbolized after being reordered in the image of information machines, it still relies on a dialectically enfolded cultural order that lies underneath.

The symbolization involved in all the above styles and activities is limited by the desymbolizing pressures that we examined earlier. As a result, resymbolization is not a realistic response. Something more is required in the form of two parallel modes of knowing and doing, not unlike what we proposed for engineering. Resymbolization is possible by making an equivalent distinction between design exemplars and analytical exemplars. Managers, like engineers, are involved in situations experienced as messy, incomplete, and contradictory. Such experiences are best dealt with on the level of symbolization. This will lead to the development of a great deal of metaconscious knowledge that may be expressed in intuitions and hunches that are capable of directing attention to issues that may be out of line or that may constitute unique opportunities. All this is embedded in experience and culture and amenable to resymbolization. The outcome may take the form of action-design exemplars for managers that are equivalent to the design exemplars for engineers. These action-design exemplars are integrated conceptions of how everything evolves in relation to everything else, how this has led to a particular issue, and how best to respond to it.

Such an action-design exemplar must then be complemented by analytical exemplars furnished by the appropriate disciplines. Once again, a synergy must be created between the knowing and doing embedded in experience and culture and the knowing and doing separated from experience and culture. Management action-design exemplars must rely on analytical exemplars for analysis and optimization.

This analysis converges with the one developed by Henry Mintzberg as to how management education might be reformed.[99] Ideally, management education should be limited to practising managers who have already developed a great deal of metaconscious knowledge to sustain the dimension that must be kept on the level of experience and culture. Doing so raises the question as to how they become practising managers in the first place. In many companies, practising managers ascend from the ranks of practising professionals, such as engineers or accountants. It may be argued that competent general managers need a much more diverse background. Can a more general preparation be furnished by business schools? This may well be possible if the curriculum is developed on two interacting levels, one related to action-design exemplars by means of action-learning and the other based on disciplines.

The initial education of general managers would have to be interrupted by many years of professional practice, when they would develop a base of metaconscious knowledge embedded in experience and culture, interacting with a base of metaconscious knowledge separated from experience and culture. They would then be ready for another phase in their professional development in which this experience would become a springboard for a more artful and skill-like approach to creating action-design exemplars; these, in turn, would be analysed and optimized by more advanced discipline-based knowledge. Classroom learning would now be closely coupled to the experience of managing.

Action-design exemplars of a more advanced kind could now be used to order experience more clearly and insightfully, followed by a more creative use of the experience. Managers may also learn how to use this experience to keep in check the action-design exemplars that are too far out to be effective or too routine to capture novelty. A tension can thus be created between art, craft, and science.

Much of what goes on in our business schools is neither elementary nor advanced management education. It is simply one-sided education based entirely on discipline-based approaches that are applied to situations dominated by one category of phenomena. It is important not to become trapped in the dualism between theoretical and practical education. An action-design exemplar may be regarded as the conscious and theoretical elaboration of practical experience. It orders, clarifies, and thus goes beyond that experience in a theoretical fashion that remains embedded in experience and culture. In this way, it provides a context within which the skilled practitioner is able to assess whether particular discipline-based tools are appropriate or whether their limitations will lead to erroneous conclusions. At the same time, action-design exemplars may be used to advance the abilities of practitioners to symbolize their experiences of managing. This can heighten their awareness of certain aspects and of the roles they play in relation to all others. It may help them become more skilled at avoiding overemphasizing or underemphasizing some aspects, by more adequately symbolizing them in relation to all others. In other words, action-design exemplars are theories in the order of what is true, no matter how highly desymbolized corporate life may be. There is no other way to overcome the limitations of discipline-based approaches. The limitations of these theories can be partially offset by the strength of the analytical exemplars used to analyse, optimize, or modify the action-design exemplars.

There are two levels of theories. The first are those of action-design exemplars that prove useful to advance the experiences of practitioners, sharpen their abilities to symbolize situations, and to go well beyond their experiences in creative and imaginative ways. The other theories are separated from experience and culture and are thus useful for examining what happens in a particular domain, in which one category of phenomena contributes a sub-function to the larger whole. The strengths and weaknesses of these two levels of theories are opposites. The one illuminates how everything evolves in relation to everything else, while the other sharpens the understanding of situations that can be understood in terms of one category of phenomena. The challenge to management education is to design a curriculum in which these two levels of theories complement one another so that the weaknesses of each become less problematic. It means that the one challenges the other and vice versa. It will introduce a certain level of discomfort that is so often experienced in real life. Everything is always open-ended, ambiguous, and incomplete. There are no right answers. The best we can do is to come up with action-design exemplars that accomplish the desired results while at the same time minimizing the harmful ones. Here we are back to the need for preventive approaches.

Once again, preventive approaches may permit managers to achieve the desired results, while steadily improving social, health, and environmental standards. In turn, by gradually raising these standards, corporations will compete in part according to their ability to make their practices more preventively oriented. Eventually free trade agreements will have to be challenged to once again allow markets to operate within democratically established standards. Preventive approaches embodied in action-design exemplars must continue to be analysed and optimized by means of analytical exemplars to ensure the optimal use of scarce resources, but no longer at the expense of human life, society, and the biosphere. In this way, managing a corporation can be part of the solution instead of the problem.

What would a more preventively oriented action-design exemplar look like? I believe there are four general principles that can guide the way. First, a greater complementarity has to be designed between the knowing and doing embedded in experience and culture and the discipline-based knowing and doing separated from experience and culture. It has created a fault line that runs through every corporation, with devastating consequences, especially in relation to poorly functioning negative feedback loops. Employees on all levels have difficulty understanding those who work on the other side of this fault line. The above issues faced by any manager are, to varying degrees, experienced by any employee who takes his or her job seriously. Making employees as self-reliant as possible eliminates much of the need to externally manage their work. As the previously mentioned Semco organization has demonstrated, this is possible to a degree that few people would have believed.

The second principle involves creating the greatest possible complementarity between people and machines. Their strengths and weaknesses are diametrically opposite, and the organization should take full advantage of this understanding. Modern organizations essentially waste the skills, motivation, and loyalty of most of their employees.

The third principle is to design organizations that give people the greatest possible control over their work. This would provide employees with opportunities to learn and expand their skill sets, which would contribute more effectively to the organization, reduce underemployment, decrease the stress that leads to physical and mental illness, and provide greater job satisfaction and thus loyalty to the corporation. It would reduce the hand-brain separation and would thus soften the previously mentioned fault line that contributes so much to the 'us-them' feelings that are so destructive.

Finally, the above three principles should be applied throughout the organization to create a complementarity between areas of redesign and their contexts. Until now, many highly creative and extremely successful initiatives have been short-lived because they were incompatible with their organizational surroundings. This fourth principle includes a careful examination of the interfaces with societies and the biosphere.

These principles for developing more viable and sustainable organizations may find a friendly home in so-called social businesses. They are designed to address the problems of the poorest portion of humanity with the understanding that investors will be entitled to recover no more than the funds they have invested; as a result, any profits earned by these social businesses will be reinvested in addressing the issues with which they struggle. It is the only way these businesses may escape from contributing to the creation of anti-economies.

In sum, we have not even begun to incorporate systematically and comprehensively the many highly successful action-design exemplars into better organizations. Our lives and future depend on it.

Resymbolizing Medicine

The effects of technique on human health may be understood in terms of the reciprocal relations we have with our physical, social, and cultural surroundings. On the physical level, we must exchange matter and energy, which our bodies can neither create nor destroy. From what we breathe, eat, and drink, we obtain what is necessary to repair and replace everything in our bodies at least once every seven years, except for our brain cells, which are only repaired. In addition, the energy required to power our bodies and brain-minds is obtained from transforming parts of these intakes. Still other intakes may result from skin contact or the penetration of the skin or body orifices. Whatever our bodies and brain-minds can no longer use must be exchanged with our surroundings.

Almost all these intakes have components that exert negative effects on our physical lives. These include a variety of chemical compounds, bacteria, viruses, microorganisms, and, almost certainly, genetically modified organisms. Their interference places demands on still other processes that are designed to protect us. If these demands do not exceed the resources of these processes, they will be successfully dealt with. If, however, they overtax these resources, we may become ill as a consequence of a deepening and prolonged struggle to regain our health. Human health on the physical level may thus be interpreted as a constant struggle between what threatens our biological life and the resources of our immune system to defend against them.

We also depend on and can be threatened by our reciprocal relations with our social surroundings. These are required for our becoming symbolic beings by acquiring a culture. Without a community, children turn out very differently, as is evident from those who have largely grown up in isolation or who have been brought up by animals. Communities provide us with acceptance, friendship, and love. Since we help to constitute these social surroundings, others look to us for social support. We place certain demands on these social surroundings as they place demands on us. If our lives are not adequately sustained, we may become lonely and depressed, and if this continues during an extended period of time, the consequences may become far worse. Since we are both individuals and unique manifestations of our society, we help to sustain and are sustained by our social surroundings.

Our reciprocal relations with our social surroundings are fragile and may be affected in many different ways. Possibly the greatest transformations in this reciprocity occurred when human beings began to live in societies, and again when these societies were transformed into mass societies as a consequence of technique. Technical mediation limits and transforms our reciprocal relations with others. As long as we have adequate resources to face the many demands that they place on us, we will live relatively happy and fulfilling lives. If, however, these social demands strain or overload our resources, we will defend ourselves in many different ways that are frequently accompanied by negative consequences for our lives. If, in the extreme, our resources are completely overwhelmed, a break-down may occur, with even more severe consequences. Once again, our health and well-being depend on this reciprocity with others; as we help to sustain them, they help to sustain us.

Human beings also depend on reciprocal relations with their spiritual surroundings. As noted, the members of a symbolic species mediate their relations with their surroundings by means of a culture. They have learned to do so by growing up in communities, which transmit a cultural order as a design for making sense of and living in the world from generation to generation. As a result, these communities create a measure of freedom with respect to surroundings that would otherwise more narrowly determine them. These new freedoms cannot be enjoyed without some protection from relativism, nihilism, and anomie. This is provided by the deepest metaconscious knowledge in the form of myths. These myths anchor the symbolic relating of everything to everything else in an ultimately unknowable reality, making it possible for communities to engage in historical journeys. However, all this cannot be accomplished without some alienation. As such, these cultures possess the very beings and lives of their members, thereby taking on a spiritual quality.

Our reciprocal dependence on our spiritual surroundings takes the form of an evolving dialectical tension between cultural unity and individual diversity. If the cultural unity imposes itself too strongly, the individual diversity may be too limited to permit the community to adapt to new and challenging conditions. In the extreme, this can result in the collapse of a culture. If the individual diversity is not sufficiently constrained, it may grow to the point that people have less and less in common, again resulting in the eventual collapse of the community. There are many developments that can upset this delicate balance, and desymbolization is what concerns us here because a weak culture can open the door to relativism, nihilism, and anomie.

The reciprocal relations with our surroundings on the physical, social, and spiritual levels are interdependent. On the level of physical exchanges, toxic compounds can affect a variety of processes in our lives, including those of the organization of our brain-minds, thus potentially affecting social and cultural exchanges. In turn, problems on the social level can lead to alcohol or substance abuse. Anomie can lead to depression and suicide. All three levels of reciprocal interdependence on our surroundings are thus dimensions of the constant tension between what keeps us healthy and what threatens our health. Health is not an absence of disease but a balance between what sustains life and what constantly undermines it.

Since technique fundamentally transforms our relations with our physical, social, and spiritual surroundings, it is having a profound effect on our well-being. Each level is being contaminated by elements and relations that are performance oriented at the expense of context compatibility. This is profoundly affecting our food webs, the fabric of social relations, and our spiritual condition. We will pay particular attention to the threat that technique poses to us as a symbolic species, as a result of desymbolization.

From the above perspective it is clear that technique presents contemporary medicine with historically unprecedented challenges. For the purpose of our analysis these may be divided into two components. The first is related to the way in which technique transforms our reciprocal relations with our surroundings. Improving performance strains and breaks the relations of the fabric that weaves everything together. This component can be significantly reduced by the use of preventive approaches. In this respect, the resymbolization of human knowing and doing in general, and the discipline-based modes in particular, will substantially contribute to genuine health care. Such health care is preferable to disease care, which is always end of pipe. In other words, the health care that would result from the transformation of technique, beginning with preventive approaches, would restore the task of medicine to a scale that would be much more manageable. What remains are other processes leading to disease but which are not (directly or indirectly) related to technique and which can therefore not be affected by its transformation. Of course, medicine will continue to face many other sources of illness.

The second component of the influence technique has on human health is through its discipline-based organization. The result has been a considerable expansion in our medical means, their growing effectiveness, and their sharply rising costs. The successes and failures of modern medicine are closely related. Simply put, in medical situations in which one category of phenomena dominates all the others, the system should perform very well, while in situations where this is not the case, it may well perform rather poorly. One category of phenomena often receives all the attention in life-threatening situations, such as heart attacks, cancer, acute appendicitis, serious gunshot wounds, or major injuries from accidents. When life hangs in the balance, most people want a lower priority to be assigned to any side effects involving other phenomena. In less acute cases, secondary categories of phenomena producing a variety of side effects receive much more consideration.

Few specialists may have the theoretical and practical background for a comprehensive assessment of all factors. Moreover, a synthesis of what various disciplines know and do must involve symbolization and hence experience and culture. We are back to the need for parallel modes of knowing and doing.

The medical system is likely to perform poorly when human health is affected in diverse ways that synergistically reinforce one another. The result is a scattered set of symptoms, none of which appears to be significant in itself or indicates the dominance of one category of phenomena. It is very difficult to treat such a set of individual symptoms without being sure as to what disease binds them together. These kinds of situations may result from several low-level interferences with different kinds of processes that interact within one or more of the physical, social, and spiritual dimensions of human health. The methodological challenges are well known. We are able to test drugs on animals, using high dosages over a short period of time. Long-term and low-dosage experiments are more difficult to conduct for a variety of reasons, including the need to control for other effects. Doing so for multiple exposures with possible positive or negative synergistic effects is virtually impossible. The widespread use of technique creates many of these kinds of situations.

We have noted that endlessly improving human life on the basis of the system of technique is likely to create many tensions within whatever it is that has been made more performing as well as between it and its surroundings. Almost all our food, drink, and air have been deeply affected by technique.[100] For example, much (if not all) of our food is produced by agribusiness. This system is organized and reorganized to improve performance as opposed to improving nutrition. One of the many reasons is the use of monocultures. They risk mining soils for a fixed set of nutrients; what is taken from the soils to ensure the greatest possible performance in the resulting crops is likely to exceed their rate of regeneration. There is some disturbing evidence to this effect,[101] but it is still largely ignored. This trend must be understood in relation to the generally poor absorption rates of nutrient supplements, which are also technically produced, with similar consequences. Despite our grocery stores being filled with all manner of vegetables, fruits, and processed foods, we may well be starving our bodies in terms of various categories of nutrients. What effect will this have on human health? In asking this question, we are back to the problem of attempting to understand the erosion of a beach in terms of individual waves. Our health can be undermined by many such effects. As we have pointed out, if fast food (of the kind produced by the McFoods of this world) were a drug, it would never have been allowed on the market, because high-dosage, short-term exposures to human beings very quickly ruin their health. Even when it is eaten only occasionally, its effects will combine with those of so many other efficiently manufactured foods that the consequences for human health are bound to be non-trivial. The difficulty in confirming this is that the threat to human health cannot be assessed by one category of assaults at a time. Hence, we must look at all the other assaults on human health in all three (physical, social, and spiritual) dimensions.

From social epidemiology we know a great deal about how the constant reorganization of human work according to the latest techniques has a devastating effect on human health. We also know that our urban habitat imposes many stressors on its inhabitants. Lifestyle factors are also important. It is somewhat ironic that we spend so much time and effort on exercising, while at the same time using every possible technique to save effort of any kind.

Technique exposes us to a great many physical, social, and spiritual assaults on our health. There is no scientific way to determine the overall effects of even a very small number of closely related effects, let alone something more ambitious. The synergistic effects of technique on the physical, social, and spiritual processes that mingle together in our lives are both positive and negative. This may lead to ways of undermining human health that are virtually impossible to know or effectively treat by means of a medical system that is discipline based. Lacking any knowledge of its own blind spots, the system, facing a situation it cannot readily classify, may well conclude that the problem must be in people's heads. Such situations undoubtedly occur from time to time, but in most cases it is more likely that we are faced with the comprehensive effects of technique on human life, society, and the biosphere. After all, any attempt to improve anything with little or no consideration of the ways in which this affects the broader context amounts to tearing the local fabric of relations in multiple ways, with far-reaching consequences. The fact that the current situation is not far worse is a tribute to a very long period of evolution during which everything evolved in relation to everything else, with the result that everything became highly adjusted to everything else in terms of sustaining and being sustained by the larger whole.

Technique has destroyed not only the traditions of human communities but also the traditions of the crafts that contributed to the sustaining of their ways of life. Traditional healing is no exception. Consider traditional Chinese medicine as an example. Possibly because of a prohibition against dissecting human cadavers, the craft of healing had to remain on the level of experience and culture. Various symptoms had to be carefully correlated with other life events, and this in the course of many generations cumulated into a very sophisticated tradition-based healing craft. As previously noted, the rapid and comprehensive changes in human life, society, and the biosphere resulting from industrialization made less effective the symbolization based on experience and culture. The resulting desymbolization made it much more difficult to correlate a wide variety of situations with human health, thereby undermining all tradition-based medicine embedded in experience and culture.

Is it possible to create a medical system that fundamentally depends on parallel modes of knowing and doing? We have noted the limitations of either mode in the face of technique. Such a system would depend on medicine as craft, science, and art. This is difficult to envisage under the conditions that are common in traditional societies, and it does not appear to be very feasible in our present world. However, a partial synergy may well be possible. It is undeniable that a number of elements of traditional Chinese medicine continue to work reasonably well in areas where the discipline-based medical system is either weak or impotent.

Can medical education be resymbolized? Can the present system benefit from the two parallel modes of knowing and doing? Is it possible to create the equivalent of action-design exemplars that integrate diagnosis and treatment and combine both modes of knowing and doing? There are no simple answers, but tackling some aspects of these issues may well point the way to further developments.

To begin with, the people who approach the current discipline-based medical system make sense of their health issues on the level of experience and culture. It is the general practitioners who must translate what these people tell them into the technical language and culture of discipline-based disease care. A variety of medical techniques are explored, the results interpreted, and other techniques prescribed. All this must be retranslated back into what patients want and need to know. In other words, the well-documented issue of patient autonomy is much more complex than it appears in the literature. The deep structural problems of this relationship cannot be adequately conceptualized, let alone dealt with, within the conceptual framework and values of mainstream medical ethics.

Moreover, the loss of autonomy with respect to the medical system is not limited to patients. General practitioners and specialists also lose their autonomy. Medical education is entirely discipline-based and hence subject to the same kinds of limitations that we have discussed for some other professions. It might even be argued that general practitioners find themselves in an extremely difficult situation. Their effectiveness depends to a considerable degree on skills they developed in their internship, but mostly after graduation. At this point, however, they no longer serve in specialized units, and the world of their practice is no longer divided according to the disciplines encountered in school. They have to establish a working relationship between their patients and the medical system that involves two parallel modes of knowing and doing.

General practitioners begin by making a diagnosis, which matches the needs of patients to the capabilities of the discipline-based system. This cannot be accomplished without two parallel modes of knowing and doing, one embedded in and the other separated from experience and culture. Patients deal with one kind, and the system with the other kind. After the system does the various diagnostic tests, the results, including the required treatments, must once again be translated back to the patients.

Consider the following example. A general practitioner has just read a journal article suggesting that patients suffering from borderline hypertension could benefit from being put on beta blockers to reduce the likelihood of their suffering a heart attack. Informing the patient of this study is difficult in itself. How significant are these findings? Will a subsequent study that adds or subtracts a few parameters come up with different results? After all, these studies can control for only a very limited number of parameters and thus depend on what are currently believed to be the most significant ones. The possibility that future studies will come up with very different conclusions cannot be ruled out. At the same time, the prescribed drugs have far-reaching consequences for human health and require other drugs to deal with the most serious side effects, while others may go untreated. In addition, the ways in which patients react to these or any other kinds of drugs vary considerably. What autonomy does the general practitioner have with respect to the medical system when advising a patient? Depending on the legal and insurance regimes, doctors may have no choice but to apply the findings of highly specialized studies out of fear that patients who are not given the drugs and who suffer a heart attack may sue for malpractice. At the same time, they may feel that in a number of situations patients may be better off without the drugs because of the many and serious side effects. If the risk of a heart attack is low, if the consequences of taking the drugs are substantial, and if there are other factors, such as the patient possibly suffering from 'white coat syndrome,' general practitioners may be reluctant to prescribe these drugs. However, because of the legal regime, they may feel compelled to do the opposite, thus contributing to a great deal of waste in the system.

All this places general practitioners who deeply care about their patients in a difficult dilemma. In order to provide them with informed advice, they must symbolize the situation in relation to a patient's life, which includes their specialized knowing and doing separated from experience and culture. Consider a general practitioner imagining what he or she would do if the patient were a spouse, son, or daughter. Having shared a great many experiences with them, the doctor would be able to more adequately symbolize the medical issue at hand in relation to the patient's life. It would no longer be possible to ignore any knowing and doing embedded in experience and culture. The risks of being sued would almost certainly be a non-issue because of the relationship. Should non-relatives be treated any differently? Nevertheless, in real situations doctors may feel they have no choice but to prescribe the drugs. They have lost their autonomy with respect to the medical system; what is technically possible must be done. Such a loss of autonomy makes the doctor-patient relationship more adversarial as a consequence of technique.

In the same way, medical specialists lose their autonomy with respect to the medical system on which they depend. All this was clearly predicted by Jacques Ellul's theory of the autonomy of technique.[102] The autonomy lost by patients, general practitioners, and specialists is gained by the medical system. The evolution of technique can generally be described in ways that require no reference to our beliefs, values, or cultures.[103] A great many medical decisions will be made in accordance with the necessities imposed by the system instead of what is best for patients, general practitioners, and specialists.

While transforming medical practice, technique is also changing human life, society, and the biosphere. We have thus far referred to the problems this creates within the physical dimension of human health in terms of toxic compounds, bacteria, viruses, microorganisms, and genetically modified organisms creeping up the food web. Biotechnology and nanotechnology will undoubtedly add to these problems. Like all other discipline-based techniques, they will open up new frontiers of pollution.

Superimposed on all this is the way technique changes people and their communities. The desymbolization of our brain-minds and cultures is weakening the metaconscious images that we have of who we are, how we are related to others, how we belong to mass societies, and how these substitutes for human communities depend on what remains of the biosphere. People who are particularly sensitive to these metaconscious images may have a feeling that they have been cast adrift, which brings a great deal of anxiety and depression. Others may metaconsciously intuit the same situation as a unique new opportunity. Having a very fuzzy metaconscious knowledge of their social selves, they may set out on a journey to 'find themselves.' They quickly learn that the Internet offers them the possibility of experimenting with different identities. Some of these appear to suit better than others, and they learn to narrow their search to certain types. In some instances, these explorations appear to lead to the most unexpected and (by traditional cultural values) even shocking results.

Carl Elliott describes a category of people who wish to become amputees and will resort to almost any means to achieve what they strongly feel they should have been all along.[104] Others discover that they must 'die to their old body' and be reborn into another by means of a complete makeover, to be achieved by plastic surgery or gender changes.[105] Still others may feel that the technical means to achieve what they desire will not be available during their lifetime. They thus join the transhumanist movement and possibly go as far as arranging to have their heads stored at very low temperatures after their deaths.[106] Some people confidently predict that soon nanotechnology will permit us to explore, repair, and recreate the human body from the molecular level up. The results will be captured in binary code and stored on the Internet as a backup 'just in case.'[107]

In some sense there is nothing new here. For some two hundred years humanity has been re-engineering itself, first by technology changing people and later by technique changing people. The term *re-engineering* is used to denote that this involved the reordering of individual and collective human life, first in the image of traditional machines and later in the image of information machines. Doing so involved the desymbolization of experience and culture along with the metaconscious knowledge developed in people's brain-minds. This opened up and closed down different forms of individual and collective human life. The more the technical order imposed itself on cultural orders by desymbolizing them, the less these orders were able to maintain a strong and healthy dialectical tension between cultural unity and individual diversity. The metaconscious images of ourselves and the world eventually became so desymbolized that people could interpret them in many different ways. The proliferation of all kinds of techniques fired their imaginations as to what was possible, ranging from the modest 'Net 2.0'[108] to the radical 'Humanity 2.0' visions of the future.[109] The re-engineering of humanity thus became cause and effect of developing the order of technique. Those who could not find themselves in these ideologies, which made them anxious and depressed, were offered antidepressants. Either way, the order of technique was stimulated by the very problems it created.

Carl Elliott observes that some of these developments appear to have a strong sexual dimension.[110] Once again, Ellul's theory of technique predicted this.[111] We have noted that people have some metaconscious knowledge of how the technical order has taken over much of their (cultural) lives, and some feel that if they want to assert themselves and exercise a measure of freedom, they must do so by means of their biological selves in a variety of ways. Their bodies can be 're-engineered' to fit the kind of self they have searched for and discovered. Adding the sexual dimension serves the sacred of transgression. These developments must be clearly distinguished from those in which the contamination of the food web disturbs the hormonal and other control structures of people's bodies, possibly, in the extreme, changing the usual gender orientation. Cultural factors have an additional influence in such cases.

All these developments have far-reaching implications. Individual and collective human life is fraught with tensions between those aspects that have been made to perform and the others that have been disordered by technique. Entirely new forms of mental illness now become possible. It is well known that some cultures, during certain times, produced unique mental illnesses that appeared out of nowhere and, sometime later, disappeared into nowhere. Carl Elliott cites examples of the 'fugue state' in nineteenth century France, and the development of multiple personalities in the United States during the 1960s and 1970s.[112] Ian Hacking refers to 'transient mental illnesses' and seeks to explain them by their fitting into an 'ecological niche' created by a particular culture during a particular time.[113] When these niches disappear, the corresponding mental illnesses vanish with them. He does not rule out other causes such as traumatic childhood events and their interference with normal development, and toxic substances affecting biological processes that may interfere with the growth of brain-minds. In all likelihood, these transient mental illnesses cannot be explained by a single cause.

These arguments parallel the ones developed here, with one fundamental difference. For some two hundred years the cultural orders of industrializing societies have failed to create the kinds of cultural niches that people may fill as they work out their intuitions of the metaconscious knowledge they have of themselves and the world. Niches refer to a specific role or function that helps to sustain, and is sustained by, an ecosystem. Today's cultural orders have been so deeply permeated and transformed by the technical order that they are no longer capable of creating such cultural niches. Equivalent niches cannot be created by the technical order since it is built up without any reference to sense. It is incapable of sustaining human life as long as we are a symbolic species. Nevertheless, transient mental illnesses can almost certainly be understood in the context of the re-engineering of individual and collective human life in the image of machines. When for some reason or another people cannot assume the new identity that is being thrust on them, they become mentally ill. This illness will take different forms in different times, which depends a great deal on the kind of culture involved and its path of desymbolization.

Consequently, the hypothesis that the development described in this work can help us to understand the forms of mental illness that come and go as a result of dislocating particular cultures in unique ways during a certain period appears sound. Since technique is fundamentally changing the social and spiritual dimensions of individual and collective human life, there will undoubtedly be significant and far-reaching health implications. It is tempting to speculate that some forms of schizophrenia may be a transient mental illness related to the effects of knowing and doing separating themselves from experience and culture, which can make human life in some socio-economic groups and cultures almost unliveable.

Individuals exploring new identities, social relations, and lives will undoubtedly find support from social workers, psychologists, psychiatrists, and other health practitioners. It is not likely that they will come up with the above kinds of explanations. They will interpret and cope with these kinds of developments within their own disciplines.

For example, the growth of multiple personalities mentioned above was largely explored in terms of possible childhood abuse. Patients were asked certain questions, which provided them with clues as to how to make sense of their condition. They began to remember episodes of abuse, and a factor of exaggeration may well have crept in to please their therapists. It is one of the risks of therapy. Here we encounter yet another instance of people changing technique and technique changing people. Patients were fitted into the taxonomy of a discipline and given the authority of science, and took this as a key to understanding who they were and what was happening in their lives. Being diagnosed as having a particular mental illness showed them that they were not the only person in this situation. They now belonged somewhere, and consciously and metaconsciously they began to act as they should. All this represents a certain medicalization of human life. In this way, disciplines help create and legitimate new forms of mental illness.

Some members of a profession may encounter patients with new and unusual symptoms. They become interested and begin to explore the situation. They seek to describe it in ways that conform to one or more of their disciplines in order to arrive at a proper diagnosis. Possible treatments are explored, and the results are published in relevant journals. Appropriate tests must be devised to measure the severity of the illness of a particular patient relative to all others belonging to the newly established category. Once a mental illness is officially diagnosed, it may gain acceptance beyond the medical profession, and public or private insurance may begin to reimburse some costs of treatment. If the potential market is large enough, pharmaceutical companies may see an opportunity for developing and marketing new drugs. In this way, the medical system and the pharmaceutical industry become agents that contribute to technique changing people. Once again, technique feeds on the very problems it helps to create. By medicalizing the health issues created by technique, we open them up to medical techniques, and this begs the question whether the problems of technique can be addressed by having more technique.

All these developments are creating a context to which the current discipline-based medical system is poorly adapted. The stress levels of many practitioners are likely to be increased. Medical ethics will be of little value in guiding their behaviour; it is almost entirely end-of-pipe and commonly grounded in philosophies that are so asocial and ahistorical that they have little sense of what is happening to humanity as a consequence of the kinds of developments that we have described in terms of technique. Being discipline-based, medical ethics has no way of knowing the limits of medical practice and will not be very helpful in preventing the medicalization of human life. What may be the consequence of adapting to technique or, in the extreme, the inability to do so is in many cases not a disease but a sociocultural disorder that may lead to physical and biological symptoms. We need to deal with technique much more directly, and this will require the resymbolization of technique with regard to human health.

Resymbolizing Legal Education

In societies whose cultural orders are dominated by a technical order, legal institutions are threatened by two considerable difficulties. How can they regulate a technical order that has been built up from discipline-based approaches? How can they cope with the loss of their symbolic support on which they have traditionally depended? We will examine these difficulties in succession. Prior to industrialization, legal institutions dealt with cultural orders that were essentially self-regulating. With the emergence of new economic orders followed by a universal technical order, the disordering effects on both the cultural and the natural orders multiplied. The law cannot rescue a society from psychopathic technical expertise any more than it can rescue human life from a psychopathic order.

For example, the preceding descriptions of engineering and management show how a wide range of dislocations are produced, of which the environmental crisis has received the most attention. For decades the law attempted to keep up by passing more and more environmental regulations, with little success. It soon became apparent that this end-of-pipe approach was ineffective and very expensive. Cost-benefit techniques were applied, with the disastrous consequences that were discussed previously. The invocation of concepts such as deregulation and self-regulation had no effect whatsoever on the functioning of the system, including the production of environmental problems. Free trade had a chilling effect on every government because of their fear of being sued by corporations claiming that environmental regulations were actually trade barriers.

Thus far, the response to global warming has taken the form of a compensatory technique. By the creation of sophisticated models of climate change, the 'efficiency' of the biosphere in handling greenhouse gases is assessed and caps are established. In order to efficiently distribute these limitations on greenhouse gas emissions, cap and trade regulatory regimes are being implemented. There is little or no consideration of whether all this makes any sense. What if these models have missed some important interactions? What if the efficient distribution of the limitations to greenhouse gas emissions are fundamentally unjust? What we are doing is to 'efficiently' manage the atmosphere just like any other resource, and by now we ought to know the consequences.

In the same vein, the law has been impotent in protecting societies from their economies' being turned into anti-economies. Financial techniques that make money with money in the most efficient manner dominate the financial sector. The Enrons and the Goldman Sachs of this world are not first and foremost the result of corruption and dishonest behaviour (although this plays a non-trivial role) but the direct consequence of technique dominating finance. Nor has the law been able to protect us from losing our social support when our communities have been steadily undermined and replaced by mass societies. It is true that the growing impotence of the law has been partly offset by integration propaganda as a technical means of creating social conformity.

The most notable exceptions to the above patterns were the creation of the precautionary and 'no regrets' principles. They made sense and were supported by a great deal of evidence. It is almost always cheaper to prevent serious and irreversible effects on the environment as opposed to creating them in the first place, waiting until a cause-effect relation has been established, and then dealing with them in an ineffective end-of pipe manner. It turns out that in a significant number of cases the use of preventive approaches will ensure that we will be better off even when serious harmful effects on the environment do not materialize in the way that was anticipated. However, preventively oriented legal principles have been unable to impose a preventive orientation on discipline-based approaches.

In sum, there are no legal remedies to a technical order that has been built up with little or no reference to sense. Its structure and evolution make this impossible. Monetarism, which tries to brush all this away by declaring current economic phenomena and trends to be 'natural,' will soon have run its course. The politicians and religious leaders who believe that there are moral and political solutions do not understand the limitations of their trade any better than do our law makers. If tomorrow we all woke up as saints but continued to make use of discipline-based approaches, very little would change.

The second problem faced by contemporary legal institutions stems from the inadequate support they derive from highly desymbolized cultures. The significance of this loss of support may be illustrated by two issues. The first is the difference between applicable and non-applicable laws in democratic societies. The second is the reason that all cultures have invented legal institutions as a response to the necessities imposed by cultural orders.

Most laws are spontaneously obeyed in democratic societies, even though the overwhelming majority of their members have neither read any of the laws nor taken courses to have them explained. Such spontaneous obedience would be incomprehensible without a close correspondence between the metaconscious values implicit in the organizations of the brain-minds of the members of a community and the explicit values embodied in its laws. When this correspondence is weak, a law risks being massively disobeyed, and judges will have no alternative but to declare the law inapplicable for the obvious and practical reason that in democratic societies it is impossible to punish or incarcerate significant portions of the population. In totalitarian societies obedience may be coerced by the use of brute force aimed at striking fear into people. As a result, the art of lawmaking is based on a clear insight into the metaconscious values of a culture and the stretching' of them in a desired direction. The scope for doing so is highly limited if the creation of inapplicable laws is to be avoided. When a culture becomes highly desymbolized, this kind of lawmaking becomes next to impossible. The technical order that now dominates most cultural orders is an order of non-sense that cannot be directed by legal sense. Moreover, many legal issues appear to be quite different when viewed in terms of a cultural order as opposed to the technical one. For example, human rights must be interpreted in terms of their meaning with respect to the technical order and not the cultural one. It is the former that is steadily undermining what little remains of freedom and democracy. Nevertheless, our politicians keep behaving as if we still lived in genuine societies with intact cultural orders.

We have argued that some categories of crime accrue no benefits to those who commit them and that these may be interpreted as a sacred transgression of technique and the nation state. The desymbolization of cultural orders has weakened all social bonds. It is no longer our communities that are vandalized, defaced with graffiti, compromised by the hacking of their information systems, or robbed of necessary income by widespread cheating on taxes when people think they can get away with it. It is the 'system' that no longer commands the respect and trust of a growing proportion of society. The applicability of a great many tax laws is restricted to situations in which people cannot get away with dodging them. There is a growing underground economy. All of this and more reflect the fact that, for many people, it is no longer *our* economy and *our* government, because they regard the system as being manifestly unjust. People metaconsciously know that it is no longer *our* system, accountable to democratic processes, because in their daily lives they experience it as an external force turned against them. Many people no longer bother to vote because they believe it will change nothing essential. Governments increasingly use the law as an organizational tool, which further aggravates the situation. Rather than recognizing this as a crisis of legitimacy and dealing with the root issues, they find 'law and order' agendas almost irresistible. All this is particularly acute in policing, which is increasingly going out of public control. Too many innocent citizens are caught up in the mistakes and brutality of the police without having any recourse to meaningful redress.

The universality of legal institutions is rooted in the role they play in stabilizing cultural orders in time, space, and the social.[114] Humanity invented cultural orders within a living world in which everything was related to everything else, with the result that nothing ever repeated itself in quite the same way. The consequences of any human actions were a function of their characteristics and of the circumstances into which they were launched. Hence, the intentions behind these activities would have a different effect in another time and place or in different social circumstances. How could people count on anything in their individual lives, and how could the cultural order of their collective life endure in the face of so much change? How could any stability and predictability be created? The answer is that this was impossible without legal institutions of some kind. They created the conditions under which one could count on the powers of nature, almost regardless of time, place, and social circumstances. They made it possible to count on others, regardless of what might happen. As a result of Western civilization, we have come to associate legal institutions with human rights, justice, and freedom. However, these legal innovations were built on very important prior developments, which may be illustrated by a few examples from earlier societies.

For a culture to sustain human life it must create a way of life that involves stable and predictable relations with local ecosystems. At one time these ecosystems were believed to be made up exclusively of living beings. Everything had a spirit, with the result that no regularity in nature could be taken for granted. Indigenous people knew that the sun would rise every morning unless the sun god decided otherwise. Early agricultural people knew very well how dependent they were on the sun, and they therefore had no choice but to intervene in this unpredictable situation. A legally binding contract had to be entered into with the sun god by means of a magical and religious ritual that would bind both parties.

Similarly, a tribe whose way of life depended on the capture, training, and barter of elephants had not simply succeeded in developing the appropriate hunting and training techniques. They had also obtained and maintained the permission of the powers of nature to appropriate elephants in order to stay alive, and this had to be assured by a suitable legal arrangement and enacted with the aid of a religion. For these reasons the creation and use of early technologies were commonly intertwined with legal, magical, and religious arrangements.

A knowledge of nature is synonymous with predictability only if nature is believed to be the equivalent of a gigantic mechanism incapable of doing anything other than following the laws inscribed in it. However, such a view of nature is barely five hundred years old. Before that time, the powers and spirits of nature had to be reckoned with, which excluded any possibility of our kind of science. Any stability could be counted on only if a legal contract with the 'powers that be' had been established.

Since natural phenomena were generally seen as spatially localized, it was commonly believed that natural powers and spirits had jurisdiction in a limited territory. Hence, to establish a particular way of life, a community had to mark out a territory and make contracts with the local gods and spirits. The rituals by which this was assured had a legal and religious character. The legal and religious institutions modelled and stabilized the relations with the natural powers, thus assuring that their experiences of the local environment would be stable and predictable, provided that the contract was sustained by the appropriate rituals. Without this, no cultural order could be established within the natural order. The gods had consented to cooperate with the human order.

These arrangements also meant that, outside a community's territory, powers would be encountered with which the people had no relationship, with the result that anything could happen. Any venture beyond the established territory required the services of a magician who was in touch with these powers and could solicit their cooperation on behalf of the community. Such a magician thus lived outside of the order of the community, which made him or her a necessity as well as a threat.

The nineteenth century interpretation of private property has greatly distorted our understanding of the challenges that had to be overcome by earlier people. The institution of private property was not first and foremost a way to protect people from theft but a way to shield them from the powers of nature reclaiming what people had appropriated from it. This included permission to capture animals for food or domestication.

In addition, these early communities were obliged to order all the social relations that were necessary for the maintenance and evolution of their ways of life within their established territories. Their members had to be able to count on others, which required that a variety of social relations be made durable. Such social arrangements could not be left to circumstance if the way of life of the community was to endure, and this was always accomplished by means of legal institutions. For example, in relationships that were later to be stabilized by the invention of the institution of marriage, both partners could change in unpredictable ways. Without the legal domination of these changes, the rearing of children, the care for elderly parents, and other social obligations could be jeopardized, and the way of life of a community could not then reliably be passed on from generation to generation. After the establishment of the institution of marriage, the partners knew what they could expect from one another, and the community knew what to expect from the couple. The relationship had now taken on a measure of predictability regardless of changes in time, place, and social circumstances. The institution limited the way in which these circumstances could affect the evolution of the relationship. The future became reliable, and sanctions could be levied against those who disturbed the culturally imposed order by divorce.

Other examples are furnished by business relations of any kind. Suppose a member of an indigenous community made a living by obtaining permission from the natural powers to appropriate horses and to train them. Other members of the community would approach this person in order to obtain a horse. Any agreement reached between them could be jeopardized by changing circumstances. In time, one of the parties could change his or her mind and argue that, because of what had happened, the horse was no longer as desirable as it had been before. Once such arrangements became backed by the equivalent of legal contracts, neither party was absolved from responsibilities because of changing circumstances. Again, the legal institution made the agreement predictable and stable over time, and the community could sanction those who changed the form of the agreement. In this way, a community could establish a cultural order for many activities by providing a system of legal models. The cultural order was legally stabilized to become reliable and predictable because individual and collective human life was no longer left to circumstance. This predictability and reliability was achieved even though in life nothing ever repeats itself in quite the same way, as a result of everything constantly adapting by evolving in relation to everything else.

The applicability or non-applicability of laws and the universality of legal institutions depended on the symbolic support of a cultural order to which they were internally and externally connected. Individual and collective human life could then evolve in an orderly fashion and not fall prey to ever-changing circumstances. Along with religious institutions, legal institutions helped to rescue human life from relativism, nihilism, and anomie. This role was well understood in ancient Greece, which recognized its dependence on law. Before that time, legal institutions helped to establish and stabilize cultural orders with little regard to their effects on the individual members of a community. This development evolved further with a very important legal innovation made by the Romans, which asserted the legal rights of citizens with respect to the state. It became one of the founding 'perfections' of Western civilization.[115] Gradually, Western civilization built on this legal innovation to develop human rights and civil liberties.

This promising legal evolution was undermined by the process of industrialization and its desymbolizing effects on cultural orders. The emergence of economic orders, followed by a universal technical order, compelled any state to take control of legal institutions in order to regulate these orders at the expense of the cultural orders. The law took on an increasingly organizational character, and as common law traditions were overwhelmed by change, legal techniques became increasingly dominant. Jacques Ellul accurately predicted these developments shortly after the Second World War.[116]

Today the law serves a double purpose. First, legal techniques play an important role in stabilizing the technical order of non-sense. Few reasonable people can find much sense when they read the contracts governing the provision of their electricity, gas, water, sewage disposal, car insurance, life insurance, communications services (regular and cell phone, television, and Internet), software licences, medical consent forms, and much more. Almost without exception, these agreements are so one-sided that they lack legitimacy, even though the courts will enforce them. The same is true for police services, which primarily serve and protect the order of non-sense and only secondarily the common good.

The second and much less important role of the law is to stabilize what little remains of the cultural order as a result of desymbolization. It must be remembered that the participants in the legal system have a discipline-based education and that, beyond their areas of expertise, they rely on highly desymbolized cultures supplemented by integration propaganda. Without the latter, the legitimacy of the legal system would be even lower than it is today, and we would likely be facing an acute legal crisis. The situation is the same as in all other spheres of human activities: will the technical order be able to outpace the many crises it creates by compensatory techniques or will it succumb to its disordering effects on cultural orders and our being a symbolic species?

This admittedly broad-brushstroke diagnosis of the law raises the following question: will the technical order rule the law or can the law rule the technical order? It is a microcosm of a much larger issue, first introduced by Jacques Ellul as the autonomy of technique.[117] As previously argued, science and technique, being discipline-based approaches to knowing and doing, have permitted contemporary societies to spectacularly increase the power of their means by sacrificing the fabrics of human lives, societies, and ecosystems. This disordering has desymbolized our minds and cultures to accommodate everything to our most powerful creations. These no longer serve us, and this is changing everything, including the law.

Consider a recent court case as an example.[118] A group of Saskatchewan organic farmers launched a class action suit against Monsanto Canada and Bayer CropScience in an attempt to recover damages suffered from the introduction of a strain of canola that was genetically modified to resist herbicides. As a result of pollen from the genetically modified canola polluting non-genetically modified canola crops, these farmers were no longer able to meet the European standards for organic products. The Canadian government had approved genetically modified canola in the mid 1990s. Initially, the Canadian Private Organic Certification Organization did not explicitly mention genetically modified organisms in their standards, but it soon followed the European precedent of forbidding them. In addition to the organic farmers' no longer being able to grow canola, there was the problem of creating a gap in the their crop rotation scheme.

The outcome of this case is entirely predictable by the conceptual framework developed in this work. The evidence presented by expert witnesses was all discipline based. To make sense of this testimony, all participants in the trial had to translate the evidence from the relevant disciplines into the world of sense. There is no scientific way of doing this, because specialists can tell us nothing reliable about the meaning and value of anything for human life, society, and the biosphere. As a result, this translation had to involve the highly desymbolized organizations of the brain-minds of the participants as well as their shared culture. Since this desymbolized culture was possessed by the technical order down to the deepest metaconscious knowledge of the sacred and myths, it is difficult to envisage how the participants might have attributed the difficulties experienced by the organic farmers to the genetically modified canola. Something that has metaconsciously been associated with the order of the greatest good known by the cultural community cannot do bad things. The cause of the problem has to be elsewhere. The lawyers of the defendants were quick to point this out. The harm to these farmers was the result of the standards set by the organic certifier that were incompatible with all plants spreading pollen and by the decision of the farmers to adhere to these standards. After all, the genetically modified canola was approved by the Canadian government and thus, presumably, safe. In other words, the blame was shifted to the organic farmers, even though international agricultural biotechnology corporations are busy modifying the ecosystems on which these farmers depend.

The expert testimony was considered by the court according to legal precedence and principles. Once again, we must be very clear on what was involved. There is no longer any legal tradition, because it, along with all other traditions, has been completely overwhelmed by the many changes associated with industrialization during the past two hundred years. There is no question of cumulatively elaborating a cultural order by means of legal institutions. Most or all of the legal principles, and the precedents on which they are based, derive from a human, social, and environmental context that no longer exists. The introduction of genetically modified organisms into the biosphere represented an experiment of unprecedented proportions. None of these kinds of organisms had participated in the process of evolution, in the course of which everything has evolved in relation to everything else as an expression of a highly shared DNA. There is therefore a reasonable, if not highly probable, chance that these genetically modified organisms constitute an entirely new form of pollution of our planet's DNA pool. The launch of genetically modified organisms into the biosphere is irreversible, and their long-term effects on all life are scientifically (that is, based on disciplines) completely unpredictable. Governments are so busy managing the technical order, and so desperate for economic growth, that with a little lobbying by industry the common good does not stand much of a chance. In this case (and in almost all similar ones), it is impossible to scientifically establish the safety or the non-safety of genetically modified organisms. Given the domination of all cultural orders by the technical order and given the almost complete lack of awareness that most experts have of the limits of their expertise (separated from experience and culture), the government made a political decision. This being the case, the government ought to have recognized that a significant portion of the nation's citizens might well disagree, and should therefore have mandated the labelling of all foods containing genetically modified organisms.

There is a broad consensus that our contemporary ways of life are unsustainable. There is an equally widespread response that essentially amounts to continuing business as usual, and agriculture is no exception. Our agribusiness systems are unsustainable from many points of view. They are mining the soils, resulting in a decline of essential nutrients in our food. They are so heavily dependent on fossil fuels that the price of food is bound to rise sharply as we deplete these fuels or limit their use to avoid global warming. They deprive rural areas of adequate numbers of jobs, compelling a massive migration to the urban centres where there is no meaningful employment, especially in the south. Their treatment of animals as resources is inhumane by the standards of any reasonable and informed people. Their monocultures are so deeply unbalancing local ecosystems as to produce more and more problems, one of which required the invention of genetically modified canola. Moreover, these systems overconsume water in a great many cases and pollute both surface waters and aquifers. Given these kinds of issues, a reasonable course of government action would be to recognize that organic farming represents a possible alternative to industrial agriculture, which is unsustainable. In other words, if we resymbolize the decisions of governments to approve genetically modified organisms and to treat their citizens as children by not permitting them to know what they are eating, we would quickly discover that they are behaving as people unwilling to consider a great many important factors. However, this is to be expected in societies dominated by a sacralized technical order.

Had the government behaved reasonably, it would have applied the precautionary principle. It is one that makes sense since it does not suffer from the limitations of discipline-based approaches. However, doing so would have made the government vulnerable to the criticism by industry that it was risking bringing all scientific and technical progress to a standstill. Industry does not want to face the fact that the agribusiness subsystem of technique feeds on its own problems by compensating for them with more herbicides, pesticides, fertilizers, and much more, thus creating a need for genetically modified organisms. These represent neither an advance in nutrition nor an advance in the sustainable growing of food, but a compensation for problems created by a technique-based approach to agriculture.

With hindsight, it is hard to believe that we did not recognize from the beginning the kinds of problems we were getting into by creating agribusiness. It represents a subsystem of technique in which the overriding criteria for decision making are output-input ratios. As a result, the system represents a highly efficient way of extracting, processing, and distributing food, but a monumental failure in ensuring that this food is compatible with our nutritional needs, the ecosystems in which it is grown, our employment needs, our energy resources, and our responsibilities to future generations and all life.

The decisions of our courts in these kinds of situations tend to be unreasonable. There no longer is a legal tradition. Many legal precedents were set under conditions that no longer exist and do not consider current factors. The legal system has essentially loosened its ties with its context, to convert what remains into resources for legal techniques and principles. The courts thus end up being blind to the limitations of discipline-based expertise and the legal precedents created with that legal expertise. Moreover, the courts are entirely blind to the fact that they no longer operate in relation to a viable cultural order, as did legal institutions in the past. The result is a deep structural bias towards adapting societies to the technical order and protecting and advancing this order even when it represents an all-consuming pursuit of efficiency. The courts appear to treat as reasonable those people who are unaware of the depth to which the organizations of their brain-minds are possessed by a technical order and who therefore identify the public good with that order.

Jennifer Chandler points to other kinds of cases exhibiting the same pattern.[119] For example, a person seeks damages as compensation for an injury received at work. There is a reasonable possibility that the use of the latest medical techniques may diminish the consequences. When plaintiffs refuse these medical techniques, their eligibility for compensation is often denied or significantly reduced. Once again, there may well be a difference of opinion between discipline-based medical expertise and the reasoned opinions of plaintiffs who do not operate on the basis of disciplines and who therefore take many more factors into account on the basis of experience and symbolization. The courts have tended to impose the most 'efficient' approach to the problem as opposed to the most reasonable one. Similarly, so-called shrink-wrap contracts, which are not voluntarily entered into by purchasers of software, have been upheld by the courts. They have sacrificed fundamental principles for the necessities imposed by contemporary systems of mass production, advertising, and consumption. Although a systematic and comprehensive investigation of the hypothesis that the law upholds the technical order at the expense of the cultural order is far from complete, the present conceptual framework predicts that this will certainly turn out to be the case.

We are thus converging towards the same kinds of implications for legal education as we encountered for that of engineers, managers, and doctors. The discipline-based approach will have to be resymbolized, which will almost certainly turn many current decisions on their heads. If we wish to have our legal systems defend the public interest, it is essential that all participants can resymbolize the legal precedents and theoretical principles in the context of our time.

Community Colleges

I would like to close this chapter with a brief comment regarding our community colleges. In the reforms proposed thus far, these colleges should become the bridge between the knowing and doing embedded in experience and culture and the knowing and doing separated from experience and culture. They must seize the fruits of the transformation of university disciplines and specialties in order to further work out the implications. There is no future in making these colleges an intellectually and professionally inferior version of our universities. There is a future, however, in engaging their students' imaginations and skills to develop the equivalents of design exemplars and action-design exemplars in every sphere of human endeavour. They could then help in the creation of a future in which the gains of efficiency will be genuine rather than having been realized by destroying the fabric of all life. They could be a vital link in creating a civilization that includes science and technique but is not enslaved to them, and instead uses genuine human values to create a liveable and sustainable future.

Epilogue: Power and Non-power

In the last two centuries Western civilization has created a system that has steadily augmented the power it has over everything, including our selves. It has expanded its reach to every aspect of human life and all parts of the globe. Its power is derived from tearing and rearranging the fabric of all life. As a result, it treats human life, society, and the biosphere as commoditized resources, and their utilization has created a variety of difficulties necessitating further technical attention. The result is a self-reinforcing development that expands the scope and power of the system.

In exercising this power, we, the people, have completely transformed all our experiences of each other and the world, and with these transformed experiences we have had to build very different lives and communities. Consequently, much as nature did in prehistory and as society did in history, technique has taken hold of humanity to the very depth of our being a symbolic species. To this, technique has added reification and globalization. Despite the general consensus that slavery is an unacceptable form of human life, our possession by technique does not appear to trouble us a great deal. We may acknowledge that our technologies are no longer appropriate and our ways of life are no longer sustainable, but we do not connect this to an exercise of power that respects no local context. Our enslavement and reification are made liveable by new generations of secular myths that have displaced their traditional precursors. We live as if there were no limits to scientific knowing, technical doing, and political organizing, with the result that we have one common future, which, being universal, respects no local culture or ecosystem. All this would be entirely satisfactory if this reflected our human values, but instead we live as if improving the outputs that can be obtained from the required inputs were the path to the only acceptable future, that of economic growth, and the only way to stay in the competitive race. We just have to accept what little this leaves for ourselves. In other words, we have constantly adjusted our values to accommodate ourselves to technique and the nation state. The power that these creations have over us depends on our willingness to live as if they represented the greatest good in an ultimately unknowable reality. Consequently, we would trust them with our lives and our world.

Constrained by our straitjacket of power, we would do well to reflect for a moment on how we and the biosphere that sustains us came into existence. My students are always stunned to learn that the efficiency of photosynthesis, which ultimately sustains all life on our planet, is estimated to be around 2 per cent over the life of plants, both in the wild and under cultivation. They are even more surprised when I explain to them that such a low efficiency continues to be a necessary prerequisite for the diversity of life that is responsible for the incredible resilience of the biosphere, which thus far has stood up to our unprecedented exercise of power over it. In this regard, the biosphere represents an approach that, in contrast to ours, keeps the use of power to a bare minimum. Imagine if plants had an almost unlimited ability to increase their power of photosynthesis. The result would have been a struggle among plants that would have led to a few coming out on top. We would not have seen the five major plant formations, each of which utilizes this low efficiency of photosynthesis to balance its energy budget according to the availability of solar energy, water, and nutrients. The DNA pool was thus able to develop its full potential for a resilient diversity based on multiple pathways for energy to pass through the system and for matter to be conserved in self-purifying closed cycles. Without this diversity of plants, the diversity of all other life forms, including our own, would be unthinkable, as well as the enormous capacity of the biosphere to sustain all that life under variable conditions.

The situation is not so different for the social ecology of a human community. If the wealth and power of an elite is permitted to grow with few constraints, any possibility of cooperation, democracy, and a public good will be steadily diminished, making a civic society impossible. We have all experienced the same kind of thing in a group. If one person is permitted to be an authoritarian leader who can impose his or her will with few constraints, the other members of the group have little or no possibility of participating in evolving and adapting the group. In both cases, the dialectical tension between individual diversity and a common unity is decisively weakened, and along with it the ability to evolve and adapt while sustaining individual life. We have shown how technique continues to create a variety of instruments for transferring wealth from the bottom to the top. It has resulted in the proliferation of 'us-them' sentiments that have all but destroyed what little remained of constraints on the system by appeals to a public good or civic society. Among the industrially advanced nations, the United States leads this destructive development in the name of freedom and democracy. Canada is not far behind.

Preventive approaches would seek to re-establish a balance between the exercise of power and the need to respect the local context, which would constrain that power. The cultural approach, based on symbolization, amounted to a strategy of making the greatest possible use of context to ensure that everything would as much as possible adapt to and evolve with everything else. The technical approach, based on maximizing efficiency, makes the least possible use of context, thereby ensuring that everything does not fit the context and cannot evolve as an integral part of it. Since, as a consequence of technique changing people, we are mesmerized by the power of the system, preventive approaches are very attractive to young people who wish to make a contribution to a more liveable and sustainable future, but not to those who feel it will threaten their hard-earned security that is vested in an autonomous discipline or specialty.

All this is further obscured by our secular political religion of democracy. It is the greatest of all political illusions. We know very well that when we wire electronic components into a circuit, it is the structure of the circuit and not the values of the people using it that will determine the function that the circuit can deliver. Similarly, our discipline-based approach to knowing and doing has 'wired' the knowledge components provided by the many disciplines and specialties into socially and historically unique patterns backed by institutions, and the outcome of the knowledge structure will not depend on the intentions or political policies of anyone, any more than does the electronic circuit. Yet we continue to hope that the election of the next mayor, premier, prime minister, or president can somehow change the structure and functioning of the system. Similarly, we expect that by our good intentions and endless talk about greener products or sustainable development, the system will somehow be transformed. In the meantime, we continue to work in our disciplinary silos that prevent us from seeing that all this talk accomplishes nothing.

To sum up, the course of our civilization in the twenty-first century will be largely shaped by the relationship between technique and culture. If the former continues to dominate the latter, all the human, social, economic, and environmental issues described in this work will continue to multiply and intensify. It is a perfect example of positive feedback. The more the ways of life of contemporary mass societies are evolved and adapted in a piecemeal fashion by discipline-based knowing and doing, the more the fabric of human life, society, and the biosphere will be torn and the more further applications of that knowing and doing will become necessary.

On the microlevel the same kinds of developments will manifest themselves. The most minute, quantitative, and statistically sound studies open the floodgates to non-sense. Examining a small number of variables independently from millions of others increases the risk that the addition of one or two more will completely change the picture. Besides, a responsible intervention in the development of technique will require large non-cumulative changes that will pull the rug out from under these kinds of methodological approaches, because everything will be changing in complex, non-linear ways involving many variables. Without being guided by attempts to symbolize the situation and dialectically connect it to everything else, the internal consistency of any study will be lacking in external consistency with human life and the world. Parallel modes of knowing and doing are always required, but to create a synergistic relation between the two would involve a complete transformation of the secular sacred and myths of our civilization.

In the meantime, the level of desymbolization continues to grow under the pressure of discipline-based knowing and doing. At some point (if this has not already happened), a tipping point may be reached where the rehabilitation of experience and culture may become impossible. At that point, our future as a symbolic species will be compromised.

This book attempts to issue a wake-up call to many of our institutions. I would like to believe that the call may be heeded by our universities since they are still somewhat in the business of weighing evidence. However, which particular discipline or specialty is really in a position to weigh this kind of evidence, which far transcends its boundaries? Which discipline or specialty can initiate meaningful changes beyond its borders? If the administration is busy paying lip service to the myths of our mass society, little leadership can be expected from that source.

Chances of significant reforms coming from other institutions are even less likely. Corporations are too busy increasing shareholder value, and governments are too tied down in managing public opinion.

Nevertheless, if our civilization is to be worthy of that name, a way will have to be found to move towards one that includes science and technique but is not enslaved to these creations. I am simply restating in a new way a very old dilemma: our 'golden calf' no longer symbolizes our use of the life-sustaining and wealth-producing capacities of the biosphere but now represents the 'fertility' of a discipline-based science and technique driving unlimited growth on a finite planet. Are we capable of understanding our frenzy around this cult in order to wake up and chart a new and more responsible course? If not, our war on ourselves and on all other forms of life will continue.

As we are creating and intensifying more and more risks, the temptation appears to be almost irresistible to ramp up the use of integration propaganda to reinforce the character of democracy as a secular political religion. One of the most successful ways of doing so has been by co-opting Christianity. People are judged and divided into the believers who belong and those who threaten everything. In this way, societies are split into those who have the truth and those who have none of it. People either believe in the democratic way of life based on economic growth, free trade, discipline-based science and technique, and the values of the day (which, in a mass society, we can no longer live) or become a threat to everything for which this stands. Throughout human history, humanity has only been able to accomplish this by invoking a religion. We must step back from this abyss and recognize that the issues we face are so complex that none of us could possibly understand them fully and completely. All of us are bound to get some of what is happening right and much of what is happening wrong. We do not need to abandon our values and commitments to recognize this. Others can help us to see what presently we do not see and to understand what presently we do not understand. By means of symbolization we make human life into a dialogue that transforms our experiences into lives and our lives into communities. To be sure, this is always anchored in myths, but without them individual and collective human life would descend into relativism, nihilism, and anomie. If we do not step back from the cult of technique and the nation state, we will be unable to continue the necessary dialogue by which we must grow our understanding of ourselves, of others, and of all life. The future of our children and our grandchildren depends on it.

Notes

Introduction

1 The references for this brief overview will be presented when the concepts and ideas are discussed in detail in later chapters.
2 Gilbert Simondon, *Du mode d'existence des objets techniques* (Paris: Aubier, 1989).
3 H.H. Gerth and C. Wright Mills, eds., *From Max Weber: Essays in Sociology* (New York: Oxford University Press, 1963); see also Rogers Brubaker, *The Limits of Rationality: An Essay on the Social and Moral Thought of Max Weber* (London: Allen and Unwin, 1984); Jacques Ellul, *The Technological Society*, trans. John Wilkinson (New York: A.A. Knopf, 1964); *The Technological System*, trans. Joachim Neugroschel (New York: Continuum, 1980).
4 Adam Smith, *The Wealth of Nations* (New York: Collier, 1902).
5 Jacques Ellul, *Les classes sociales*, ed. Michel Hourcade, Jean-Pierre Jezequel, and Gerard Paul (Bordeaux: Institut d'Etudes Politiques, Université de Bordeaux, 1998); *La pensée Marxiste*, ed. Michel Hourcade, Jean-Pierre Jezequel, and Gerard Paul (Paris: La Table Ronde, 2003).
6 Gerth and Mills, *From Max Weber*; Brubaker, *The Limits of Rationality.*
7 John Kenneth Galbraith, *The New Industrial State* (New York: New American Library, 1985).
8 Jacques Ellul, *Perspectives on Our Age: Jacques Ellul Speaks on His Life and Work* (Toronto: House of Anansi Press, 2004, rev. ed.); W.H. Vanderburg, *Living in the Labyrinth of Technology* (Toronto: University of Toronto Press, 2005).
9 Nelly Viallaniex, *Kierkegaard et la parole de Dieu* (Paris: Champion, 1977).

1 Symbolization: Getting in Touch with Ourselves and the World

1 The concept of our bodies working in the background is based on Hubert Dreyfus, *On the Internet* (New York: Routledge, 2001) and *What Computers Still Can't Do* (Cambridge: MIT Press 1992). I have adapted it to our lives working in the background via the organizations of our brain-minds.
2 Timothy D. Wilson, *Strangers to Ourselves: Discovering the Adaptive Unconscious* (Cambridge, MA: Belknap Press of Harvard University Press, 2002).
3 It is important for the reader to understand the conditions under which this work emerged. After completing a doctorate in mechanical engineering, I continued my studies of technology via the social sciences. My principal interest was the 'culture' of engineering, and it quickly became evident that this required an understanding of how babies and children acquired a culture and how these cultures worked in their lives. Coming as an outsider, I naively assumed that all this could not work without a great deal of what today is called neural plasticity. This concept has found widespread acceptance, but the implications for psychology, social psychology, sociology, and anthropology are far from developed. As best as I can understand it, on the one hand the works emphasizing neural plasticity do not adequately develop this concept in relation to our being a symbolic species. On the other hand, works that start from the premise that we are a symbolic species are far from fully developing our dependence on neural plasticity. If we grow our brain-minds by means of processes of differentiation and integration, each development in the lives of babies and children builds on previous ones and prepares for others. Symbolization ensures that all these developments are related to all the others. The concept of neural plasticity is implicit in my first book, which seeks to understand how babies and children develop the organizations of their brain-minds and how the cultures of their communities evolve. After reading a first draft, my post-doctoral mentor warned me that the book would not have much success because it was too far ahead of the thinking of that time. This turned out to be true, and most current studies of neural plasticity and our being a symbolic species have ignored it. The study was published under the title of *The Growth of Minds and Cultures: A Unified Theory of the Structure of Human Experience* (Toronto: University of Toronto Press, 1985).
4 Michael Polanyi, *Personal Knowledge: Towards a Post-critical Philosophy* (New York: Routledge, 1998).
5 Thomas S. Kuhn, *The Structure of Scientific Revolutions*, 2nd. ed. (Chicago: University of Chicago Press, 1970).

6 For a discussion of the model of human skill acquisition developed by Stuart Dreyfus, see the special issue entitled 'Human Expertise in the Age of the Computer' of the *Bulletin of Science Technology & Society* 24, no. 3 (June 2004): 175–274.
7 Vanderburg, *The Growth of Minds and Cultures.*
8 T.W. Deacon, *The Symbolic Species: The Co-evolution of Language and the Brain* (New York: W.W. Norton, 1998).
9 For complete details, the reader is referred to Vanderburg, *The Growth of Minds and Cultures.*
10 William E. Connolly, 'Essentially Contested Concepts,' in *The Terms of Political Discourse* (Princeton, NJ: Princeton University Press, 1984), 10–44.
11 L. Wittgenstein, *Philosophical Investigations,* trans. G.E.M. Anscombe (Oxford: Blackwell Publishing, 1953).
12 Deacon, *The Symbolic Species.*
13 W.H. Beston, W. Heron, and T.H. Scott, 'Effects of Decreased Variation in the Sensory Environment,' *Canadian Psychology* 8 (1954): 70. J.P. Zubeck, D. Pushkar, W. Sansom, and J. Gowing, 'Perceptual Changes after Prolonged Sensory Isolation (Darkness and Silence),' *Canadian Journal of Psychology* 15 (1961): 83.
14 Robert Karasek and Töres Theorell, *Healthy Work: Stress, Productivity, and the Reconstruction of Working Life* (New York: Basic Books, 1990).
15 Helen Keller, *The Story of My Life* (New York: Norton, 2003).
16 Deacon, *The Symbolic Species.*
17 R. Shayna Rosenbaum, Stefan Köhler, Daniel L. Schachter, Morris Moscovith, Robyn Westmacott, Sandra B. Black, Fuquiang Gao, and Endel Tulving, 'The Case of K.C.: Contribution of a Memory-Impaired Person to Memory Theory,' *Neuropsychologia* 43 (2005): 989–1021.
18 Jacques Ellul, *Propaganda: The Formation of Men's Attitudes,* trans. Konrad Kellen and Jean Lerner (New York: Vintage Books, 1973); Jacques Ellul, 'De la signification des relations publiques dans la société technicienne: Un cas de passage de l'information à la propagande,' *L'Année sociologique,* 1963: 69–152.
19 Vanderburg, *Living in the Labyrinth of Technology.*
20 A. Peterson, 'The Philosophy of Niels Bohr,' *Bulletin of the Atomic Scientists* 19 (Sept. 1963): 8–14.
21 A. Peterson, *Quantum Physics and the Philosophical Tradition* (Cambridge, MA: MIT Press, 1968), 188.
22 'Autobiographical Note,' in *Albert Einstein: Philosopher-Scientist*, ed. P.A. Schlipp (Evanston, IL: Library of Living Philosophers, 1949).

23 W. Heisenberg, *Physics and Beyond: Encounters and Conversations* (New York: Harper and Row, 1971), 63.
24 W.M. Cook, 'Ability of Children in Colour Discrimination,' *Child Development* 2 (1931): 303; P. Malrieu, 'Quelques problèmes de la vision des couleurs chez l'enfant,' *Journal de Psychologie* 52 (1955): 222; J. André, *Études sur les termes de couleur dans la langue latine* (Paris: Klinsksieck, 1949), 427.
25 Claude Lévi-Strauss, *La pensée sauvage* (Paris: Plon, 1962).
26 Jacques Ellul, *The Humiliation of the Word*, trans. Joyce Main Hanks (Grand Rapids, MI: Eerdmans, 1985).
27 Maurice Merleau-Ponty, *Phenomenology of Perception* (New York: Routledge and K. Paul, 1962).
28 Kuhn, *The Structure of Scientific Revolutions.*
29 J.S. Bruner and Leo Postman, 'On the Perception of Incongruity: A Paradigm,' *Journal of Personality* 18 (1949): 206–23.
30 Oliver Sacks, 'Phantom Fingers: The Case of the One-Armed Pianist,' in *Musicophilia: Tales of Music and the Brain* (New York: Alfred A. Knopf, 2007).
31 See especially Jacques Ellul, *The New Demons,* trans. C. Edward Hopkin (New York: Seabury, 1975); Vanderburg, *The Growth of Minds and Cultures.* These ideas arose from intuitions that I developed, from sitting in on his lectures during my five years of post-doctoral work, regarding the concept of culture implicit in his sociological and historical analyses.
32 M. Eliade, *A History of Religious Ideas* (Chicago: University of Chicago Press, 1968); *The Sacred and the Profane*, trans. W. Trask (New York: Harper and Rowe, 1961).
33 P. Ricoeur, *The Symbolism of Evil,* trans. E. Buchanan (New York: Harper and Rowe, 1967).
34 Raimundo Panikkar, *Myth, Faith and Hermeneutics: Cross-Cultural Studies* (New York: Paulist Press, 1979).
35 Arnold Toynbee, *A Study of History,* abridgement of vols. 1–10 by D.C. Somervell (London: Oxford University Press, 1946).
36 Kuhn, *The Structure of Scientific Revolutions.*
37 Georges Devereux, *From Anxiety to Method in the Behavioral Sciences* (New York: Humanities Press, 1967).
38 Ludwig Feuerbach, *Lectures on the Essence of Religion*, trans. Ralph Manheim (New York: Harper and Row, 1967).
39 Emile Durkheim, *On Suicide,* trans. Robin Buss (New York: Penguin, 2006).
40 Ellul, *Perspectives on Our Age*; Ellul, *The New Demons*; Vanderburg, *Living in the Labyrinth of Technology.*
41 Toynbee, *A Study of History*; Joseph A. Tainter, *The Collapse of Complex Societies* (Cambridge: Cambridge University Press, 1988).

42 Vanderburg, *Living in the Labyrinth of Technology*.

43 Jacques Ellul, *On Freedom, Love, and Power*, comp., ed., and trans. Willem H. Vanderburg. (Toronto: University of Toronto Press, 2010)

44 Ellul, *The Humiliation of the Word*.

45 In spite of the lip service that our civilization pays to freedom, exercising it is almost always extraordinarily difficult. It is easy to affirm this and other values, but it is far more difficult to examine how our conformity may contribute to their violation and to stand up to do something about it. This is especially true when the consequences of non-conformity are very serious, as in Nazi Germany under Hitler, fascist Italy under Mussolini, communist Russia especially under Stalin, and American democracy under McCarthy. The great secular political religions of the twentieth century (communism, national socialism and democracy) performed the same role as traditional religions did in the past, and their 'heretics' were treated in much the same way. The works of Robert J. Lifton and Erich Fromm also come to mind. Today there exists a widespread belief that all living systems are in decline, leading to the need for a new kind of secular magic that proclaims every change as bringing us greater sustainability. Very few of us can be counted as exercising our freedom in accordance with our values and thus living as non-conformists. To the extent that we do, we experience the extraordinary constraints technique imposes on our lives. The exercise of our freedom can have no meaning other than in relation to what alienates us. Resisting the alienation of a past era is wonderfully easy, politically correct, and self-justifying. If democracy sets us free, we had better know from what it liberates us. Our values, like our experiences, are dialectically structured. Unless we know what alienates our lives, we will have no idea how we may be set free.

2 Desymbolization: Losing Touch with Ourselves and the World

1 Ellul, *On Freedom, Love and Power*.

2 Dreyfus, *What Computers Still Can't Do*.

3 I am relying on the excellent interpretation advanced by Hubert Dreyfus, *On the Internet*. See especially chapter 4.

4 Dreyfus, *On the Internet*.

5 Ellul, *Propaganda*.

6 Dreyfus, *On the Internet*.

7 Hubert Dreyfus, Stuart E. Dreyfus, and Tom Athanasiou, *Mind over Machine: The Power of Human Intuitions and Expertise in the Era of the Computer* (New York: Free Press, 1986).

8 Willem H. Vanderburg, *The Labyrinth of Technology* (Toronto: University of Toronto Press, 2000). See especially chapter 3.
9 Ellul, *The Technological System*; Vanderburg, *Living in the Labyrinth of Technology*, see especially chapter 10.
10 Vanderburg, *The Labyrinth of Technology.* See especially chapters 4 and 5.
11 Ellul, *Propaganda.* Ellul's interpretation of human life in a mass society is at the heart of my disagreements with postmodernist views regarding the way audiences relate to television as well as to other technologies. These include the *reception theory* of Wolfgang Iser, the *reader response theory* of Stanley Fish, and what Michael de Certeau and John Fiske referred to as *resistant activity*. I will argue in the present and following chapters that, although people continue to form their own meanings, they now do so within the confines of a highly desymbolized mass society, and this changes everything. For postmodernist interpretations, see Michael de Certeau, *The Practice of Everyday Life*, trans. Steven Rendall (Berkeley: University of California Press, 1984); Stanley Fish, *Is There a Text in This Class? The Authority of Interpretive Communities* (Cambridge, MA: Harvard University Press, 1980); John Fiske, *Television Culture*, 2nd ed. (New York: Routledge, 2010); Wolfgang Iser, *The Act of Reading: A Theory of Aesthetic Response* (Baltimore: Johns Hopkins University Press, 1978).
12 David Riesman, *The Lonely Crowd: A Study of Changing American Character* (New Haven, CT: Yale University Press, 1969); Vanderburg, *Living in the Labyrinth of Technology.*
13 Jerry Mander, *Four Arguments for the Elimination of Television* (New York: Morrow, 1978).
14 Neil Postman, *Amusing Ourselves to Death: Public Discourse in the Age of Showbusiness* (New York: Penguin Books, 1988).
15 Marshall McLuhan, *The Global Village: Transformations in World Life and Media in the 21st Century* (New York: Oxford University Press, 1989).
16 Guy Debord, *The Society of the Spectacle* (New York: Zone, 1994).
17 Lewis Mumford, *The Myth of the Machine; Technics and Human Development* (New York: Harcourt, Brace, and World, 1967).
18 Cornelius Castoriadis, *The Imaginary Institution of Society* (Cambridge, MA: MIT Press, 1998).
19 Roger Caillois, *Le mythe et l'homme* (Paris: Gallimard, 1972).
20 Ellul, *The Technological System*; Vanderburg, *The Growth of Minds and Cultures.*
21 William E. Connolly, 'Essentially Contested Concepts,' in *The Terms of Political Discourse* (Princeton, NJ: Princeton University Press, 1984), 10–44.
22 Vanderburg, *The Labyrinth of Technology.* See especially chapter 4.

23 Ellul, *The Humiliation of the Word.*
24 Ibid.
25 Ibid.
26 Joel Bakan, *The Corporation: The Pathological Pursuit of Profit and Power* (Toronto: Penguin Canada, 2004).
27 McLuhan, *The Global Village.*
28 Sherry Turkle, *The Second Self: Computers and the Human Spirit* (New York: Simon and Schuster, 1984); Sherry Turkle, *Living on the Screen: Identity in the Age of the Internet* (New York: Simon and Schuster, 1995).
29 Craig Brod, *Technostress: The Human Cost of the Computer Revolution* (Reading, MA: Addison-Wesley, 1984).
30 Karasek and Theorell, *Healthy Work.*
31 Tim Blackmore, *War X: Human Extensions in Battlespace* (Toronto: University of Toronto Press, 2005).
32 Vanderburg, *Living in the Labyrinth of Technology.*
33 Benson Snyder, 'Literacy and Numeracy: Two Ways of Knowing,' *Daedalus* 119 (1990): 233–56.
34 Ibid.
35 Vanderburg, *Living in the Labyrinth of Technology.* See especially chapter 10.
36 Don Tapscott, *Grown Up Digital; How the Net Generation is Changing Your World* (New York: McGraw-Hill, 2009); Matthew Fraser, *Throwing Sheep in the Boardroom: How Online Social Networking Will Transform Your Life, Work and World* (Hoboken, NJ: Wiley, 2008). For more critical assessments, see Jane M. Healy, *Endangered Minds: How Computers Affect Our Children's Minds – For Better and Worse* (New York: Simon and Schuster, 1998); Mark Bauerlein, *The Dumbest Generation: How the Digital Age Stupefies Young Americans and Jeopardizes Our Future (or Don't Trust Anyone Under 30)* (New York: Jeremy P. Tarcher / Penguin, 2009); Jaron Lanier, *You Are Not a Gadget* (New York: Alfred A. Knopf, 2010); Lee Siegel, *Against the Machine: How the Web is Reshaping Culture and Commerce and Why It Matters* (New York: Spiegel and Grau, 2008).
37 Vanderburg, *Living in the Labyrinth of Technology.*
38 Ibid.
39 Kuhn, *The Structure of Scientific Revolutions.*
40 Stephen Hawking, *A Briefer History of Time* (London: Bantam, 2005).
41 Ibid.
42 David Bohm, *Wholeness and the Implicate Order* (New York: Routledge, 2002).
43 Bernard D'Espagnat, *In Search of Reality* (New York: Springer-Verlag, 1983).

44 Many studies have implied this kind of connection. See, for example, Mary Midgley, *Science as Salvation: A Modern Myth and Its Meaning* (New York: Routledge, 1992). We will return to this subject.
45 M. McCloskey, 'Intuitive Physics,' *Scientific American* 248 (April 1983): 122–30. For evidence of an 'intuitive arithmetic,' see J. Lave, 'The Values of Quantification,' in *Power, Action and Belief*, ed. J. Law (London: Routledge and Kegan Paul, 1986), 88–111.
46 McCloskey, 'Intuitive Physics.'
47 Sheila Tobias, *They're Not Dumb, They're Different: Stalking the Second Tier* (Tucson, AZ: Research Corporation Foundation for the Advancement of Science, 1990).
48 Vanderburg, *Living in the Labyrinth of Technology.*
49 Ibid.
50 Devereux, *From Anxiety to Method in the Behavioral Sciences.*

3 Colliding Orders and the Triumph of the Real

1 Henry Hodges, *Technology in the Ancient World* (London: Allen Lane, 1970).
2 Sigfried Giedion, *Mechanization Takes Command: A Contribution to Anonymous History* (New York: Oxford University Press, 1955); Hodges, *Technology in the Ancient World.*
3 Ellul, *The Humiliation of the Word.*
4 Vanderburg, *Living in the Labyrinth of Technology.*
5 E.P. Thompson, *The Making of the English Working Class* (Harmondsworth, UK: Penguin Books, 1976).
6 Vanderburg, *The Labyrinth of Technology.*
7 Vanderburg, *Living in the Labyrinth of Technology*; Jacques Ellul, *Changer de révolution: L'inéluctable prolétariat* (Paris: Editions Seuil, 1982).
8 Smith, *The Wealth of Nations.*
9 Karasek and Theorell, *Healthy Work.*
10 Smith, *The Wealth of Nations.*
11 Galbraith, *The New Industrial State.*
12 David C. Korten, *When Corporations Rule the World* (West Hartford, CT: Kumarian Press, 1995).
13 Karl Polanyi, *The Great Transformation* (New York: Farrar and Rinehart, 1944).
14 Georg Simmel, *Ville et modernité*, sous la direction de Jean Rémy (Paris: L'Harmattan, 1995); 'The Metropolis and Mental Life,' in *Cities and Society*, ed. P.K. Hatt and A.J. Reiss (New York: Free Press, 1957), 32–57.

15 Willem H. Vanderburg, 'Technology and the Law: Who Rules?' *Bulletin of Science, Technology & Society* 27, no. 4 (2007): 322–32; Vanderburg, *The Growth of Minds and Cultures*.
16 Vanderburg, *Living in the Labyrinth of Technology*.
17 Ellul, *Les classes sociales; La pensée Marxiste*.
18 Ibid.
19 Jacques Ellul, *Métamorphose du bourgeois* (Paris: Calmann-Levy, 1967).
20 Ellul, *The New Demons*; Richard Stivers, *Evil in Modern Myth and Ritual* (Athens, GA: University of Georgia Press, 1982); Vanderburg, *Living in the Labyrinth of Technology*.
21 Vanderburg, *Living in the Labyrinth of Technology*.
22 Ludwig Feuerbach, *Lectures on the Essence of Religion*, trans. Ralph Manheim (New York: Harper and Row, 1967).
23 Lewis Mumford, *The Myth of the Machine: The Pentagon of Power* (New York: Harcourt, Brace and Jovanovich, 1967).
24 Max Weber, *The Protestant Ethic and the Spirit of Capitalism*, trans. Talcott Parsons (New York: C. Scribner, 1985).
25 Giedion, *Mechanization Takes Command*; Jacques Ellul, 'Remarks on Technology and Art,' *Bulletin of Science, Technology & Society* 21, no. 1 (Feb. 2001): 26–37.
26 Raymond Williams, *New Keywords: A Revised Vocabulary of Culture and Society*, ed. Tony Bennett, Lawrence Grossberg, and Meaghan Morris (Malden, MA: Blackwell Publishing, 2005).

4 The Triumph of the Technical Order over Cultural Orders

1 The original five-stage acquisition model was published in Dreyfus and Dreyfus, with Athanasiou, *Mind over Machine*. For a partial update and a range of applications, the reader is referred to a special issue of the *Bulletin of Science, Technology & Society* (June 2004). The interpretation of this theory and its applications by Harry Collins and Robert Evans, *Rethinking Expertise* (Chicago: University of Chicago Press, 2007), is very different from my own. So is their understanding of the concept of tacit knowledge developed by Michael Polanyi, *Personal Knowledge*. My interpretation, *The Growth of Minds and Cultures*, of how babies and children learn to skilfully cope with the world converges with those of Stuart and Hubert Dreyfus as well as that of Michael Polanyi.
2 Dreyfus and Dreyfus, *Mind over Machine*, 32.
3 Hubert Dreyfus and Stuart Dreyfus, 'What Is Morality? A Phenomenological Account of the Development of Ethical Expertise,' in *Universalism vs.*

Communitarianism: Contemporary Debates in Ethics, ed. David M. Rasmussen (Cambridge, MA: MIT Press, 1990), 237–64.

4 George Sturt, *The Wheelwright's Shop* (Cambridge: Cambridge University Press, 1963).

5 Shlomo Avineri, *The Social and Political Thought of Karl Marx* (Cambridge: Cambridge University Press, 1968).

6 Gerth and Mills, *From Max Weber*; see also Brubaker, *The Limits of Rationality.*

7 Ludwig Wittgenstein, *Philosophical Investigations,* trans. G.E.M. Anscombe (Oxford: Basil Blackwell, 1967).

8 Vanderburg, *Living in the Labyrinth of Technology.*

9 Brubaker, *The Limits of Rationality.*

10 Vanderburg, *Living in the Labyrinth of Technology.*

11 Ibid.

12 Devereux, *From Anxiety to Method in the Behavioral Sciences.*

13 Werner Heisenberg, *The Physical Principles of the Quantum Theory*, trans. Carl Eckart and Frank C. Hoyt (New York: Dover Publications, 1930).

14 McCloskey, 'Intuitive Physics,' 122–30. For evidence of an 'intuitive arithmetic,' see Lave, 'The Values of Quantification,' 88–111.

15 Gilbert Simondon, *Du mode d'existence des objets techniques.*

16 Willem H. Vanderburg, 'Rethinking Engineering Design and Decision Making in Response to Economic, Social and Environmental Crises, *Bulletin of Science, Technology & Society* 29, no. 5 (October 2009): 421–32.

17 Christian Berggren, 'The Fate of Branch-Plants – Performance versus Power,' in *Enriching Production: Perspectives on Volvo's Uddevalla Plant as an Alternative to Lean Production,* ed. Ake Sandberg, 107–8; Christian Berggren, *The Volvo Experience: Alternatives to Lean Production in the Swedish Auto Industry* (Basingstoke, UK: Macmillan, 1994); Ricardo Semler, *Maverick: The Success Story Behind the World's Most Unusual Workplace* (New York: Warner Books, 1993).

18 Willem H. Vanderburg, 'Can the University Escape from the Labyrinth of Technology? Part 1: Rethinking the Intellectual and Professional Division of Labour and Its Knowledge Infrastructure,' *Bulletin of Science, Technology & Society* 26, no. 3 (June 2006): 171–7; 'Part 2: Intellectual Map Making and the Tension between Breadth and Depth,' 178–88; 'Part 3: A Strategy for Transforming the Profession,' 189–203; 'Part 4: Extending the Strategy to Medicine, the Social Sciences and the University,' 204–16.

19 Jacques Ellul, *The Technological Society*, trans. John Wilkinson (New York: Vintage Books, 1964).

20 Andrew Freeman, 'Turning Digits into Dollars: A Survey of Technology in Finance,' *The Economist*, 26 Oct. 1996, 3–16; Rob Norton, 'Which Offices or Stores Really Perform Best,' *Fortune*, 31 Oct. 1994, 38.

21 M.F. Elliott-Jones, *Input-Output Analysis: A Nontechnical Description* (New York: Conference Board, 1971); Michael L. Lahr and Erik Dietzenbacher, eds., *Input-Output Analysis: Frontiers and Extensions* (New York: Palgrave, 2001).
22 Henry Mintzberg, *Managers, Not MBAs: A Hard Look at the Soft Practice of Managing and Management Development* (San Francisco: Berrett-Koehler Publishers, 2004).
23 Ivar Sonbo Kristansen and Gavin Mooney, eds., *Evidence-Based Medicine: In Its Place* (New York: Routledge, 2004); Scott Weingarten and Jack Tinker, eds., *Evidence-Based Medicine: A Critical Assessment* (London: Royal Society of Medicine Press, 2002).
24 Russell L. Ackoff, 'The Future of Operational Research Is Past,' *Journal of the Operational Research Society* 30, no. 2 (1979): 93–104.
25 Jacques Ellul, *La technique, ou, L'enjeu du siècle* (Paris: Economica, 1990).
26 Ellul, *The Technological Society*.
27 Willem H. Vanderburg, 'The Anti-economy Hypothesis' (parts 1–3), *Bulletin of Science, Technology & Society* 29, no. 1 (Feb. 2009): 48–56.
28 Galbraith, *The New Industrial State*.
29 Vanderburg, *Living in the Labyrinth of Technology*.
30 Galbraith, *The New Industrial State*.
31 Vanderburg, *The Labyrinth of Technology*.
32 Ralph Estes, *Tyranny of the Bottom Line: Why Corporations Make Good People Do Bad Things* (San Francisco: Berrett-Koehler Publishers, 1996).
33 Korten, *When Corporations Rule the World*.
34 S. Anderson and J. Cavanagh, *Top 200: The Rise of Corporate Global Power* (Institute for Policy Studies, 2000).
35 Vanderburg, *The Labyrinth of Technology*, chapter 3.
36 Ibid.
37 Vanderburg, 'Can the University Escape from the Labyrinth of Technology?', parts 1–4.
38 Herman E. Daly and John B. Cobb Jr., *For the Common Good: Redirecting the Economy toward Community, the Environment, and a Sustainable Future* (Boston: Beacon Press, 1989).
39 Ibid.
40 Ibid.
41 C. Cobb, T. Halstead, and J. Rowe, 'If the GDP Is Up, Why Is America Down?' *Atlantic Monthly*, October 1995: 59–78; and *The Genuine Progress Indicator: Summary of Data and Methodology* (Washington, DC: Redefining Progress, 1995).
42 B.R. Allenby and D.J. Richards, eds., introduction to *The Greening of Industrial Ecosystems* (Washington, DC: National Academy Press, 1994).

43 Karasek and Theorell, *Healthy Work*.
44 Vanderburg, *The Labyrinth of Technology*, chapter 7.
45 Vanderburg, 'The Anti-economy Hypothesis,' parts 1–3.
46 J. Bogle, *The Battle for the Soul of Capitalism* (New Haven, CT: Yale University Press, 2005); J. Buchan, *Frozen Desire: The Meaning of Money* (New York: Farrar, Straus and Giroux, 1997); and B. Lietaer, *The Future of Money: A New Way to Create Wealth, Work and a Wiser World* (New York, Random House, 2001).
47 Vanderburg, *Living in the Labyrinth of Technology*.
48 Ellul, *The Technological System*.
49 Vanderburg, 'The Anti-economy Hypothesis,' parts 1–3.
50 Lewis Mumford, *The Myth of the Machine: The Pentagon of Power* (New York: Harcourt, Brace and Jovanovich, 1970).
51 Turkle, *The Second Self*.
52 Galbraith, *The New Industrial State*.
53 Brubaker, *The Limits of Rationality*.
54 Ellul, *The New Demons*; Vanderburg, *Living in the Labyrinth of Technology*.
55 Ellul, *La technique, ou, L'enjeu du siècle*.
56 Jack P. Manno, *Privileged Goods: Commoditization and Its Impact on Environment and Society* (Boca Raton, FL: Lewis Publishers, 2000).
57 Michael Ignatieff, *The Rights Revolution* (CBC Massey Lecture) (Toronto: House of Anansi Press, 2000).
58 Ellul, *On Freedom, Love, and Power*.
59 Ellul, *Perspectives on Our Age*; and Vanderburg, *Living in the Labyrinth of Technology*.
60 Ellul, *The New Demons*.
61 See for example Thomas L. Friedman, *The Lexus and the Olive Tree* (New York: Farrar, Straus and Giroux, 1999), and *The World Is Flat: A Brief History of the Twenty-First Century* (New York: Farrar, Straus and Giroux, 2006).
62 Bakan, *The Corporation*.
63 Jacques Ellul, *L'empire du non-sens: l'art et la société technicienne*, (Paris: Presses universitaires de France, 1980); Willem H. Vanderburg, 'Comments on the Empire of Non-Sense: Art in a Technique-Dominated Society,' *Bulletin of Science, Technology & Society*, Vol. 21, No. 1, (Feb. 2001): 38–54.

5 Desymbolization and Resymbolization

1 Terry Eagleton, *The Illusions of Post-modernism* (Oxford: Blackwell Publishers, 1996).
2 Devereux, *From Anxiety to Method in the Behavioral Sciences*.

3 Herman Daly, 'Incorporating Values in a Bottom-Line Ecological Economy,' *Bulletin of Science, Technology & Society* 29, no. 5 (Oct. 2009): 349–57.
4 Wassily Leontieff, 'Letters: Academic Economics,' *Science*, 217: 104–5.
5 Galbraith, *The New Industrial State*.
6 Vanderburg, *Living in the Labyrinth of Technology*.
7 Herman Daly, *Beyond Growth: The Economics of Sustainable Development* (Boston: Beacon, 1996).
8 Jacques Ellul, *The Political Illusion*, trans. Konrad Kellen (New York: Knopf, 1967).
9 Peter Victor, *Managing without Growth: Slower by Design, Not Disaster* (Cheltenham, UK: Edward Elgar Publishing, 2008).
10 Robert Skidelsky, *John Maynard Keynes, 1883–1946: Economist, Philosopher, Statesman* (London: Pan Books, 2004).
11 Possibly the most persuasive advocate of all of this is Thomas L. Friedman. See especially his books entitled *The Lexus and the Olive Tree* and *The World Is Flat*.
12 Galbraith, *The New Industrial State*.
13 R. Richta, *Civilization at the Crossroads: Social and Human Implications of the Scientific and Technological Revolution* (Prague: International Arts and Science Press, 1969).
14 Vanderburg, 'The Anti-economy Hypothesis,' parts 1–3, 48–56.
15 Lietaer, *The Future of Money*.
16 R.T. Naylor, *Wages of Crime: Black Markets, Illegal Finance, and the Underworld Economy* (Montreal: McGill-Queens University Press, 2002).
17 J. Bogle, *The Battle for the Soul of Capitalism* (New Haven, CT: Yale University Press, 2005).
18 E. Janszen, 'The Next Bubble: Priming the Markets for Tomorrow's Big Crash,' *Harper's Magazine*, February 2008, 39–45.
19 Shoshana Zuboff and James Maxmin, *The Support Economy: Why Corporations Are Failing Individuals in the Next Episode of Capitalism* (New York: Viking, 2002).
20 Jacques Ellul, *The Politics of God and the Politics of Man*, trans. Geoffrey W. Bromiley (Grand Rapids, MI: Eerdmans, 1972).
21 Jeff Sharlet, *The Family: The Secret Fundamentalism at the Heart of American Power* (New York: Harper Perennial, 2009).
22 Ellul, *On Freedom, Love, and Power*.
23 Jacques Ellul already warned of this danger almost half a century ago in his book *The Technological Society*.
24 William Cronon, *Nature's Metropolis, Chicago and the Great West* (New York: W.W. Norton, 1992).

25 Estes, *The Tyranny of the Bottom Line.*
26 D.C. Korten, *The Post-corporate World: Life after Capitalism* (San Francisco: Berrett-Koehler, 2000).
27 Estes, *The Tyranny of the Bottom Line.*
28 Daly and Cobb, *For the Common Good*; Cobb, Halstead, and Rowe, 'If the GDP Is Up, Why Is America Down?' 59–78; and *The Genuine Progress Indicator.*
29 David K. Johnston, *Free Lunch: How the Wealthiest Americans Enrich Themselves at Government Expense and Stick You with the Bill* (New York: Penguin Group, 2007).
30 Ibid.
31 Ibid.
32 Ibid.
33 Ibid.
34 Ibid.
35 Ibid.
36 Ibid.
37 Ibid.
38 Weber, *The Protestant Ethic and the Spirit of Capitalism.*
39 Vanderburg, *Living in the Labyrinth of Technology.*
40 Vanderburg, 'Can the University Escape from the Labyrinth of Technology? Part 1: Rethinking the Intellectual and Professional Division of Labour and Its Knowledge Infrastructure,' 171–7; 'Part 2: Intellectual Map Making and the Tension between Breadth and Depth,' 178–88; 'Part 3: A Strategy for Transforming the Profession,' 189–203; 'Part 4: Extending the Strategy to Medicine, the Social Sciences and the University,' 204–16.
41 Vanderburg, 'The Anti-economy Hypothesis,' parts 1–3.
42 Polanyi, *The Great Transformation.*
43 Daly, *Beyond Growth.*
44 Ibid.
45 Ibid.
46 Ellul, *The Technological Society.*
47 Vanderburg, *The Growth of Minds and Cultures.*
48 I.L. Horowitz, 'Big Five and Little Five: Measuring Revolutions in Social Science,' *Society*, March/April 2006, 9–12.
49 Ibid.
50 Vanderburg, *Living in the Labyrinth of Technology.*
51 Shulamit Reinharz, *On Becoming a Social Scientist: From Survey Research and Participant Observation to Experimental Analysis* (New Brunswick, NJ: Transaction Books, 1984).

52 Devereux, *From Anxiety to Method in the Behavioral Sciences*.
53 Ellul, *The New Demons*; Vanderburg, *Living in the Labyrinth of Technology*.
54 Ellul, 'Remarks on Technology and Art,' 26–37.
55 Giedion, *Mechanization Takes Command*.
56 Jacques Ellul, *L'empire du non-sens: L'art et la société technicienne* (Paris: Presses Universitaires de France, 1980); W.H. Vanderburg, 'Comments on the Empire of Non-sense: Art in a Technique-Dominated Society,' *Bulletin of Science, Technology & Society* 21, no. 1 (Feb. 2001): 38–54.
57 John Horgan, *The End of Science* (Reading, MA: Addison-Wesley, 1996); D'Espagnat, *In Search of Reality*.
58 Ellul, *L'empire du non-sens*.
59 Daniel Bell, *The Coming of the Post-industrial Society: A Venture in Social Forecasting* (New York: Basic Books, 1976); Alain Touraine, *The Post-industrial Society; Tomorrow's Social History: Classes, Conflicts and Culture in the Programmed Society*, trans. Leonard F.X. Mayhew (New York: Random House, 1971); R. Richta, *Civilization at the Crossroads: Social and Human Implications of the Scientific and Technological Revolution* (Prague: International Arts and Science Press, 1969); McLuhan, *The Global Village*; Jean Baudrillard, *The Consumer Society: Myths and Structures* (London: Sage, 1998); Günter Friedrichs and Adam Schaff, eds., *Microelectronics and Society: For Better or for Worse: A Report to the Club of Rome* (New York: Pergamon Press, 1982); Zbigniew Brzezinski, *America in the Technotronic Age* (New York: Wiley, 1972); Debord, *The Society of the Spectacle*; Mumford, *The Myth of the Machine*; Galbraith, *The New Industrial State*.
60 Karasek and Theorell, *Healthy Work*.
61 Vanderburg, *The Labyrinth of Technology*. See especially chapter 11.
62 Michael Pollen, *In Defense of Food: An Eater's Manifesto* (New York: Penguin Books, 2009).
63 Raymond Kurzweil, *The Age of Spiritual Machines: When Computers Exceed Human Intelligence* (New York: Viking, 1999); *The Singularity Is Near: When Humans Transcend Biology* (New York: Viking, 2005).
64 Dreyfus, *On the Internet*; Dreyfus, *What Computers Still Can't Do*.
65 Edward Tenner, *Why Things Bite Back: Technology and the Revenge of Unintended Consequences* (New York: Knopf, 1996); Chellis Glendinning, *When Technology Wounds: The Human Consequences of Progress* (New York: William Morrow, 1990); Carolyn Marvin, *When Old Technologies Were New: Thinking about Electric Communications in the Late Nineteenth Century* (New York: Oxford University Press, 1988).
66 Abraham Maslow, *The Psychology of Science: A Reconnaissance* (New York: Harper and Row, 1966).

67 The original results were published years after the study had been submitted to an engineering education journal that eventually published it with trivial modifications. Since that publication in 1994, I have continued to monitor the evolution of undergraduate engineering education, and I found no reason to revise the scores when I included them in *The Labyrinth of Technology*. The results were sent to the deans of all the reputable engineering schools in North America (except Mexico), the Canadian Engineering Accreditation Board, the (U.S.) Accreditation Board for Engineering and Technology, the president of the Professional Engineers of Ontario, and the Canadian Academy of Engineering. Not a single response was received, other than a request from a Canadian university for some photocopies. In the meantime, officials and spokespersons of the profession continue to claim that the profession is engaged in an ongoing process of protecting the public interest above everything else, as required by law. The time has surely come for society to revoke the privilege of the profession to regulate itself, for failure to protect the public interest, except in the narrowest interpretation of that term.

68 Benson R. Snyder, *The Hidden Curriculum* (New York: Knopf, 1971).

69 Vanderburg, *The Labyrinth of Technology*.

70 Bakan, *The Corporation*.

71 Johnston, *Free Lunch*.

72 What I have referred to as metaconscious knowledge is comparable to Michael Polanyi, *Personal Knowledge: Towards a Post-critical Philosophy* (London: Routledge, 1998). Similarly, the disciplinary matrices of Khun, *The Structure of Scientific Revolutions*, are, for the larger part, metaconscious.

73 Some engineering faculty members have treated this as a lack of human factors engineering in the curriculum, as if the addition of one more discipline could affect the kinds of structural problems to which I have referred.

74 Vanderburg, 'Rethinking Engineering Design and Decision Making,' 421–32.

75 Berggren, 'The Fate of Branch-Plants,' 107–8; Berggren, *The Volvo Experience*.

76 Semler, *Maverick*.

77 Karasek and Theorell, *Healthy Work*.

78 Vanderburg, *The Labyrinth of Technology*.

79 Ibid.

80 Johnston, *Free Lunch*.

81 Mintzberg, *Managers Not MBAs*.

82 Warren G. Benis and James O'Toole, 'How Business Schools Lost Their Way,' *Harvard Business Review*, May 2005, 96–104.

83 Henry Mintzberg, *The Rise and Fall of Strategic Planning: Reconceiving Roles for Planning, Plans, Planners* (Toronto: Maxwell Macmillan Canada, 1994).
84 Mintzberg, *Managers Not MBAs.*
85 From its very beginning, the limitations of operations research and its failure in most situations have been pointed out in the literature. See for example Russell L. Ackoff, 'The Future of Operational Research Is Past,' *Journal of Operational Research Society* 30, no. 2 (1979): 93–104; 'Resurrecting the Future of Operational Research,' *Journal of Operational Research Society* 30, no. 3 (1979): 189–99.
86 Patrick J. Wright, *On a Clear Day You Can See General Motors: John Z. De Lorean's Look inside the Automotive Giant* (London: Sidgwick and Jackson, 1979).
87 Gerth and Mills, *From Max Weber*; see also Brubaker, *The Limits of Rationality.*
88 Galbraith, *The New Industrial State.*
89 Vanderburg, *Living in the Labyrinth of Technology.*
90 John Ralston Saul, *Voltaire's Bastards: The Dictatorship of Reason in the West* (New York: Vintage Books, 1993).
91 Dreyfus and Dreyfus, with Athanasiou, *Mind over Machine.*
92 I believe that a great deal of Henry Mintzberg's diagnosis and prescription of what ails management science and business administration may be interpreted as a particular instance of the more general patterns described in this work. His explanation of the heroic style of management is described in *Managers Not MBAs.*
93 Benson Snyder, 'Literacy and Numeracy: Two Ways of Knowing,' *Daedalus* 119 (1990): 233–56.
94 Turkle, *The Second Self.*
95 Vanderburg, *Living in the Labyrinth of Technology.*
96 Martin Shane, 'The Duty to Prevent Emotional Harm at Work: Arguments from Science and Law, Implications for Policy and Practice,' *Bulletin of Science, Technology & Society* 24, no. 5 (Aug. 2004): 305–15; Schumpeter [pseud.], 'Overstretched,' *The Economist* 395, no. 8683 (20 May 2010): 72.
97 Mintzberg, *Managers Not MBAs*, 154.
98 Barry James, 'Executives' Warning on Raw Capitalism,' *International Herald Tribune*, 15 October 1996, 23.
99 Mintzberg, *Managers Not MBAs.*
100 Very few studies have attempted to assess the implications of the overall effect that the contamination of our food, drink, and air has on human beings. This is rather surprising because human health does not come in distinct and separate components. It is ultimately the overall effect that,

in relation to our resources, helps us to understand what sustains and threatens life and, hence, to understand health and disease. For a rare study see Eric Skjei and M. Donald Whorton, *Of Mice and Molecules* (New York: Dial Press, 1983).

101 Pollan, *In Defense of Food*. See especially part 2.

102 Ellul, *The Technological Society*. See also Bjorn Hofmann, 'The Myth of Technology in Health Care,' *Science and Engineering Ethics* 81, no. 1 (March 2002): 17–29; Eileen Gambrill, 'Evidence-Informed Practice: Antidote to Propaganda in the Helping Professions?' *Research on Social Work Practice* 20, no. 3 (Jan. 2010): 302–20.

103 Ellul, *The Technological System*.

104 Carl Elliott, 'A New Way to Be Mad,' *Atlantic Monthly*, December 2000, 72–84.

105 Ibid.

106 Ibid.

107 Eric Drexler, *The Engines of Creation* (London: Fourth Estate, 1990); Hans Moravec, *Mind Children: The Future of Robot and Human Intelligence* (Cambridge, MA: Harvard University Press, 1988).

108 Tapscott, *Grown Up Digital*.

109 Carl Elliott, 'Humanity 2.0,' *Wilson Quarterly*, September 2003, 13–20.

110 Elliott, 'A New Way to Be Mad.'

111 Jean-Luc Porquet, *Jacques Ellul, l'homme qui avait (presque) tout prévu* (Paris: Cherche Midi, 2003).

112 Elliott, 'A New Way to Be Mad.'

113 Ian Hacking, *Rewriting the Soul: Multiple Personality and the Sciences of Memory* (Princeton, NJ: Princeton University Press, 1998); *Mad Travelers* (Charlottesville: University Press of Virginia, 1998).

114 With his permission, I previously reported the answers that Jacques Ellul gave in a doctoral course regarding the universality of legal institutions. It first appeared in *The Growth of Minds and Cultures*, for which he wrote the foreword. I have also referred to his arguments in 'Technology and the Law: Who Rules?' 322–32.

115 Jacques Ellul, *Histoire des institutions* (Paris: Presses universitaires de France, 1970).

116 Ellul, *The Technological Society*.

117 Ibid.

118 I am relying on various interpretations of this case, presented in two special issues of the *Bulletin of Science, Technology & Society*, guest-edited by Jennifer Chandler. I am also relying on her subsequent article showing

that this case may well be an instance of a more general pattern that she is exploring. See Jennifer Chandler, 'The Autonomy of Technology: Do Courts Control Technology or Do They Just Legitimize Its Social Acceptance?' *Bulletin of Science, Technology & Society* 27, no. 5 (Oct. 2007): 339–48.

119 Chandler, 'The Autonomy of Technology.'

Index